AF783350

MECHANISMS OF VIRAL PATHOGENESIS

DEVELOPMENTS IN MOLECULAR VIROLOGY

Yechiel Becker, Series Editor
Julia Hadar, Managing Editor

Volume Editors: A. Kohn and P. Fuchs
Israel Institute for Biological Research
Ness-Ziona, Israel

Ilana Turner, Technical Editor

Other books in the series:
Becker, Y. (ed.) *Herpesvirus DNA* (1981)
Becker, Y. (ed.), *Replication of Viral and Cellular Genomes* (1984)

MECHANISMS OF VIRAL PATHOGENESIS

From Gene to Pathogen

Proceedings of 28th OHOLO Conference, held at Zichron Ya'acov, Israel, March 20–23, 1983

Edited by

A. Kohn
P. Fuchs

Martinus Nijhoff Publishing
a member of the Kluwer Academic Publishers Group
Boston/The Hague/Dordrecht/Lancaster

Distributors for North America:
Kluwer Academic Publishers
190 Old Derby Street
Hingham, MA 02043

Distributors Outside North America:
Kluwer Academic Publishers Group
Distribution Centre
P.O. Box 322
3300 AH Dordrecht
The Netherlands

Library of Congress Cataloging in Publication Data

Oholo Biological Conference (28th : 1983 : Zikhron
Ya'akov, Israel)
Mechanisms of viral pathogenesis, from gene to
pathogen.

(Developments in molecular virology)
Includes indexes.
1. Host-virus relationships—Congresses. 2. Virus
diseases—Congresses. I. Kohn, Alexander. II. Fuchs, P.
III. Title. IV. Series.
QR482.037 1983 616'.0194 83-17408
ISBN 978-94-011-8543-1 ISBN 978-94-011-9297-2 (eBook)
DOI 10.1007/978-94-011-9297-2

Conference Organizing Committees

Organizing Committee (28th Conference)

A. Kohn
P. Fuchs } Chairmen
Israel Institute for Biological Research, Ness-Ziona, Israel

Y. Becker, Hebrew University—Hadassah Medical School, Jerusalem, Israel
E. Winocour, Weizmann Institute of Science, Rehovot, Israel
A. Yaniv, Tel-Aviv University, Ramat-Aviv, Israel

Permanent Committee

S. Gitter, Tel-Aviv University Sackler School of Medicine, Ramat-Aviv, Israel
N. Grossowicz, Hebrew University, Jerusalem, Israel
I. Hertman, Israel Institute for Biological Research, Ness-Ziona, Israel
A. Keynan, Hebrew University, Jerusalem, Israel
M. A. Klingberg, Israel Institute for Biological Research, Ness-Ziona and Tel-Aviv University Sackler School of Medicine, Ramat-Aviv, Israel
A. Kohn, Israel Institute for Biological Research, Ness-Ziona and Tel-Aviv University Sackler School of Medicine, Ramat-Aviv, Israel
M. Sela, Weizmann Institute of Science, Rehovot, Israel
G. Simon, Israel Institute for Biological Research, Ness-Ziona, Israel

I. Turner, Secretary
M. Navon, Technical Organizer
Israel Institute for Biological Research, Ness-Ziona, Israel

Oholo Conferences

1956 Bacterial Genetics (not published)

1957 Tissue Cultures in Virological Research (not published)

1958 Inborn and Acquired Resistance to Infection in Animal (not published)

1959 Experimental Approach to Mental Diseases (not published

1960 Cryptobiotic Stages in Biological Systems; Eds. N. Grossowicz, S. Hestrin and A. Keynan (Elsevier, Amsterdam 1961)

1961 Virus—Cell Relationships (Israel Institute for Biological Research, Ness-Ziona)

1962 Biological Synthesis and Function of Nucleic Acids (Israel Institute for Biological Research, Ness-Ziona)

1963 Cellular Control Mechanisms of Macromolecular Synthesis (Israel Institute for Biological Research, Ness-Ziona)

1964 Molecular Aspects of Immunology (Israel Institute for Biological Research, Ness-Ziona)

1965 Cell Surfaces (Israel Institute for Biological Research, Ness-Ziona)

1966 Chemistry and Biology of Psychotropic Agents (not published)

1967 Structure and Mode of Action of Enzymes (Israel Institute for Biological Research, Ness-Ziona)

1968 Growth and Differentiation of Cells *in vitro* (Israel Institute for Biological Research, Ness-Ziona)

1969 Behavior of Animal Cells in Culture (Israel Institute for Biological Research, Ness-Ziona)

1970 Microbial Toxins (Israel Institute for Biological Research, Ness-Ziona)

1971 Interaction of Chemical Agents with Cholinergic Mechanisms (Israel Institute for Biological Research, Ness-Ziona)

1972 Immunity in Viral and Rickettsial Diseases; Eds. A. Kohn and M.A. Klingberg (Plenum Press, New York, London 1973)

1973 Control of Gene Expression; Eds. A. Kohn and A. Shatkay, in: Advances in Experimental Medicine and Biology (Plenum Press, New York, London 1974)

1974 Sensory Physiology and Behavior; Eds. R. Galun, P. Hillman, I. Parnas and R. Werman, in: Advances in Behavioral Biology (Plenum Press, New York, London 1975)

1975 Air Pollution and the Lung; Eds. E.F. Aharonson, A. Ben-David and M.A. Klingberg (John Wiley, New York, Toronto 1976)

1976 Host Parasite Relationships in Systemic Mycoses; Eds. A. Ben-David, A. Beemer, M.A. Klingberg and E.S. Kuttin (Karger, Basel 1977)

1977 Skin: Drug Application and Evaluation of Environmental Hazards; Eds. G.A. Simon, Z. Paster, M.A. Klingberg and M. Kaye (Karger, Basel 1978)

1978 Extrachromosomal Inheritance in Bacteria; Eds. I. Hertman, A. Shafferman, A. Cohen and S.R. Smith (Karger, Basel 1979)

1979 I. Neurobiology of Cholinergic and Adrenergic Transmitters; Eds. E. Heldman, A Levy, Y. Gutman and Z. Vogel (Karger, Basel 1980)

II. Endogenous Peptides and Centrally Acting Drugs; Eds. A. Levy, E. Heldman, Z. Vogel and Y. Gutman (Karger, Basel 1980)

1980 New Developments with Human and Veterinary Vaccines; Eds. A. Mizrahi, I. Hertman, M.A. Klingberg and A. Kohn (Alan R. Liss, New York 1981)

1981 Chemical Approaches to Understanding Enzyme Catalysis; Eds. B.S. Green, Y. Ashani and D. Chipman (Elsevier, Amsterdam 1982)

1982 Behavioral Models and the Analysis of Drug Action; Eds. M.Y. Spiegelstein and A. Levy (Elsevier, Amsterdam 1982)

1983 Mechanisms of Viral Pathogenesis (From Gene to Pathogen); Eds. A. Kohn and P. Fuchs (Martinus Nijhoff Publishing, Boston)

Forthcoming Conference:

1984 Boundary Layer Structure—Modelling and Application to Air Pollution and Wind Energy

Contributing Authors

S.A. Aaronson
National Cancer Institute
Bethesda, Maryland, USA

R. Abulafia
Department of Genetics
Weizmann Institute of Science
Rehovot, Israel

H. Akashi
University of Alabama
Birmingham, Alabama, USA

Y. Aloni
Department of Genetics
Weizmann Institute of Science
Rehovot, Israel

E. Antmann
Institute for Virus Research
German Cancer Research Center
Heidelberg, Germany

Y. Asher
Department of Molecular Virology
The Hebrew University
Jerusalem, Israel

M. Aubert
Centre National d'Etudes sur la Rage
Nancy Malzeville, France

A. Barasofsky
Department of Molecular Virology
The Hebrew University
Jerusalem, Israel

P.N. Barrett
Institute for Virology and Immunobiology
University of Wuerzburg
Wuerzburg, Germany

R. Barzilay
Israel Institute for Biological Research
Ness-Ziona, Israel

B. Beaty
University of Alabama
Birmingham, Alabama, USA

Y. Becker
Department of Molecular Virology
The Hebrew University
Jerusalem, Israel

J.R. Bell
California Institute of Technology
Pasadena, California, USA

E. Ben-Asher
Department of Genetics
Weizmann Institute of Science
Rehovot, Israel

T. Ben-Hur
Department of Molecular Virology
The Hebrew University
Jerusalem, Israel

D. Ben-Nathan
Israel Institute for Biological Research
Ness-Ziona, Israel

A. Ben-Zeev
Department of Genetics
Weizmann Institute of Science
Rehovot, Israel

K. Berns
University of Florida
College of Medicine
Gainesville, Florida, USA

D.H.L. Bishop
University of Alabama
Birmingham, Alabama, USA

E. Canaani
Department of Chemical Immunology
Weizmann Institute of Science,
Rehovot, Israel

M.J. Carter
Institute for Virology and Immunobiology
University of Wuerzburg
Wuerzburg, Germany

P.W. Choppin
The Rockefeller University
New York, New York, USA

W. Clark
University of Alabama
Birmingham, Alabama, USA

C.A. Clifford
Department of Immunology and
Medical Microbiology
University of Florida
Gainsville, Florida, USA

J. Cohen
Department of Chemical
Immunology
Weizmann Institute of Science
Rehovot, Israel

P. Coulon
Université Paris-Sud
Orsay Cedex, France

R.S. Daniels
National Institute for Medical
Research
London, England

M. Dolivo
Institute of Physiology
Lausanne, Switzerland

A.R. Douglas
National Institute for Medical
Research
London, England

O. Dreazen
Department of Chemical
Immunology
Weizmann Institute of Science
Rehovot, Israel

R.M. Elliott
MRC Virology Unit
Institute of Virology
Glasgow, Scotland

A. Flamand
Université Paris-Sud
Orsay Cedex, France

J.B. Flanegan
Department of Immunology and
Medical Microbiology
University of Florida
Gainesville, Florida, USA

F. Fuller
University of Alabama
Birmingham, Alabama, USA

D.H. Gilden
Department of Molecular Virology
The Hebrew University
Jerusalem, Israel

D. Givol
Department of Chemical
Immunology
Weizmann Institute of Science
Rehobot, Israel

B.W. Grinnell
Department of Microbiology
University of Virginia
Charlottesville, Virginia, USA

P. Gruss
Department of Microbiology
University of Heidelberg
Heidelberg, Germany

E. Guetta
Department of Biology
Ben-Gurion University of the Negev
Beer Sheba, Israel

J. Hadar
Department of Molecular Biology
The Hebrew University
Jerusalem, Israel

N. Hay
Department of Genetics
Weizmann Institute of Science
Rehovot, Israel

M. Horowitz
Department of Chemical
Immunology
Weizmann Institute of Science
Rehovot, Israel

E.B. Jakobovits
Department of Genetics
Weizmann Institute of Science
Rehovot, Israel

R. Kaempfer
Department of Molecular Virology
The Hebrew University
Jerusalem, Israel

A. Keysary
Israel Institute for Biological Research
Ness-Ziona, Israel

R.M. Kinney
Centers for Disease Control
Fort Collins, Colorado, USA

M. Kita
Institut Pasteur
Paris Cedex, France

A. Klar
Department of Chemical Immunology
Weizmann Institute of Science
Rehovot, Israel

H.-D. Klenk
Institute for Virology
Justus-Liebig University
Giessen, Germany

S. Knoller
Department of Molecular Virology
The Hebrew University
Jerusalem, Israel

M. Knossow
Department of Biochemistry and Molecular Biology
Harvard University
Cambridge, Massachusetts, USA

K. Koschel
Institute for Virology and Immunobiology
University of Wuerzburg
Wuerzburg, Germany

K. Kristensson
Huddinge Hospital
Huddinge, Sweden

P. Kucera
Institute of Physiology
Lausanne, Switzerland

O. Laub
Department of Genetics
Weizmann Institute of Science
Rehovot, Israel

J.F. Lees
MRC Virology Unit
Institute of Virology
Glasgow, Scotland

J.H. Mathews
Centers for Disease Control
Fort Collins, Colorado, USA

J. McGowan
Department of Microbiology
University of Virginia
Charlottesville, Virginia, USA

E. Norrby
Karolinska Institute
Stockholm, Sweden

R. Oren
Israel Institute for Biological Research
Ness-Ziona, Israel

C. Örvell
National Bacteriological Laboratory
Stockholm, Sweden

J.C. Paulson
Dept. of Biological Chemistry
UCLA School of Medicine
Los Angeles, California, USA

P. Pfeiffer
Department of Genetics
Weizmann Institute of Science
Rehovot, Israel

D. Portnoi
Institut Pasteur
Paris Cedex, France

C.R. Pringle
MRC Virology Unit
Institute of Virology
Glasgow, Scotland

R. Pruzan
Department of Genetics
Weizmann Institute of Science
Rehovot, Israel

D. Ram
Dept. of Chemical Immunology
Weizmann Institute of Science
Rehovot, Israel

G. Rechavi
Dept. of Chemical Immunology
Weizmann Institute of Science
Rehovot, Israel

G.N. Rogers
Dept. of Biological Chemistry
UCLA School of Medicine
Los Angeles, California, USA

J.T. Roehrig
Centers for Disease Control
Fort Collins, Colorado, USA

P. Rollin
Institut Pasteur
Paris Cedex, France

D. Ron
Department of Biology
Ben-Gurion University of the Negev
Beer Sheba, Israel

H. Rosen
Dept. of Molecular Virology
The Hebrew University
Jerusalem, Israel

R. Rott
Institute for Virology
Justus-Liebig University
Giessen, Germany

G. Sauer
Institute for Virus Research
German Cancer Research Center
Heidelberg, Germany

A. Scemama
Department of Molecular Virology
The Hebrew University
Jerusalem, Israel

C. Scholtissek
Institute for Virology
Justus-Liebig University
Giessen, Germany

G. di Segni
Department of Molecular Virology
The Hebrew University
Jerusalem, Israel

H. Sheshberadaran
Karolinska Institute
Stockholm, Sweden

R.E. Shope
University of Alabama
Birmingham, Alabama, USA

Y. Shtram
Department of Molecular Virology
The Hebrew University
Jerusalem, Israel

J.J. Skehel
National Inst. for Medical Research
London, England

H. Skolnik-David
Department of Genetics
Weizmann Institute of Science
Rehovot Israel

J.H. Strauss
California Institute of Technology
Pasadena, California, USA

E. Tabor
Department of Molecular Virology
The Hebrew University
Jerusalem, Israel

J. Tal
Department of Biology
Ben-Gurion University of the Negev
Beer Sheba, Israel

Y. Teitz
Sackler School of Medicine
Tel-Aviv University
Ramat-Aviv, Israel

V. Ter Meulen
Inst. for Virology & Immunobiology
University of Wuerzburg
Wuerzburg, German

D.W. Trent
Centers for Disease Control
Fort Collins, Colorado, USA

R. Wagner
Department of Microbiology
The University of Virginia
Charlotteville, Virginia, USA

G.E. Wartret
University of Alabama
Birmingham, Alabama, USA

R. Watanabe
Inst. for Virology & Immunobiology
University of Wuerzburg
Wuerzburg, Germany

M.D. Waterfield
Imperial Cancer Research Fund
London, England

H. Wege
Inst. for Virology & Immunobiology
University of Wuerzburg
Wuerzburg, Germany

D.C. Wiley
Department of Biochemistry and
Molecular Biology
Harvard University
Cambridge, Massachusetts, USA

M. Willcocks
Inst. for Virology & Immunobiology
University of Wuerzburg
Wuerzburg, Germany

I.A. Wilson
Department of Biochemistry and
Molecular Biology
Harvard University
Cambridge, Massachusetts, USA

WELCOMING ADDRESS

Ladies and Gentlemen

The OHOLO conferences were initiated 28 years ago by the Israel Institute for Biological Research, and take their name from the site of the first meeting on the shores of Lake Kinnereth. The purpose of these meetings is, as it was at their initiation, "to foster interdisciplinary communication between scientists in Israel, and to provide added stimulus by the participation of invited scientists from abroad".

I thought at first that as the OHOLO conferences are so well established, there is no need for formal introduction to the present one. However, going through the topics of the next one (which is mathematics and energy), it occurred to me that the present conference is of a singular nature. It is the last conference of a continuum of 27 meetings dealing with various aspects of biochemistry, biology and molecular genetics.

It is not by chance that the topics of our conferences followed outstanding achievements in Biology. We started with bacterial genetics (1956), followed by biological synthesis and function of nucleic acids (1962), cellular control mechanisms of macromolecular synthesis (1963), molecular aspects of immunology (1964), strategies for the control of gene expression (1973), extrachromosomal inheritance in bacteria (1978) and finally to this year's topic—viral pathogenesis—or more appropriately named "from gene to pathogen."

Our present topic represents the peak of scientific endeavour which started with the initiation of the study of bacterial genetics at Cold Spring Harbor, and which achieved a profound understanding of basic molecular processes underlying the inheritance of genes and gene expression. Great names in molecular biology were connected with our conferences: Lwoff, Luria, Monod, Benzer, Spiegelman, S. Cohen, Gorini, Levinthal, Magasanik, Kabat and others. This year's conference is certainly in line with the tradition of the past. There is no doubt that the progress in molecular genetics represents the most dramatic scientific breakthrough in our century and every one who witnessed the flowering of this discipline, and had even a modest part in it, may be proud of this victory of the human mind.

It is with this sense of deep appreciation that I welcome the participants of the 28th OHOLO conference and wish our guests a very pleasant time in Zichron Ya'acov. Thank you.

I. Hertman, Director
Israel Institute for Biological Research
Ness-Ziona, Israel

PREFACE AND ACKNOWLEDGMENTS

This conference is devoted to the elucidation of the various molecular mechanisms determining the virulence or persistence and latency of animal viruses.

The concept of virulence entails not only the infecting microorganism but also the host which is affected by it. The host may be a cell or it may be an organism. The term "virulence" is associated with those properties of the virus which lead to the death of the cells or to disease symptoms in an organism.

Since virulence is defined as a product of interaction of virus and the host, a statement about the level of virulence of any particular strain of virus can be made only if the species and age of the host, the route of inoculation and the dose are kept constant, so that this property of "virulence" can in fact be attributed to the genomic structure and the molecular expression of it in a virus.

When viruses cause disease by multiplication in a target tissue or organ (e.g. CNS, lungs, liver) the virulence is a function of virus multiplication and the extent of cell damage. Thus poliovirus is not considered virulent as long as it infects only the cells lining the intestinal tract, but when it has the ability to invade the central nervous system it is virulent. Many viruses (e.g. arboviruses) may be virulent to newborn animals, but not to adults in the same species.

On the cellular level the first requirement for the ability of a virus to infect a cell is the necessity in a cell to have proper receptors for binding the virus. Poliovirus is not virulent in chicken cells because it cannot attach to it and infect it. The opportunity for expression of virulence is thus first determined by the chemistry of virus:cell encounter. This entails cell receptors as well as the virus attachment proteins (VAP). The structure of these viral antigens is genomically determined and therefore can be affected by mutations. In addition, there are posttranslational events, such as e.g. the cleavage of hemagglutinins in myxoviruses and of F proteins in paramyxoviruses which are essential in establishing a productive infection (Choppin, Rott, Trent).

In viruses with divided genomes such as myxo, bunya and orbiviruses the reshuffling of the genomic pieces in cells infected by more than one strain of virus may result in a change in virulence to a particular host. This reassortment, which can be experimentally manipulated, permits the analysis of progeny virus and the assignment of virulence to certain genes. So, for instance, the work of Bishop already indicated that the medium-sized RNA species of the bunyavirus genome that codes for glycoproteins determines neurovirulence in the California group of virus and viscerotropism in group C bunyavirus. He has also shown that the L-RNA gene products (transcriptase and replicase) can mitigate the virulence that is prescribed by M-RNA gene products. Reassortant viruses with a defective L-RNA are not transmitted by arthropod vectors. Genetic analysis of reassortment of influenza and bunyavirus points to a polygenic nature of pathogenicity of these viruses. In the paramyxoviruses the normal expression of M protein is involved in persistent infection, as e.g. in SSPE induced by measles virus (Norrby, Carter).

The host contribution to virulence of myxo and paramyxoviruses involves the proteolytic cleavage of HA or F proteins. Thus cells (or animal species) that do not possess the proper protease, would not be infected by these viruses (Akov).

In rabies virus Coulon and Flamand identified the antigenic site on the glycoprotein G which is associated with virulence of this virus. Mutants which are not neutralized by certain monoclonal antibodies directed to different epitopes on G, have a significantly reduced pathogenicity in mice. Thus in this virus the virulence seems to be associated with a particular antigenic structure on the G-protein.

The examples depicted till now are connected with the effects of structural changes in some viral proteins which determine virulence either per se or as a result of the structure being differently affected by host cell enzymes.

Another type of virulence depends on the regulatory functions of the viral or host genomes. Wagner shows that the inhibition of cellular RNA synthesis by vesicular stomatitis virus depends on the presence and role of a 47 nucleotide leader sequence present at the 3′ end of the viral genome, and the presence in that leader squence of TAATA-like and consensus-like nucleotide sequences.

This type of regulation of virulence is also encountered in herpes virus which needs a thymidine kinase for the phosphorylation of its nucleosides (Becker). The presence and activity of this enzyme may determine whether the virus will be actively multiplying in the infected cells or establish a latent infection, or be reactivated from its latent state. Mutants of HSV which lack the TK gene can infect epidermal cells and even proceed to nerve cells in ganglia, but they are unable to multiply there and to proceed to brain so as to cause encephalitis. The virus can multiply, however, in host cells which provide their kinase for the synthesis of viral components.

Finally one should perhaps discuss in this context the concept of oncogenes and the fact that normal animal genes become expressed as a result of finding themselves in the vicinity or association with promoter regions supplied by oncogenic viruses, or by rearrangement of the gene (Canaani, Aaronson).

Other host specific regulatory factors which may affect the expression of a gene are the attenuators and the enhancers, which have been by now quite well molecularly defined and are discussed in this conference in the context of SV40 gene expression (Aloni, Gruss) and retrovirus pathogenesis.

Research on virulence is still in its beginning stages. Many individual elements concerning either the virus or the host contributing to virulence have been studied, often in detail, but the overall picture—the orchestration effect of all contributing factors—still eludes us and remains a challenge for future research.

ACKNOWLEDGMENTS

The Conference was organized and sponsored by the Israel Institute for Biological Research, Ness-Ziona, Israel, affiliated to the Tel-Aviv University Sackler School of Medicine.

The financial support given to the conference by the following institutions is gratefully acknowledged:

The British Council (Israel and UK)

Ministry of Energy and Infrastructure
National Council for Research & Development
Jerusalem, Israel

Office of US Naval Research
Branch Office, London, UK

Weizmann Institute of Science
Rehovot, Israel

CONTENTS

1

ATTENUATION IN THE CONTROL OF GENE EXPRESSION IN ANIMAL VIRUSES

YOSEF ALONI, NISSIM HAY, HAGIT SKOLNIK-DAVID, PIERRE PFEIFFER, RIVKA ABULAFIA, RON PRUZAN, EDNA BEN-ASHER, EDWARD B. JAKOBOVITS, ORGAD LAUB and AVRI BEN-ZE'EV
Department of Genetics, Weizmann Institute of Science, Rehovot 76100, Israel

SUMMARY

Viral structural proteins self-assemble to produce the capsid of the virion. For an efficient self-assembly process the structural proteins should be synthesized in optimal amounts. This could be accomplished by a mechanism which somehow couples transcription in the nucleus to translation in the cytoplasm. A mechanism of gene regulation which couples transcription and translation, termed "attenuation" exists in procaryotes. This control is manifested through the synthesis of a small "leader" peptide. Successful synthesis of this peptide leads to transcription-termination, whereas in the absence of its synthesis, the RNA polymerase is allowed to continue transcription through the structural genes that follow the DNA sequence coding for the leader peptide. In the present communication we summarize the available information concerning transcription termination and attenuation in eucaryotes and we show that a mechanism resembling attenuation in procaryotes regulates the production of VP_1, VP_2 and VP_3 in SV40 and gene expression in other animal viruses. In analogy to procaryotes the leader protein of SV40 late RNA ("agnoprotein") stabilizes the RNA conformation, stem-and-loop structure followed by us, which leads to transcription-termination, while deficiency of the agnoprotein leads to stabilization of an alternative RNA conformation which allows the RNA polymerase to continue transcription through the structural genes. These observations show that RNA polymerase II responds to a transcription-termination signal similar to that to which the procaryotic polymerase responds and they are included in a model in which a feedback control mechanism regulates the transcription of the viral mRNAs in the nucleus and the translation of their encoded proteins in the cytoplasm. The model has striking similarities to the attenuation model in amino acid biosynthetic operons of bacteria suggesting that SV40 has exploited a procaryotic control mechanism and adjusted it to the eucaryotic environment.

PRELUDE

In comparison to procaryotes the formation of mRNA in eucaryotes is a more complex process (1). In addition to regulation at the initiation step, there is regulation of RNA capping, processing, polyadenylation and splicing in the nucleus, mRNA transport from the nucleus and mRNA stability and frequency of translation in the cytoplasm. In eucaryotes transcription initiation is widely believed to be the major controlling step in mRNA formation and little is known about the specificity of transcription termination. In procaryotes transcription termination is believed to be a second major level regulating the formation of mRNA (2-10). In the Introduction we summarize the available information concerning transcription termination and attenuation in eucaryotes, in the Results we present our studies indicating that transcription termination and primarily attenuation could be a major controlling process in the formation of mRNA in viruses and cells and in the Discussion we present a detailed model in which attenuation and mRNA modulation in a feedback control mechanism quantitatively regulate SV40 gene expression.

INTRODUCTION

Transcription-termination in procaryotes

Many regions specifying termination of transcription in procaryotes have now been analyzed. Two especially striking features for rho-independent termination of transcription in vitro are: a GC-rich region of dyad symmetry that invariably precedes the termination site and a stretch of uridine residues located near the 3' terminus of the transcript. Other features also appear to influence termination, such as the distribution of AT versus GC base pairs, distal sequences and termination and antitermination proteins (2-10). The mechanism of transcription termination is thought to involve formation of a stable stem-and-loop structure in the RNA and deceleration of the elongation rate within the stem. Release of the RNA from the template occurs during the synthesis of the uridylic acid residues (2-10). Martin and Tinoco (11) have shown that the rU-dA duplex formed at the termination site is exceptionally unstable and could provide the major driving force for termination and transcription release. In termination sites where one or more of these elements are absent, transcription-termination is dependent on the termination factor rho.

Transcription-termination in eucaryotes

In eucaryotes, transcription is carried out by three RNA polymerases; RNA polymerase I carries out the transcription of the ribosomal RNA genes, RNA polymerase II carries out the transcription of genes coding for proteins and for capped snRNAs U_1 to U_6. RNA polymerase III carries out transcription of the genes of the small molecular weight RNAs such as t-RNA 5S r-RNA, and VA RNA in adenovirus. Most of the studies concerning transcription termination in eucaryotes were performed on RNA polymerase III and RNA polymerase I and very little is known about termination of RNA polymerase II transcription.

Termination of transcription by RNA polymerase III. RNA polymerase III appears to recognize a short sequence at the end of the gene rather than the structure of the RNA which is formed as it is transcribed (12-14). This sequence requires four or more T-residues and perhaps no more than additional residues at each side of the T-cluster. This latter requirement may either be two GC residues or the absence of an A cluster directly adjacent to or within one nucleotide of the T cluster (14). A typical termination signal would be GCTTTTGC (15). More distal residues have not been found to influence termination. The dyad symmetries found at the 3' end of RNA polymerase III transcript differ from a typical termination site in that they are not immediately adjacent to the string of T residues where transcription terminates. The dyad symmetry appears not to be directly involved in the mechanism of transcription termination (15). One exception is the Drosophila 5S RNA where the dyad symmetry is immediately adjacent to four U-residues (16).

Termination of transcription by RNA polymerase I. There is a relatively high degree of homology within the mature rRNA sequences of various eucaryotes. However, right at the putative transcription termination sites the homology is limited with the exception of yeast and Tetrahymena (17-19). In both organisms transcription termination occurs at a (A+T)-rich region within a stretch of T-residues. However, the termination signal is lacking both a GC-rich region preceding the T-cluster and a dyad symmetry that can form a stable stem-and-loop structure. The termination region of the yeast rRNA operon contains an extended dyad symmetry that can form a stable hairpin structure but it is located downstream from the termination site (18). A somewhat less extended dyad symmetry but at the same position towards the gene can be observed near the rRNA termination site of Xenopus laevis (17). This structural similarity between two terminators of RNA polymerase I suggests that sequences beyond the site of termination are involved in the termination process and that this region may be recognized by a termination factor. A eucaryotic rho-like factor has indeed been implicated in transcription termination of rRNA of Tetrahymena (20,21). It is interesting to note that the sequence GCTTTTGTC found at the termination site of rRNA operon in Xenopus (17) is similar to that of the 5S rRNA (see above).

The 3' terminal nucleotides of the mouse rRNA operon can form a stable stem-and-loop structure but it is not followed by a cluster of T-residues (22). The 3' end of 28S rRNA is located about 30 nucleotides upstream and at this region there is homology with 28S rRNAs of yeast Drosophila and Xenopus (17-22). The possibility that the primary transcript terminates further downstream could not be ruled out completely, since demonstration of the 45S rRNA being the primary transcript is impossible in any way at the present time. Further accumulation of data concerning the transcription termination site of rRNA genes of eucaryotes is required for determining the specific signals of termination.

Termination of transcription by RNA polymerase II. Transcription by RNA polymerase II usually procedes beyond the 3' end of the mature mRNA. This is then followed by a processing step and polyadenylation. The polyadenylation signal AAUAAA, resides about 25 nucleotides upstream from the polyadenylation site. The signal is usually not present in histone mRNAs (23).

Transcription termination of the β-globin mRNA was found to occur within a region 1400±100 nucleotides downstream from the poly(A) site (24). The resolution of these experiments does not allow a conclusion whether or not the termination is at a precise site, but does show that termination occurs within a specific region.

Late in Adeno 2 infection, the primary transcript covers 30,000 nucleotides, but it apparently stops within the terminal 650 bases without reaching the end of the genome (25,26).

Birnstiel and his colleagues have shown that histone mRNAs which usually lack the polyA tail at their 3' end, terminate in an RNA hairpin structure. However, this structure is not followed by a string of uridylic acid residues (23,27). This group has recently isolated a rho-like factor and they have postulated that this regulatory factor can interact with the palindrome structure leading to transcription termination (28). This model resembles the rho-dependent termination in procaryotes (2-10). Clarification of the exact mode of action of the rho-like factor awaits its complete purification and utilization of an in vitro reconstituted transcriptional system. It should be mentioned, however, that it is difficult to prove conclusively that the 3' ends of histone mRNA arise directly from transcription termination and not from processing of a longer precursor molecule.

Some of the snRNA are also transcribed by RNA polymerase II. It was noticed that a stretch of U-residues is not found at the 3' end of these RNA molecules with the exception of U_6. The lack of a putative termination signal is consistent with the suggestion that the snRNAs are transcribed as part of a larger primary transcript (29).

Attenuation

In procaryotes transcription termination sites are also located within the operon. The internal termination sites cause premature termination of the transcripts and quantitatively regulate the level of gene expression, by selectively reducing the transcription of distal portions of the operon. This mechanism of regulation has been termed attenuation (10). In contrast to termination at the end of the operon, attenuation is not a complete termination process. It can be overcome by regulatory proteins as in bacteriophage λ or in response to changes in physiological conditions, as in the case of amino acid operons in bacteria (2-10). The sequences at the attenuation sites share common features with those of the termination sites. The DNA immediately preceding the site of attenuation is GC-rich with dyad symmetry and the 3' terminus of the transcript contains a series of uridine residues.

In eucaryotes, an abundant population of promoter-proximal RNA chains has been observed in whole nuclear RNA (30,31), in the β-globin gene (65,66), in adenovirus type 2 (25,32,33) and in simian virus 40 (34-41). On the basis of these results and DNA sequence data (67,77,78,100) it has been suggested that a premature termination process resembling attenuation in procaryotes also occurs in eucaryotes (33,37-39,67,77,78, 101). Moreover, these studies have shown that the adenosine analog 5,6-dichloro-1-β-ribofuranosylbenzimidazole (DRB) enhances premature termination (25,30,35,37,38,42,66) but its mode of action is not completely understood.

In Adeno virus infected cells there are two observations that may indicate regulation of gene expression by an attenuation mechanism: (i) Controlled premature termination occurs during Ad-2 infection (131-134). It has been shown that the promoter at 16.4 m.u. produces RNA beginning at the same cap site both early and late in infection. From the late transcripts five 3' co-terminal groups of mRNA are formed (1), but early in infection only the L1 poly(A) site is found in mRNA. Early transcripts seem to terminate around m.u. 50 whereas the late transcription continues around m.u. 98 (1). This would be equivalent to a read-through process in an attenuation mechanism. (ii) The E1-A specific gene product facilitates expression of E1-B, E2, E3, E4 and L1 mRNAs (1,135). It was demonstrated that the E1-A specific product functioned at the level of transcription initiation, but beyond that the mechanism is not yet understood. The most attractive hypothesis at present suggests that the E1-A gene product inhibits the function of a cellular repressor. It remains to be distinguished whether the mechanism of E1-A gene product is a result of direct activation of transcription or inhibition of repressor (1,135). The E1-A gene product has the characteristic of an antiattenuator factor, resembling the N gene product of λ (137). In the λ system, N gene product-mediated antitermination of transcription enables RNA polymerase to transcribe λ genes distal to the various terminators, and is essential for the growth of the phage.

In retroviruses it has been shown that the LTR contains sequences that code for both the initiation and termination of RNA transcription. In these sequences the promoter region is immediately adjacent to the terminator signals. The termination signal must be ignored by RNA polymerase II during transcription of the LTR at the 5' end of the RNA. However, the enzyme would have to recognize such signals at the end of the transcript, 5400 nucleotides downstream. A model resembling an attenuation mechanism in procaryotes (78) has been suggested in which readthrough at the 5' end or termination at the 3' end are dependent on alternative RNA secondary structures.

In all the above studies the use of the term "attenuation" closely follows the original usage in procaryotic systems however there is no experimental basis to suggest that the mechanism of attenuation in eucaryotes is similar to that in procaryotes. Furthermore, there is no experimental evidence that DRB enhances premature termination at the attenuation site.

In the present communication, we shall show that SV40 has exploited a mechanism resembling attenuation in procaryotes for quantitatively regulating its gene expression. We suggest that a similar mechanism regulates the expression of other viral and cellular genes.

SV40 as a model system

The use of animal viruses as model systems for probing the complexities of molecular control mechanisms has been particularly fruitful. It is generally felt that an understanding of genetic regulation in viruses will provide insight into similar regulatory processes in eucaryotic cells. The molecular biology of SV40 has been under intensive investigation for a number of years, and these studies have provided considerable information regarding the regulation of gene expression, in particular, transcriptional and posttranscriptional processing of mRNA (43-47).

SV40 provides several unique advantages as a model system for such studies. These include the following: 1) The viral genome is a small circular molecule (M.W. 3.4×10^6) that contains genetic information for only six proteins and may be additional small polypeptides (104). 2) The DNA can be obtained in large quantities, which is imperative for many experiments in molecular biology. 3) The same RNA polymerase (polymerase II) transcribes both viral and cellular RNA. 4) The viral and cellular RNAs undergo similar posttranscriptional modification (e.g. capping at the 5' terminus, polyadenylation at the 3' terminus, internal methylation and splicing). 5) A number of mutants and hybrid viruses are available for study. 6) Transcriptional complexes are easy to obtain. 7) The entire nucleotide sequence of the genome of this virus and the localization of the major and minor initiation sites for transcription have been determined (46,47).

The SV40 genome is comprised of early and late genes that are localized in symmetrical halves of the viral DNA. The segment between 0.67 and 0.17 on the map is transcribed in counterclockwise direction prior to the onset of viral DNA replication, and codes for the early viral proteins. The second segment (from 0.67 to 0.17) is transcribed in abundance after initiation of viral DNA replication in a clockwise direction. It encodes the information for the late proteins: VP_1, VP_2 and VP_3. The capsid proteins have been mapped approximately between 0.95 and 0.16, 0.76 and 0.97, and 0.83 and 0.97 respectively (see Fig. 1) (46,47).

At late times after infection of monkey cells with SV40, RNA transcripts are initiated at major and minor sites (48-50). The RNA polymerase initiating transcription at the major initiation site (residue 243) synthesizes a primary transcript (51). As a consequence of different RNA splicing, two major viral mRNAs (16S and 19S mRNA) (52,53) as well as minor mRNA species are produced. The minor species differ from the major ones in their leader sequences (46,47,49). The major 16S and 19S RNAs share common 5' (residue 243) and 3' (residue 2592) ends. However the major 16S mRNA is characterized by an intron which extends inclusively from residue 445 to residue 1380 (Fig. 2) while the major 19S mRNA is characterized by an intron which extends inclusively from residue 292 to residue 475 (Fig. 3). The 16S mRNA encodes information for two proteins: the leader protein known as "agnoprotein" (54) identified in SV40 infected cells (37,55-57) and the capsid protein VP_1 (Fig. 2). The 16S mRNA is apparently the in vivo message of both proteins (37,57).

The most abundant group of 19S mRNA is characterized by an intron of 184 nucleotides (46,49). As a result of the splicing process 152 of the 202 nucleotides of the 16S leader are missing. Consequently, the leader of the 19S mRNA contains an open reading frame for a protein of 29 amino acids (49). This protein shares the first 13 amino acids with the agnoprotein and overlaps with the AUG of VP_2 (Fig. 3). We have designated this postulated (49) but as yet undiscovered protein as $agnoprotein^{29}$ (39).

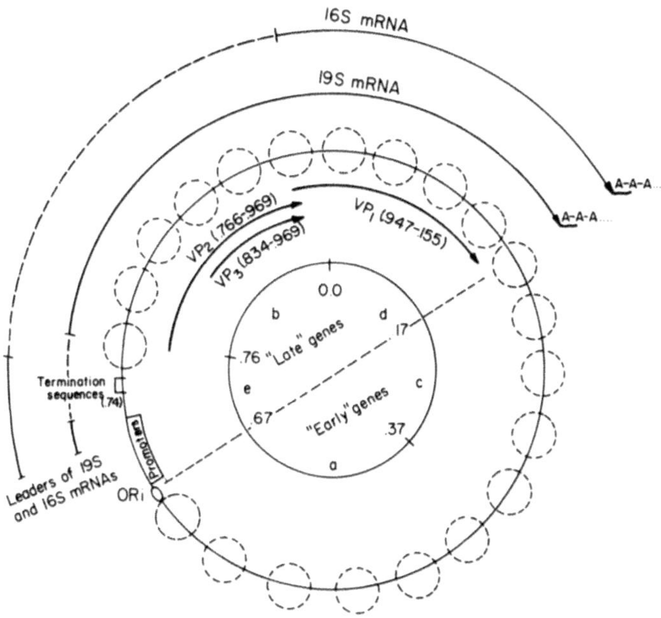

Fig. 1. Transcriptional map of SV40 indicating the five Eco RI; HpaI and BgI restriction fragments of SV40 genome in the central circle. Arrows on RNAs indicate 3' termini, poly(A) tails, and direction of transcription. Decimal numbers represent map units of the SV40 genome. Dashed lines (---) indicate sequences spliced out of 16S and 19S mRNAs. The location of the coding regions are indicated by heavy lines. The small dashed circles denote the distribution of nucleosomes about the SV40 minichromosome with an exposed region that contains the origin of replication (ORi) promoters for late transcription and transcription termination sequences (46). Only the major leaders of 19S and 16S mRNAs are represented; for more details see ref. 46.

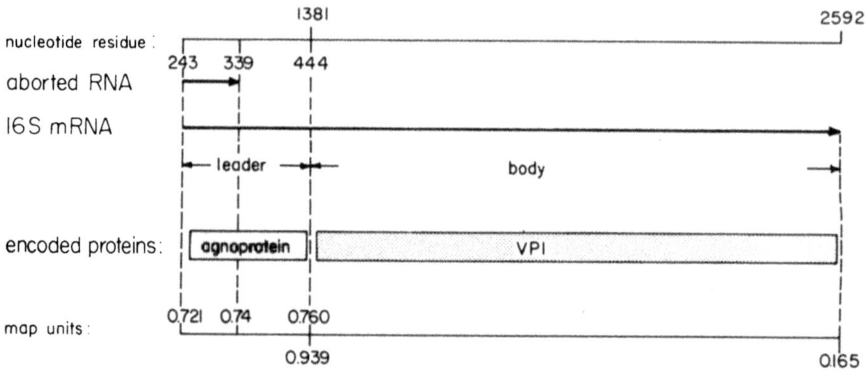

Fig. 2. Map coordinates of the aborted RNA and of the 16S mRNA and its encoded proteins. Nucleotide residue numbers in all the figures refer to the wild-type SV40 sequence by Reddy et al. (68). The figure is from Hay et al. (37).

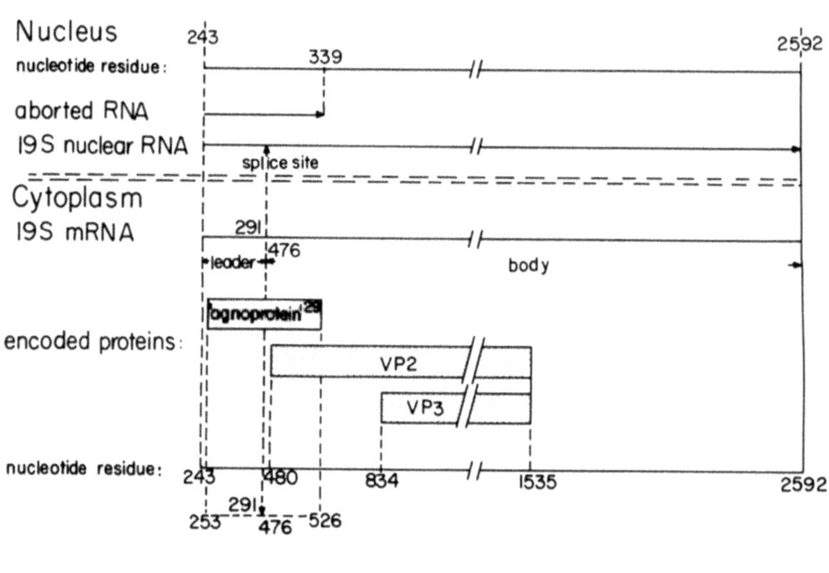

Fig. 3. Map coordinates of the aborted RNA, of the 19S nuclear RNA and of the 19S mRNA and its encoded proteins.

SV40 mRNA is polycistronic. In principle, eucaryotic mRNA could be polycistronic if a mechanism controlling ribosome binding to the initiation codon existed. Kozak (58,59) has proposed a "scanning model" for selection of the initiation sites by eucaryotic ribosomes, which suggests that the 40S ribosome binds to the 5' end of the mRNA and advances towards the 3' end until it encounters the first AUG codon. It has been speculated that secondary structure in and around the initiator AUG codon could be a factor that sequesters the initiation codon from the ribosome. This would lead to the selection by the ribosome of an AUG located further downstream in the genome. Modulation of RNA secondary structure could then be a mechanism controlling the selection of the starting site for translation (37,39,60,61).

RESULTS

Attenuation in the control of SV40 gene expression

Rationale. The product of encapsidation, the virion, is composed of SV40 DNA and cellular histones surrounded by a capsid consisting of viral proteins, VP_1, VP_2 and VP_3 in the proportion 420/36/96, respectively, per capsid (62,63). The simplest encapsidation mechanism requires self-assembly of viral proteins to form capsids (39,64). We assume that a controlled production of the structural proteins VP_1, VP_2 and VP_3 is critical for an efficient self-assembly process (39). This is true for other viruses as well. We suggest that attenuation and modulation of RNA secondary structure are the mechanisms that quantitatively regulate the production of the capsid proteins (37,39). We describe below experiments that support this suggestion.

The promoter-proximal SV40 nascent RNA terminates in vitro at a transcription termination signal

Experiments performed in the last few years have shown an abundant population of promoter-proximal nascent RNA chains on SV40 transcriptional complexes (VTC), minichromosomes and isolated nuclei (34-41). These three systems are capable of elongating the preinitiated viral RNA in vitro. We adopted these systems in order to verify whether these short nascent viral RNAs are of discrete lengths and whether DRB enhances premature termination at physiological sites.

In the following experiment VTC were prepared from untreated and DRB-treated cultures and incubated in vitro for 5 min in the presence of α-^{32}P-UTP. The briefly elongated RNAs were purified and analyzed by gel electrophoresis. Fig. 4, lanes A and B show one major band in a position corresponding to a length of 93-95 nucleotides as well as other minor bands in the RNA preparations of both untreated and DRB-treated cultures (37). Longer viral RNA is absent in the preparation of the DRB-treated culture, as shown by the fact that the region above the major band has no radioactivity. Identical bands were obtained when the viral RNA was purified from minichromosomes and isolated nuclei (38,41).

To locate the RNA in the major band on the physical map of SV40 DNA, the RNA was eluted from the gel and hybridized to two sets of restriction fragments. The map locations of the restriction fragments of the two sets are shown in part E of Fig. 4. Lane C of Fig. 4 shows that hybridization occurred in one set solely with fragment e (0.67-0.76 map units) and lane D of Fig. 4 shows that hybridization occurred in the second set solely with fragment d (0.73-0.82 map units). On the basis of these results, we conclude that the labeled RNA in the major band originates from a region of the genome spanning 0.73-0.76 map units (nucleotides 267-440) (68).

More direct evidence for the position of the 93-95 nucleotide RNA was obtained by the experiment illustrated in Fig. 5. The 93-95 nucleotide RNA was recovered from a gel similar to that shown in Lane A of Fig.4 and was hybridized with the "late" strand of a fragment spanning from the Hpa II site (nucleotide 267) to the Bam H1 site. The RNA-DNA hybrid was treated with RNase to remove the nonhybridized portion of the RNA, and the length of the RNase-resistant portion was determined, after denaturation, by gel electrophoresis (37). Fig. 5 shows that the RNase treatment reduced the length of the RNA from 93-95 nucleotides to 69-71 nucleotides. The minor bands may result from incomplete digestion of the RNA. We conclude that the 5' end of the original RNA is at nucleotide 243 and the 3' end at nucleotides 335-337. We have designated this RNA, attenuated RNA. Further evidence confirming the position of the attenuated RNA comes from a fingerprint analysis of RNase T1 digest of this RNA, followed by nearest-neighbor analysis of each spot (40). The latter analysis also allowed for the identification of the in vivo pause sites

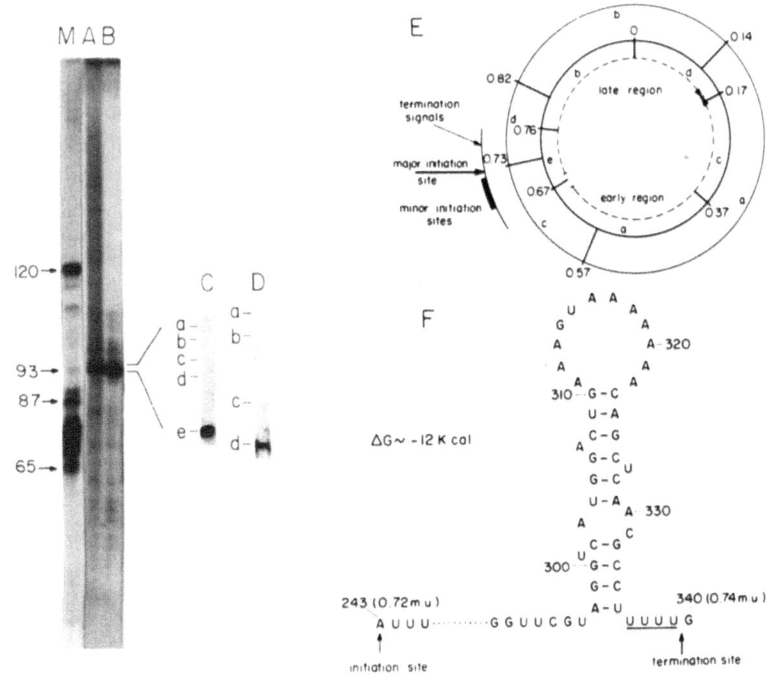

Fig. 4. Size analysis of viral RNA of untreated and DRB-treated cultures and the predicted secondary structure of the attenuator region. VTC were extracted from untreated and DRB-treated infected cultures by the Sarkosyl extraction method (113) and incubated in vitro in the presence of α-^{32}P-UTP for 5 min, and ^{32}P-RNAs were extracted. The labeled RNAs were denatured and subjected to electrophoresis on a 12% acrylamide gel in 7 M urea. (Lane A) ^{32}P-RNA of untreated cultures; (Lane B) ^{32}P-RNA of DRB-treated cultures; (Lane M) length markers of E. coli RNA (a gift from M. David). The RNA in the major bands in lanes A and B (93-95 nucleotides in length) was eluted from the gel by electrophoresis (115) and hybridized to two sets of restriction fragments obtained by digestion of SV40 DNA with Eco. R1, Hpa 1 and Bgl 1 (Lane C) and with Taq 1, Hpa 11 and Bam H1 (Lane D). The map locations of the fragments in lane C are shown in the inner circle and those of lane D in the outer circle in (E). (F) Secondary structure at the 3' end of the aborted RNA. The ΔG was calculated as described (116). Nucleotide residue numbers refer to the wild-type SV40 sequence of Reddy et al. (68).

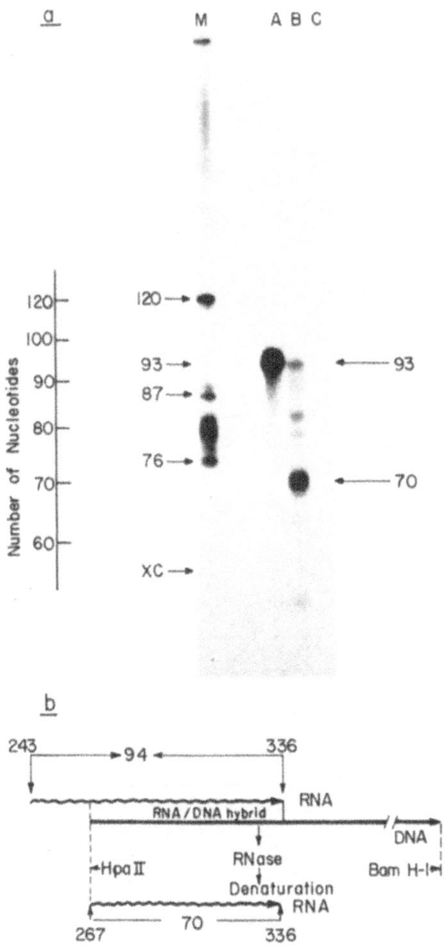

Fig. 5. Mapping of the 93-95 nucleotide labeled RNA by RNAase digestion of RNA-DNA hybrid. (a) Shows an autoradiograph of 12% acrylamide gel in 7 M urea as in Fig.4 (Lane A). The 93-95 nucleotide labeled RNA annealed with the L strand of the 0.73-0.14 map unit DNA fragment with no RNAase treatment. (Lane B) Same as in lane A, but with RNAase treatment. (Lane M) Length markers of E. coli RNA. (b) Diagram showing the plan of the experiment and the deduced topography of the labeled RNA with the lengths of the RNA before and after RNAase treatment. Nucleotide residue numbers refer to the wild-type SV40 sequence of Reddy et al. (68).

The locations of the in vivo pause and in vitro attenuation sites: GC-rich dyad symmetry region in the process of pausing and attenuation

When the labeled RNA present in the major band is digested with RNase T1 and fingerprinted (69) an oligonucleotide pattern characteristic of a transcript from the region between nucleotides 243 and 335-337 can be predicted. However, in an actual experiment, only those oligonucleotides which were elongated in vitro should produce radioactive spots, because the in vivo synthesized oligonucleotides are unlabeled. The junctions between labeled and unlabeled oligonucleotides define the in vivo pause sites of RNA polymerase molecules.

Fig. 6 shows the RNase T1 fingerprints of the labeled RNA present in the major band (see Fig. 4) (40). The fingerprint of 4 and 10 min labeled RNAs display 10 spots of which 3 and 3' are not well resolved. In further experiments they were analyzed together.

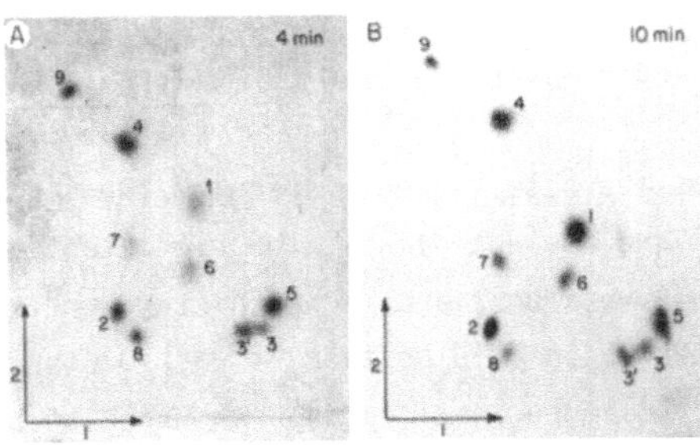

Fig. 6. Fingerprint analysis of 4 and 10 min major labeled RNA band as in Fig. 4. The ^{32}P-labeled RNAs recovered from the gel were digested completion with RNase T1 and the products were separated in two dimensions (40) 1- electrophoresis; 2 - chromatography.

To match the oligonucleotides of each spot with the predicted T1 oligonucleotides, each spot of Fig. 6 was eluted and digested to completion with RNase T2. The digestion results in the transfer of the labeled phosphate to the 5' nearest nucleotide. The labeled nucleotides produced were then identified by paper electrophoresis (Fig. 7).

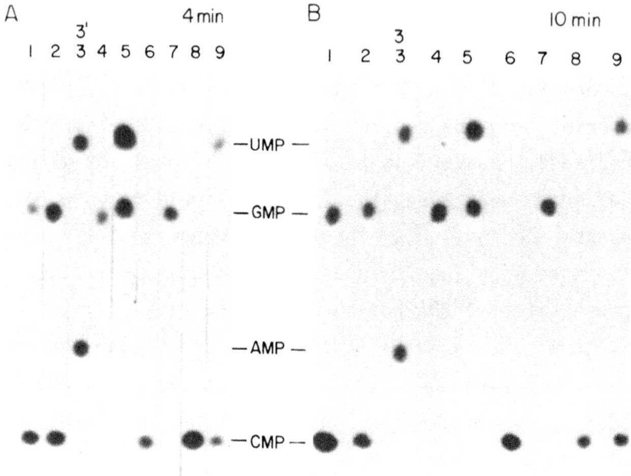

Fig. 7. Nearest-neighbor analysis by RNase T2 digestion of the various oligonucleotides in Fig. 6A and B. T1 oligonucleotides were recovered from the fingerprints of Fig. 6A and B and totally digested with RNase T2. The products were separated by high voltage electrophoresis as detailed by Skolnik-David and Aloni (40).

Fig. 8 shows the base sequence spanning the major initiation site (nucleotide 243) and nucleotide 351, the predicted RNase T1 cleavage sites (following G residues) and the predicted labeled nucleotides obtained after the RNase T2 digestion of each of the oligonucleotides. Note that several oligonucleotides are not expected to be labeled. Ten consecutive oligonucleotides spanning nucleotide 259 to nucleotide 336 were identified. It is evident that spot 1 contains the two oligonucleotides CUG (nucleotides 259-261 and 268-270) because the ratio of radioactivity between CMP and GMP in spot 1 is 2:1 (see Fig.7). It should be noted that some polymerase molecules may have initiated elongation 13 nucleotides downstream from the transcription initiation site at nucleotide 256 but we are unable to resolve it, because nucleotides 256-258 cannot be identified by our analyses.

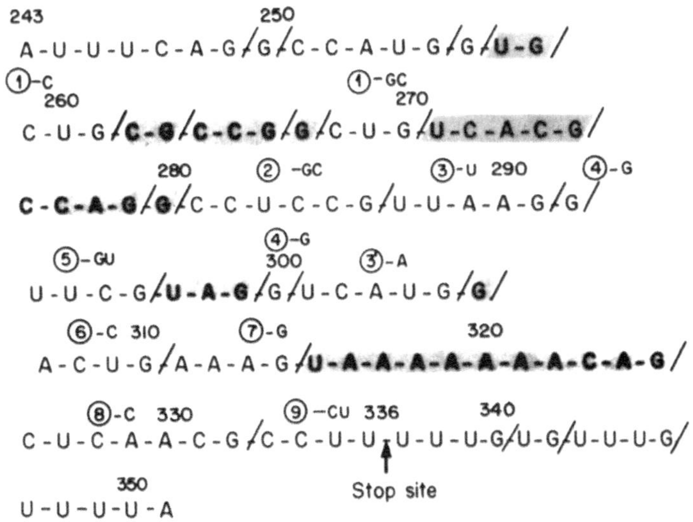

Fig. 8. The oligonucleotides obtained after the RNase T1 digestion and the labeled nucleotides in each oligomer obtained following the RNase T2 digestion. Nucleotide residue numbers refer to wild-type SV-40 sequence of Reddy et al. (68). The slash indicates the RNase T1 cleavage site (following G). The nucleotides above each oligonucleotide are those which are labeled following the RNase T2 digestion. The shaded oligonucleotides are not labeled in the present analysis.

In none of our fingerprint analyses have we ever observed an additional spot that could correspond to the oligonucleotide CCAUG (nucleotides 251 - 255) (see Fig. 8). It appears, therefore that in vivo the first pause site of active RNA polymerase molecules in VTC is located 13 - 16 nucleotides downstream from the transcription initiation site. It is noteworthy that spot 2 consistently contains about two to three times more radioactivity than spot 1, in spite of the fact that spot 2 has only two labeled phosphates as compared to three in spot 1 (see Fig. 8). Based on this observation we conclude that the highest concentration of RNA polymerase molecules on VTC is in the vicinity of the oligonucleotide CCUCCG (281 - 286) of spot 2 (40). This could be the location of an in vivo pause site. If this is an exclusive pause site then the occurrence of RNA polymerase molecules upstream from it could reflect a situation in which two or more RNA polymerase molecules have initiated transcription on the same VTC and they are physically blocked by the enzyme present at the pause site. Alternatively, it is possible that RNA polymerase molecules also pause in vivo at the open promoter complex (70,71).

The precise 3' end of the RNA in the major band (the in vitro attenuation site) maps at nucleotide 336. This conclusion is based on the 1:1 ratio of radioactivity between CMP and UMP in spot 9 (Fig. 7 and see Fig. 8). The RNA in the major band is therefore 94 nucleotides long (nucleotides 243-336). This confirms the above estimation.

SV40 DNA sequences spanning nucleotides 243 and 336 possess dyad symmetries sufficient for the formation of stable hairpin conformation. This conformation is designated "attenuation". The pair of inverted repeat sequences 1+2; 3+4 of the "attenuation" conformation is shown in Fig. 9.

The oligonucleotide of spot 2, which we have suggested above to be the position of a pause site of RNA polymerase molecules in vivo, is at the end of a GC-rich region of dyad symmetry (1+2 in Fig. 9). The major in vitro attenuation site is at the second uridine residue that follows a GC-rich region of dyad symmetry (3+4 in Fig. 9). Based on these observations, we conclude that the eucaryotic RNA polymerase II when transcribing SV40 DNA responds to signals for pausing and attenuation, similar to those in procaryotes (2,72-75).

It has been suggested that in the trp leader region the function of the early (1+2) hairpin is to retard the polymerase in order to permit the translating ribosome to catch up and remain coupled to the transcription apparatus (73,74). This of course does not apply to the SV40 system where transcription and translation are uncoupled processes. However there are several other potential functions for the pause in the first opening of the DNA helix and at the 1+2 hairpin structure. One of them is to help mediate interactions between RNA polymerase and other proteins such as "attenuator" and "antiattenuator" factors. Another function is to allow capping and methylation of the cap at the 5' end of the RNA transcripts. In HeLa cells and adenovirus 2-infected cells, the prematurely terminated transcripts (i.e. attenuated RNAs) are already capped and methylated (31,33) but the RNA at the pause sites has not been analyzed. In this regard it is interesting to note that Lycan and Danna (76) have found in nascent SV40 RNA, in addition to capped ends, some unprocessed pppA ends, and undermethylated caps.

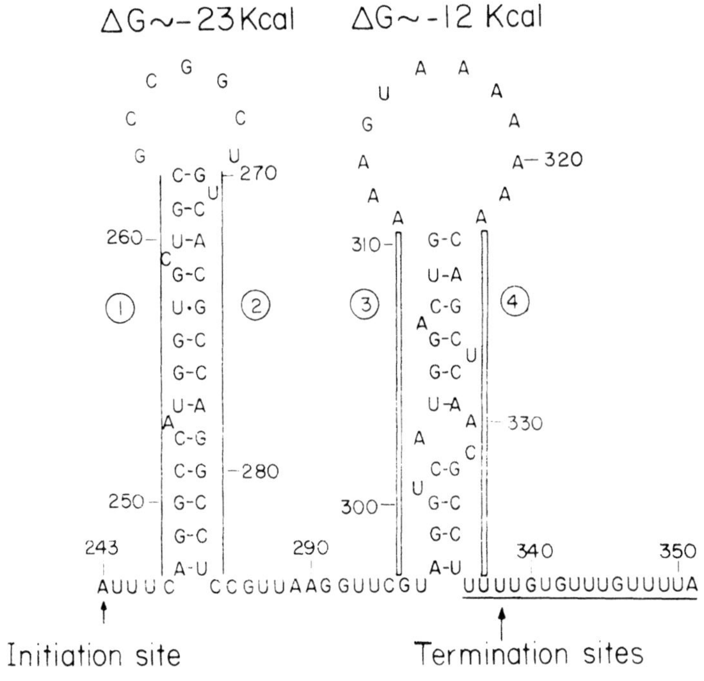

Fig. 9. The "attenuation" conformation of the attenuated RNA. Taken from Hay et al. (37). The ΔG was calculated as described (116). Nucleotide residue number refers to the wild-type SV40 sequence of Reddy et al. (68).

The SV40 minichromosome system. SV40 DNA is found in infected cells in the form of a minichromosome. It possesses a beaded structure composed of cellular histones and supercoiled viral DNA in a molecular complex which is very similar to that of cellular chromatin (79-81). This has made SV40 an attractive model system in which to study the organization and expression of eucaryotic chromatin. The chromatin structure of the 5'-flanking region of several actively transcribed cellular and viral genes is known to be different from the bulk of chromatin. Such regions are generally hypersensitive to DNase I digestion (82-90). Consistent with this, is the observation that ~ 25% of the SV40 minichromosomes display, precisely within the DNase I sensitive region a stretch of DNA not contained within a typical nucleosome structure (92-93). Transcription occurs on SV40 minichromosomes (35,94), and the late promoters map at the exposed region (34,46,95,96). The possible role of the exposed region in directing the specificity of transcription initiation of the late genes is suggested by the observation that E. coli RNA polymerase and the eucaryotic polymerase II initiate transcription primarily within this region in vitro (96,97). It is assumed that the higher order structure of the actively transcribed minichromosomes is involved at other levels of regulation during transcription (89,90). Below, we characterize a system of actively transcribing SV40 minichromosomes and describe the in vitro conditions for premature termination which can be followed by the release of the attenuated RNA (41).

Elongating viral RNA is associated with SV40 minichromosomes while prematurely terminated RNA detaches from the template

To determine the relationship between the viral template responsible for transcription and the RNA made at various salt concentrations, SV40-infected cells were labeled with (^{3}H)thymidine before extraction of the nuclei. The labeled nuclear extract was incubated for synthesis of ^{32}P-labeled RNA either at moderate or at high salts for varying times. The reactions were stopped by the addition of EDTA to 10 mM and the samples were loaded onto sucrose gradients for separation of the major viral nucleoprotein pools.

As shown in Fig. 10, the extracts incubated at moderate salts (100 mM NaCl; 30 mM ammonium sulfate) contained two major peaks of viral nucleoprotein of which the 250S peak comprised mature and immature virions and the 75S peak, the minichromosome pool of replicating and transcribing molecules (91).

RNA synthesis for 5 min revealed two major populations of which one was bound to the 75S structures reflecting attachment of RNA to the minichromosome template and a second population was present at the top of the gradient in a non-bound state. Increasing the time of incubation to 20 min or 60 min led to almost total release of RNA from the 75S peak and to its accumulation at the top of the gradient as a template-free fraction. This was in contrast to the results obtained at high salt concentrations (500 mM NaCl; 300 mM ammonium sulfate) where no major fraction of unbound RNA was obtained. Rather, increasing the time of synthesis led to accumulation of RNA into the minichromosome peak as a species attached to its template. The fraction of RNA running ahead of the minichromosome peak suggests the presence of long RNA chains.

It should be noted, that the minichromosome peak displayed in the sucrose gradient profile at high salt was shifted to a 55S position in the gradient. This, as previously reported, is due to the loss of histone H1 from the molecules which leads to a reduction in the compaction of the minichromosomes (98,99). Minichromosome-associated and unbound RNAs were purified and hybridized to Southern blots of the five restriction fragments shown in Fig. 4. Fig. 11 shows that the RNA associated with the minichromosome after 60 min incubation at moderate salt (100 mM NaCl) hybridized primarily with fragment 'e' and to a lesser extent with fragment 'b' which lies immediately downstream, reflecting poor growth of RNA chains. Labeled RNA found at the top of the gradient, on the contrary, hybridized exclusively with fragment 'e'. The situation is even more pronounced at high salt (300 mM ammonium sulfate), where minichromosome-associated RNA hybridized to the 'late" fragments (b,d and e) as well as to the 'early' fragments (a and c). This result does not discriminate between hybridization to the 'late' or 'early' strands but indicates, on the other hand, distribution of RNA polymerase molecules over almost all the viral genomes. In contrast, RNA found at the top of the gradient hybridized primarily with the promoter-proximal 'e' fragment. This indicates that SV40-specific RNA found detached from the template both after incubation at low and high salt does not originate from random degradation or accidental release of growing RNA chains, but rather points to the existence of a well defined class of RNA molecules prematurely terminated and released from their template.

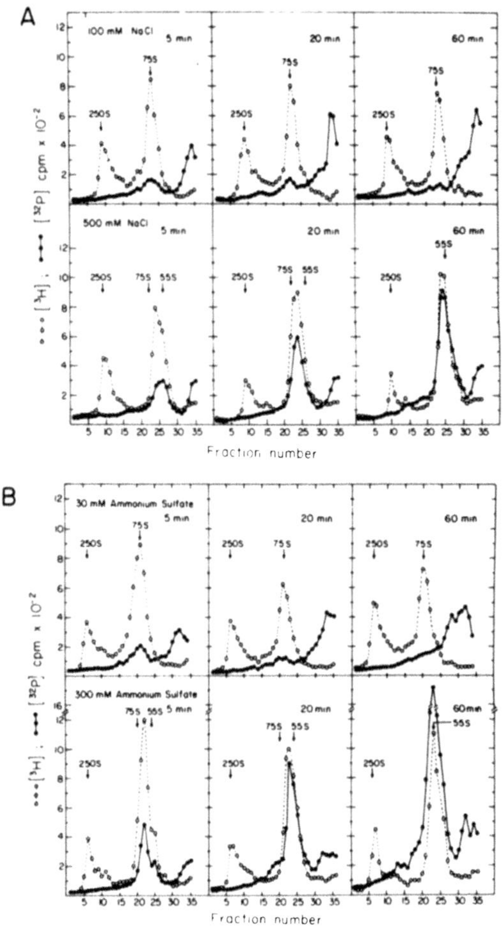

Fig. 10. Sucrose gradient analysis of transcription mixtures programmed by nuclear extracts from SV40-infected cells. Transcription mixtures were adjusted to the desired salt concentration (A-NaCl; B-ammonium sulfate) and supplemented with NTPs, $MnCl_2$ and 100 μCi (α-^{32}P)UTP as described (41). After 5, 20 and 60 min total RNA was prepared, and layered onto precooled 5-30% sucrose gradients supplemented with 200 mM NaCl. After centrifugation for 105 min in a Beckman SW41 rotor at 4°C and 32,000 r.p.m. the gradients were collected from the bottom and the acid-precipitable radioactivity was determined. The fractions containing the minichromosome-associated and free RNA (top of the gradient) were pooled and the RNA extracted for further analysis. For details see (41).

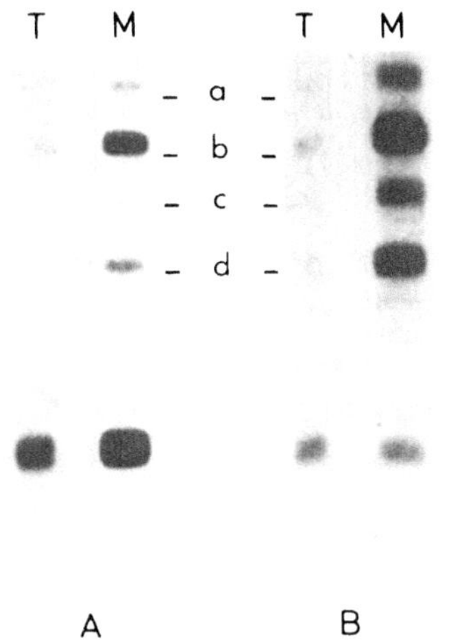

Fig.11. Hybridization pattern of the minichromosome-associated and free RNA to Southern blots of the SV40 DNA restriction fragments depicted in Fig.4. RNA synthesized by nuclear extracts at 100 mM NaCl (A) or 300 mM ammonium sulfate (B). T: free RNA remaining at the top of the gradient; M: minichromosome-associated RNA.

The prematurely terminated RNA can be released from the template as discrete species

To establish the characteristics of the viral RNA released from the template, (^{3}H)thymidine-labeled nucleoproteins were extracted and incubated for 10 min and 60 min at moderate salt concentrations (30 mM ammonium sulfate). The reactions were stopped with 10 mM EDTA and sedimented on a sucrose gradient from which the region of the 75S peak and the top of each gradient were pooled. The labeled RNA was loaded onto a denaturing acrylamide gel. As shown in Fig. 12, the small amount of RNA associated with the 75S peak both at 10 and 60 min was heterogenous in length with no major bands visible. The RNA derived from the released fraction, for both times of incorporation, showed a major band of 94 nucleotides in length as well as other bands. The other bands were not analysed since they were not consistantly observed.

The 94 nucleotide band of RNA was excised from the gel, eluted electrophoretically and hybridized to a Southern blot of SV40 DNA fragments as described in Fig.4. The purified RNA hybridized exclusively to the viral DNA fragment spanning 0.67-0.76 map units (Fig. 12B). To establish the viral origin of the 94 nucleotide band, the above experiments were repeated on sucrose gradient purified minichromosomes followed by the hybridization elution procedure. Fig. 12C Lane A shows that after 60 min incubation at moderate salt concentrations (100 mM NaCl) a major viral RNA band of 93-95 nucleotides is found at the top of the gradient. It is interesting to note the occurrence of an additional reproducible band of ~98 nucleotides present above the major band. A similar band is also found following similar analyses of Sarkosyl extracted VTC or isolated nuclei.

The 94 nucleotide viral RNA band has the same fingerprint as that found for the 94 nucleotide band produced in the VTC transcribing system (see Fig. 6). The viral RNA in the major band spans, therefore, nucleotides 243-336. Thus the actively transcribing minichromosomes provide a system in which the attenuated RNA can be released from the template. The present observation supports our previous suggestion that the eucaryotic RNA polymerase II, like the procaryotic enzyme, recognizes a hairpin structure followed by a stretch of uridylic acid residues, present at the 3' end of the transcript, as a signal of transcription termination. This is the first demonstration of in vitro specific termination of polymerase II transcription that is followed by the release of the RNA transcript (41). That premature termination occurred primarily at a moderate salt concentration could be explained by the nature of nucleic acid interaction. If the mechanism of premature termination is similar to the mechanism of transcription termination in procaryotes we can predict that transcription slows down in the GC-rich region (see Fig. 9). The rU-dA duplex that is subsequently formed is exceptionally unstable at moderate salt concentrations. The unstable interaction could provide a major driving force for termination and release of the transcript (11). Increased salt concentrations result on the other hand, in a partial stabilization of the rU-dA duplex and also stimulate the rate of synthesis and the enzyme then crosses the critical region at an increased rate. Another possibility is that the production and release of the attenuated RNA is dependent upon regulatory factors that are salt sensitive (Hay and Aloni, submitted for publication).

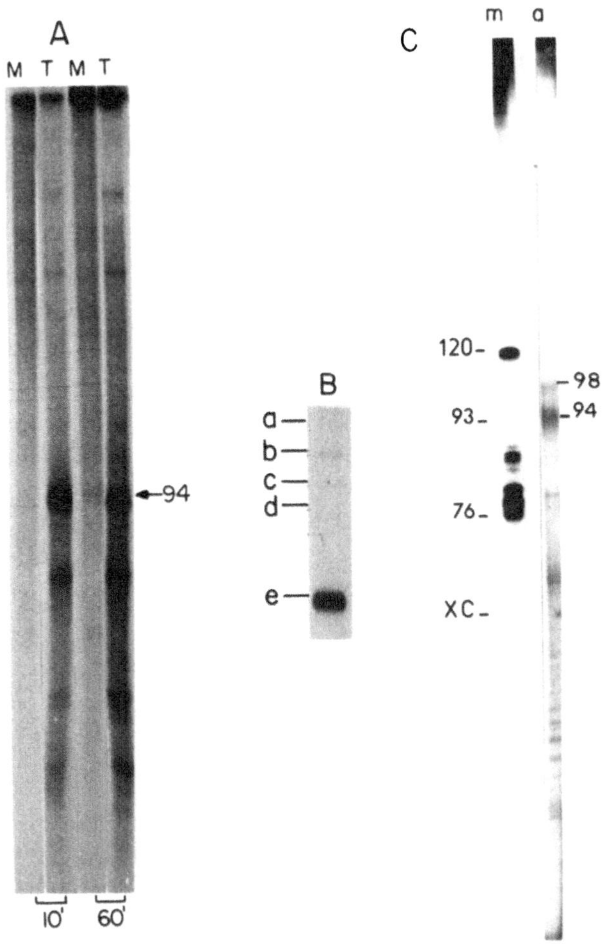

Fig. 12. Acrylamide gel electrophoresis of minichromosome bound and released RNA. Nuclear extracts were incubated for RNA synthesis in the presence of 30 mM ammonium sulfate for 10 or 60 min. The reactions were stopped, the mixtures loaded onto sucrose gradients, and sedimented as in Fig.10. (A): the regions of the 75S minichromosome peaks and the tops of the gradients were pooled, RNA purified and loaded onto the gel. M: minichromosome-bound RNA: T: released RNA. (B) The band of 94 nucleotides was eluted and hybridized to a Southern blot as in Fig. 4. (C) Purified minichromosomes were incubated for RNA synthesis in the presence of 100 mM NaCl for 60 min. The mixture was loaded onto a sucrose gradient and sedimented as in Fig. 10. The top fractions of the gradient were pooled, RNA purified and viral RNA selected by the hybridization elution procedure and loaded onto the gel. a: released viral RNA; m: E. coli t-RNA length markers; xc: xylene-cyanol marker.

The leader protein (agnoprotein) enhances premature termination. The late leader region of SV40 encodes a 7,900 Mr (61 amino acid) protein (see Fig.2). We have postulated that the leader protein in SV40 like the leader peptide in the biosynthetic operons in bacteria has a function in the attenuation mechanism (37,39). As an approach to defining this function we have determined the following: (i) its subcellular distribution in SV40-infected cells and (ii) the correlation between its synthesis, at various times after infection and the level of premature termination using wild type virus and an insertion mutant Δ79 (a generous gift of Dr. G. Khoury). Δ79 is a viable insertion mutant in which two nucleotides were inserted at the Hpa II side (55). Consequently, whereas the RNA secondary structure transcribed from the promoter-proximal region is maintained, this mutant is unable to synthesize an authentic leader protein. The results and conclusions of these experiments are as follows: (i) The predominant localization of the agnoprotein is in the cytosol. However, a minor detectable fraction of agnoprotein can be found associated with the nuclear matrix and in association with the viral minichromosome (Ben-Ze'ev, Skolnik-David and Aloni, unpublished results). An indirect immunofluorescence method with antibodies raised against purified agnoprotein confirmed the biochemical fractionation procedure. In Fig. 13 it can be seen that rabbit antiserum against the agnoprotein showed an intensive granular fluorescence (Fig.13A). The same immune antiserum was non-reactive with mock-infected cells (Fig. 13B). Fig. 13C shows a similar experiment with antibodies raised against SV40 T-Ag. It is evident that the pattern of fluorescence in Fig. 13A indicates the dominant localization to be the perinuclear and cytoplasmic regions with a granular appearance. Detectable fluorescence was also observed in the nucleus. In contrast, Fig. 13C shows an absolute nuclear localization of the SV40 large T-Ag in the same set of cells. Similar results were obtained by Nomura et al. (57). (ii) Fig. 14 shows the time course of appearance of the agnoprotein. It is interesting to note that the synthesis of the structural proteins VP_1, VP_2 and VP_3 is not increased from 42 h to 50 h post infection, rather on the contrary it appears that there is even a somewhat reduced synthesis of these viral proteins. In contrast, the agnoprotein is synthesized at the highest rate at the latest time after infection (50 h). Pulse-chase experiments performed at 50 h post-infection indicate a half life of 2-3 h for the agnoprotein (37,55). The short half life of the agnoprotein and its accumulation at the latest time after infection indicate a potential in vivo regulatory function for the leader protein which is most pronounced at the latest time after infection. Indeed, a direct correlation between the accumulation of the agnoprotein and the production of the 94 nucleotide attenuated RNA in isolated nuclei was observed. Moreover, cells infected with the viable insertion mutant Δ79 showed no agnoprotein and almost no premature termination (Hay and Aloni, submitted for publication). Based on these results we conclude that the function of the agnoprotein is to stabilize the viral RNA conformation that leads to premature termination. In addition, a potential salt-sensitive rho-like factor may mediate the release of the attenuated RNA from its template (N. Hay and Y.Aloni, unpublished results).

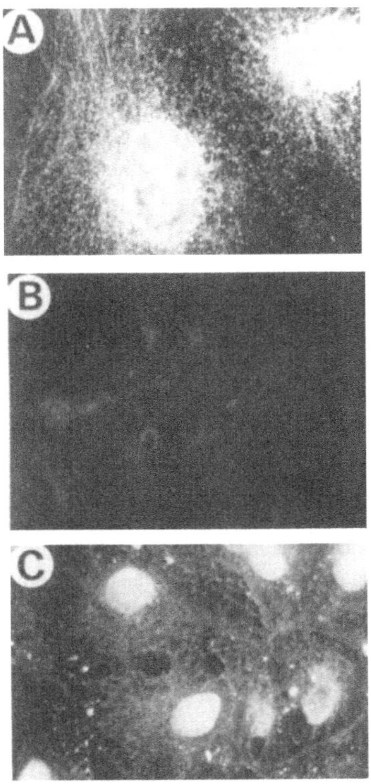

Fig. 13. Indirect immunofluorescent staining of SV40 infected cells with antibodies against agnoprotein and T-antigen. A, SV40 infected BSC-1 cells were stained with antibody against agnoprotein, that was raised by injecting agnoprotein from SDS gels into a rabbit. The second antibody was rhodaminated goat anti rabbit antibody. Before staining, the cells were fixed with 3.7% formaldehyde in PBS and permeabilized with 1% Triton X-100 in PBS for 10 min. B, Uninfected BSC-1 cells stained as above. C, SV40 infected BSC-1 cells as in A, were stained with anti T antibody and rhodaminated goat anti rabbit antibody as in A.

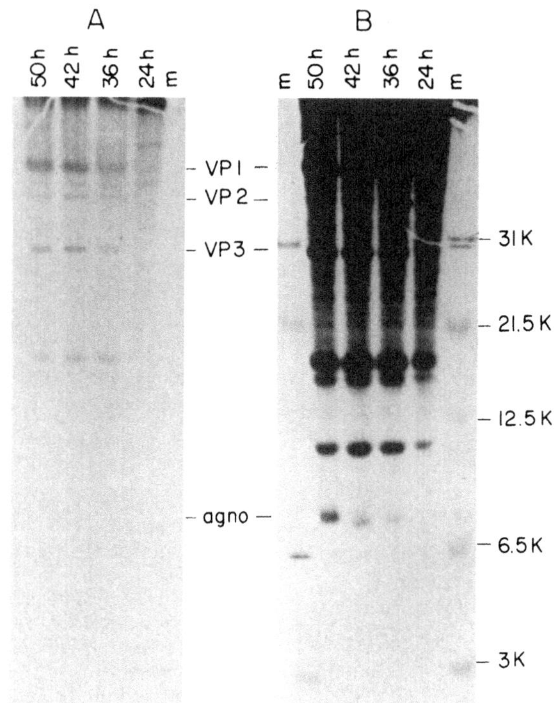

Fig. 14. Time course of the synthesis of the leader protein (agnoprotein) of SV40. BSC-1 cells ($0.5x10^6$ cells) were infected with wt SV40. At various times after infection cells were labeled for 3 h with ^{14}C-arginine, collected in 0.1 ml sample buffer (10 mM sodium phosphate pH 7.2, 7 M urea, 1% SDS, 1% β-mercaptoethanol 0.01% bromophenol blue) lysed and analyzed on a 15% polyacrylamide gel containing 0.1% SDS in 6 M urea. The gel was fluorographed and dried. The autoradiogram was obtained by exposing the dried gel to Kodak XR-2 film at -80°C. m - designate molecular weight markers.

The attenuator RNA is not associated with the nuclear matrix: association of nascent RNA with the nuclear matrix pertains to the anti-attenuation mechanism. The structural framework of the nucleus, known as the nuclear matrix, is obtained when nuclei are depleted of their membranes, soluble molecules and chromatin by subsequent treatments with detergents, nucleases and high salt (117-119). Several reports have demonstrated the association of newly synthesized RNA with the nuclear matrix as well as RNA processing products (120-129), presumably via two proteins of Mr 41,500 and 43,000 (130).

The present study addressed the question of whether the nuclear matrix is actively involved in the attenuation mechanism in SV40 (R.Abulafia, A. Ben-Ze'ev, N. Hay and Y. Aloni, submitted for publication). Isolated nuclei were prepared from SV40 infected cells, pretreated with or without DRB or proflavine, and were incubated in the presence of α-^{32}P-UTP for 5 min. Nuclei, DNase plus salt soluble fractions and nuclear matrices were prepared and ^{32}P-labeled RNAs were extracted. One portion of each ^{32}P-RNA preparation was hybridized to restriction fragments as outlined in Fig.4. The remaining portion of each preparation was used to select the viral RNA by hybridization to, and elution from, SV40 DNA on filters. The purified RNAs were then subjected to gel electrophoresis.

Fig.15a shows that total nuclear viral RNA hybridized with all the restriction fragments, but with a relative enrichment of hybridization with the promoter-proximal fragment e and its adjacent fragment b. The hybridization with fragment e is more pronounced with the RNA derived from the DNase and salt soluble fraction as compared with that of the matrix fraction. These results indicate that the soluble fraction is enriched with the promoter proximal viral RNA species. The gel electrophoresis analyses in Fig.15b show that the enrichment of the soluble fraction with the promoter-proximal viral RNA species is, at least in part, due to the existence of the 94 nucleotide attenuator RNA. The attenuator RNA is found among total nuclear RNA and in the soluble fraction but is not detected in the matrix fraction.

The enrichment in the accumulation of the attenuator RNA is even more pronounced in preparations of ^{32}P-RNA synthesized in nuclei from the DRB pretreated cells. Fig.15a shows that in all three preparations (nuclear, matrix and soluble) there is an almost exclusive signal with fragment e. However, the gel analysis (Fig.15b) shows that the attenuator RNA is detectable only in the nuclear and soluble preparations and not in the matrix fraction. These observations suggest that the matrix bound viral RNA is heterogenous in its length and ir presumably spans the entire length of fragment e. It can also be seen in Figs. 15a and 15b, that when the infected cells are pretreated with proflavine, the in vitro synthesized ^{32}P-labeled RNA does not hybridize with restriction fragment e and there is no detectable 94 nucleotide band on the gel. This is in agreement with our observation indicating that pretreatment of the infected cells with the intercalating drug, proflavine, causes readthrough (37).

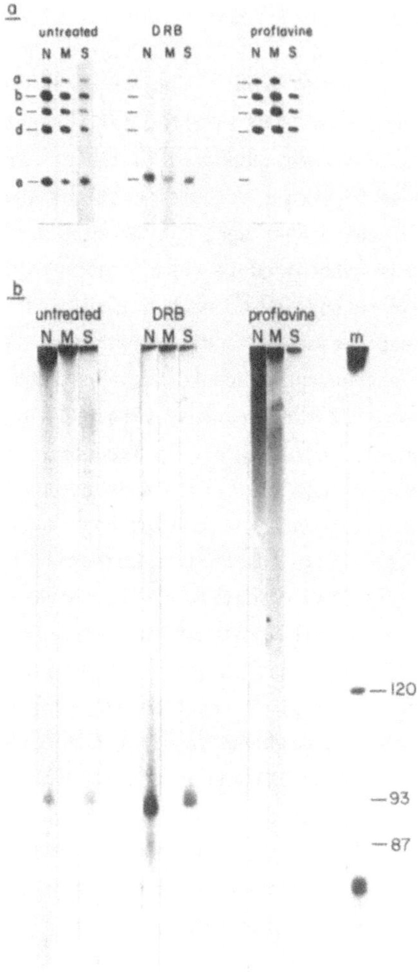

Fig. 15. A 94 nucleotide long SV40 RNA that hybridizes to the SV40 promoter proximal DNA (0.67-0.76 m.u.) is not associated with the nuclear matrix. a. - Isolated nuclei (N) prepared from an equal number of untreated, DRB treated (75 µM, 40 min) or proflavine treated (80 mM, 2 min) SV40 infected cells were incubated with α-^{32}P-UTP as described in Fig.4. The nuclear matrix fraction (M) and a DNase plus salt soluble fraction (S) were prepared and ^{32}P-labeled RNA was extracted and hybridized to the five DNA restriction fragments as described in Fig. 4. b. - An equal portion of each RNA sample obtained as described in a was hybridized to, and eluted from, SV40 DNA filters and the SV40 specific RNA molecules were analyzed by gel electrophoresis in 12% acrylamide -7M urea gels as in Fig. 4. m, ^{32}P-labeled tRNA markers.

We conclude that the attachment of the nascent viral RNA to the nuclear matrix pertains to the antiattenuation mechanism (see Discussion).

Is premature-termination a general mechanism in the regulation of gene expression in animal viruses. As an approach to answer this question we have scrutinized the DNA sequence of promoter-proximal regions of several animal viruses, currently under investigation in our laboratory. These analyses revealed the existence of dyad symmetries in the DNA with the potential of forming stem-and-loop structures in the RNAs that are immediately followed by U-residues. In all these cases alternative RNA conformation are also feasible. Fig. 16 represents these transcription-termination signals in comparison with those of the attenuation sites of the amino acid biosynthetic operons of bacteria. It is apparent that the eucaryotic stem-and-loop structures are less stable than the procaryotic ones and in addition they include unpaired bases. It is conceivable that the unpaired bases provide the recognition specificity for the attenuator and/or antiattenuator factors or for the rho-like termination factor.

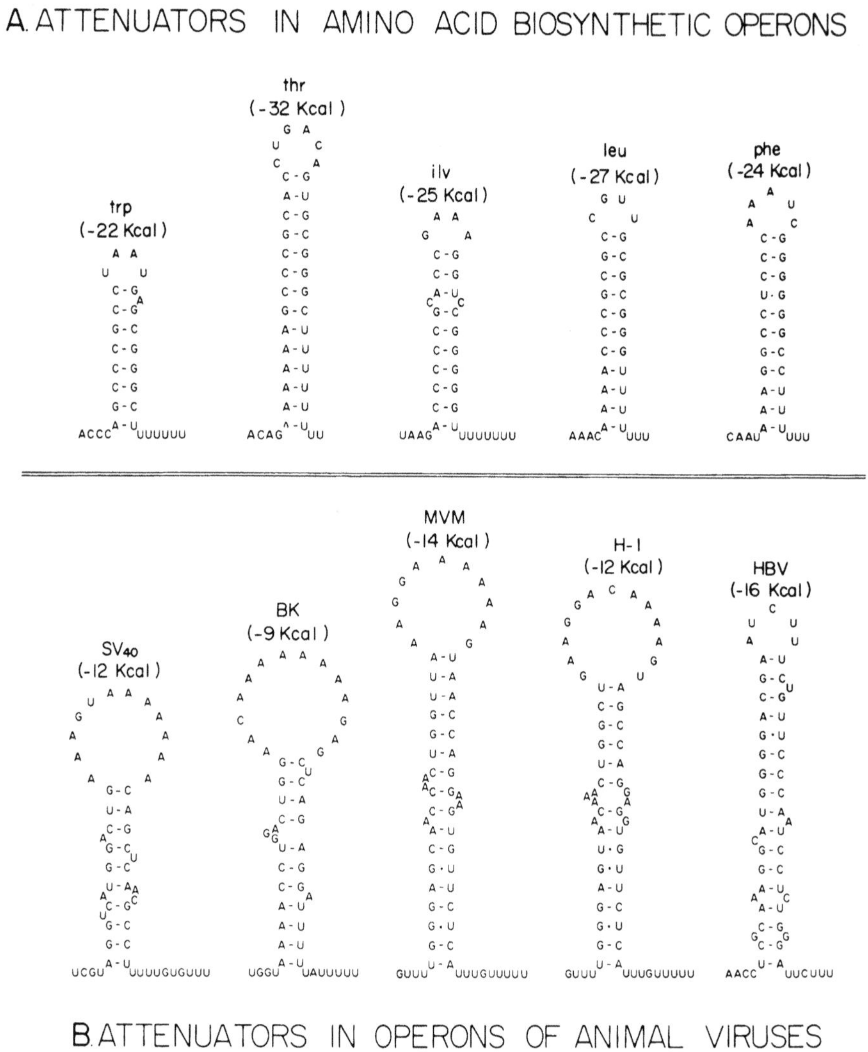

Fig. 16. Comparison between the structure of the attenuators of the amino acid biosynthetic operons of bacteria and those of animal viruses. The structure of the attenuators of the amino acid biosynthetic operons of bacteria were adopted from Kolter and Yanofsky (102). The structure of the attenuators of the animal viruses suggested here are based on the nucleotide sequence of MVM (138), H-1 (139), HBV (140) and BK (141).

DISCUSSION

A model for quantitative regulation of SV40 gene expression by attentuation and nRNA modulation in a feedback control mechanism

We have shown in the Results that viral RNA initiated at nucleotide 243 terminates in vitro 94 nucleotides downstream, at a typical procaryotic transcription termination structure, suggesting that an attenuation mechanism, resembling attenuation in procaryotes, may regulate SV40 late transcription. The same 94 nucleotides comprise the 5' end of a 202 nucleotide leader (nucleotides 243-444) that is spliced to the body (nucleotides 1381-2592) of the 16S mRNA that encodes for the structural protein VP_1 (see Fig.2).

The structural relationship of the 16S leader to the leader sequences of procaryotic mRNAs is striking and raises the possibility that, like them, it also participates in an attenuation mechanism by influencing the expression of its downstream structural gene VP_1. The 16S leader sequence, like those of procaryotes, encodes information for a small protein known as agnoprotein (54) recently discovered in several laboratories (37,55-57 and see Figs. 13 and 14). In in vitro translation systems, the agnoprotein appears to be translated from the 16S poly $(A)^+$ RNA species (37, 104).

We have noticed two pairs of inverted complementary repeat sequences in the attenuated RNA and at the 5' end of the major 16S mRNA (37). These two pairs are designated 1+2 and 3+4 in Fig.17. In the nucleus, conformation 3+4 serves as a typical transcription termination structure that has been implicated in the process of premature termination (2-10,102). When conformation 1+2 is first produced during transcription, the production of the competing conformation (2+3) (Fig.17B) is prevented. Thus the base pairing that occurs between 1 and 2 appears to be most important in the stabilization of the base pairing in the terminator region 3+4. A striking observation is that in conformation 1+2, 3+4, the agnogene AUG and the adjacent 4 nucleotides are sequestered in the stem 1+2, leaving only 5 nucleotides for ribosome binding to the 16S mRNA in the cytoplasm. A second AUG (nucleotides 303-305) is sequestered in the stem of 3+4. We predict that when conformation 1+2, 3+4 prevails at the 5' end of the 16S mRNA, the AUG initiation codon for translation of VP_1 (nucleotides 1423-1426) is the first to be encountered by the ribosome, and thus VP_1 is synthesized. Conversely, when conformation 2+3 prevails at the 5' end of the 16S mRNA, the agnogene AUG is accessible for ribosome binding and the agnoprotein is translated. We speculate that modulation between the

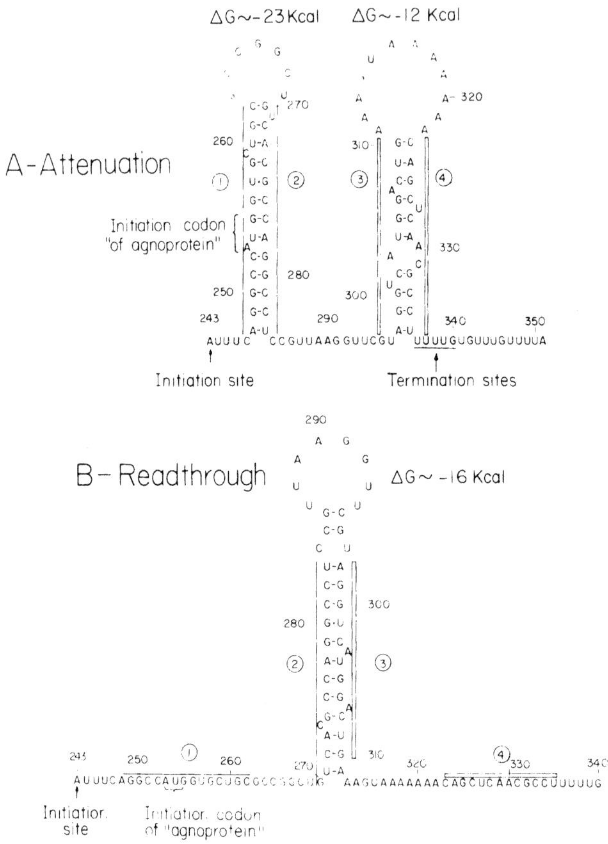

Fig. 17. Schema of alternative conformation in the attenuated RNA and at the 5' end of the 16S mRNA. (A) Attenuation conformation showing sequestration of initiation codon of agnoprotein in the stem 1+2, and a typical termination signal in 3+4. (B) Readthrough conformation in which the initiation codon of agnoprotein is available for ribosome attachment. For details see text. The ΔG were calculated as described (16).

two conformations is a fundamental element in a feedback control mechanism that regulates the amount of 16S mRNA synthesized in the nucleus and the amounts of VP_1 produced in the cytoplasm. For such control at the attenuator site, a deficiency or surplus of VP_1 must be sensed and communicated to the transcribing RNA polymerase molecule. Our prediction is that the concentration of the agnoprotein in the cell is the distinctive feature communicated to the transcribing polymerase in the regulation of termination at the attenuator. According to this suggestion, the amount of agnoprotein in the cell determines, at least in part, the equilibrium between the alternative conformations of the 94 nucleotides at the 5' end of the transcript. The suggestion that the agnoprotein is a molecule that transfers information from the cytoplasm to the nucleus is supported by the observations that it can be isolated from both the cytoplasm and the nucleus (37, 56, 57 and see Fig.13), that it can be found in association with SV40 minichromosomes (37, 57) and that it is a nucleic-acid-binding protein (55,56). An interesting correlation exists between our observation that in vitro attenuation depends on low salt concentrations and readthrough on high salt concentrations (see Fig.10) and the observation of Jay et al. (56) that under the same low salt concentrations the agnoprotein binds to nucleic acids while under high salt concentrations it elutes from them. Moreover, the short half-life (2-3 hrs) of the agnoprotein (37,55) is suitable for that of a regulatory protein.

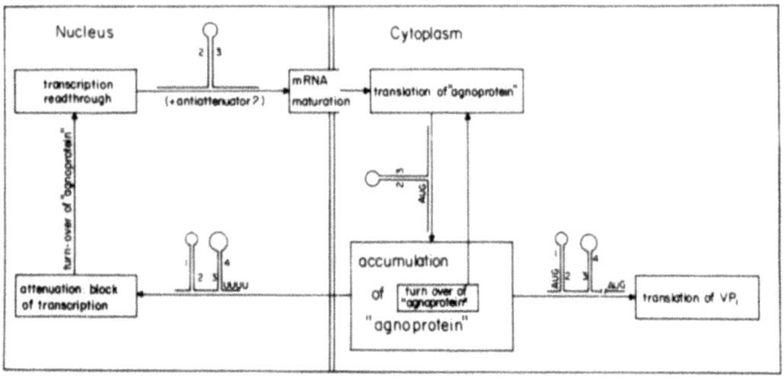

Fig. 18. A model for quantitative regulation of the synthesis of VP_1 by attenuation and mRNA modulation in a feedback control mechanism. Image 1+2, 3+4 is (A) of Fig. 17, and image 2+3 is (B) of Fig. 17. For details see text and (37,39).

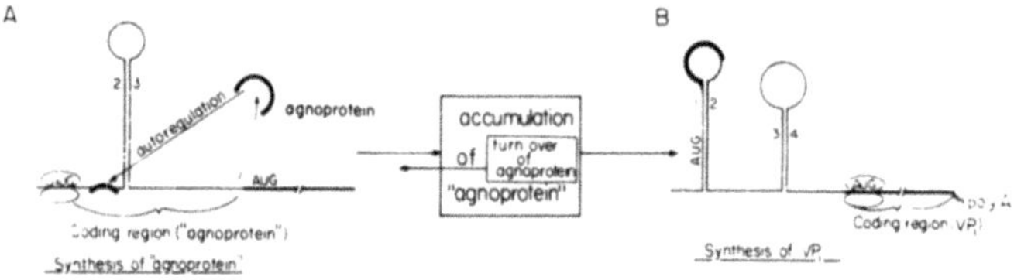

Fig. 19. Agnoprotein regulates the conformations at the 5' end of the 16S mRNA. For details see text and (37,39).

The details of the model are illustrated in Fig. 18 (37). During transcription of the 5' end of the transcripts, initating at nucleotide 243, a certain fraction, arising from the readthrough 2+3 conformation, is synthesized into a complete primary transcript. We speculate that an antiattenuator factor is needed to facilitate the 2+3 conformation. The primary RNA transcripts are processed to mature 16S $poly(A)^+$ mRNA and are transported to the cytoplasm. In this conformation, the accessible agnogene AUG is encountered by the ribosome, and 61 triplets are translated to the agnoprotein. Once a critical concentration of the agnoprotein is attained, it serves as a repressor of its own synthesis by binding to its mRNA and stabilizing conformation 1+2, 3+4 (see Fig.19). As a result of this RNA folding, the ribosome encounters the initiation codon for the translation of VP_1, and the structural protein is synthesized. At the same time, the agnoprotein is transported to the nucleus and stabilizes conformation 1+2, 3+4, leading to premature termination (Fig.18). Consequently, the VP_1 is synthesized in the cytoplasm, transcription in the nucleus is aborted. Reinitiation of the process occurs once the concentration of the agnoprotein is reduced below a critical level.

Quantitative regulation of VP_2 and VP_3 by attenuation and mRNA modulation in a feedback control mechanism

The most abundant group of 19S mRNA (the mRNA of VP_2 is characterized by a gap of 184 nucleotides which extends exclusively from residue 292 to residue 475 (see Fig.3). Thus as a result of the splicing process 152 of the 202 nucleotides of the 16S leader are missing, consequently the 19S mRNA cannot encode for the agnoprotein. Instead, in the leader of 19S mRNA there is an open reading frame for the synthesis of a 29 amino acid protein (103). This protein shares the first 13 amino acids with the agnoprotein and overlaps with the AUG of VP_2 (Fig.20). Hence a small overlapping region of the SV40 genome can potentially code for the synthesis of two different proteins in two reading frames (see Fig. 3) a situation that finds precedent in the SV40 genome in the C terminus of VP_2/VP_3 and the N terminus of VP_1. We have designated this postulated (39,103) but as yet undiscovered protein, $agnoprotein^{29}$ (39). As a result of the splicing process the nucleotides comprising the 3+4 stem-and-loop structure of Fig.17 are also removed. As a result in the 19S mRNA it is not possible to form the same alternative conformations as were suggested for the 16S mRNA in the regulation of the synthesis of VP_1. Since these alternative conformations (see Fig.17 are fundamental elements in our model (37) substitute sequences must exist if the synthesis of VP_2 and VP_3 are regulated by a feedback mechanism similar to that regulating the synthesis of VP_1.

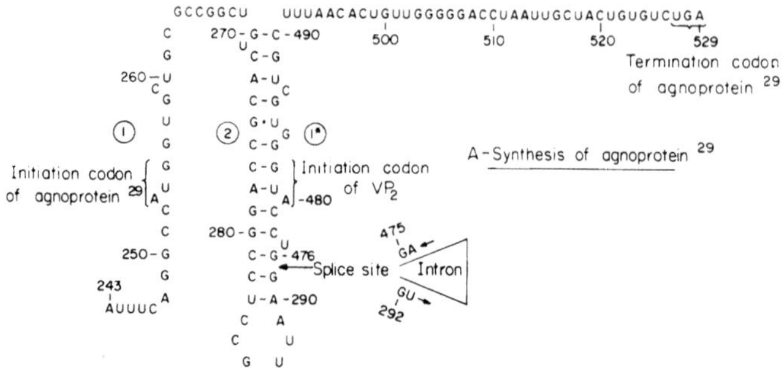

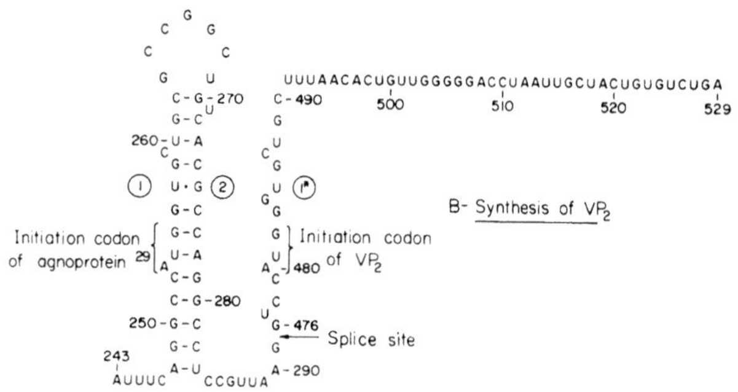

Fig. 20. Schema of alternative conformation at the 5' end of the 19S mRNA. A - conformation showing sequestration of initiation codon of VP_2 while the initiation codon of agnoprotein[29] is available for ribosome attachment. The sequence shows the entire reading frame of agnoprotein[29]. B - Conformation showing sequestration of initiation codon of agnoprotein[29], while the initiation codon of VP_2 is available for ribosome attachment.

Indeed, 17 out of 18 nucleotides preceding and including the AUG of the agnoprotein are identical to the nucleotides preceding the AUG corresponding to the initiation codon for VP_2 (105) (see strand 1 and 1* in Fig. 20). Consequently two alternative conformations can be predicted for the sequences at the 5' end of the 19S mRNA (see Fig. 20). In conformation A, nucleotides in strand 2 base-pair with those in strand 1*. In conformation B, the nucleotides in strand 2 base-pair with those in strand 1. In conformation A, the agnogene[29] AUG is available for ribosome binding and the VP_2 AUG is sequestered in the stem 2+1*. Conversely, in conformation B the agnogene[29] AUG is sequestered in the stem 1+2 and the VP_2 AUG is available for ribosome binding.

Similar to the regulation of synthesis of VP_1 (37) we speculate that modulation between the two alternative conformations is a fundamental element in a feedback control mechanism that regulates the amount of 19S mRNA synthesized in the nucleus and the amount of VP_2 produced in the cytoplasm. At the level of transcription the regulation between attenuation and read-through is based cn the same alternative conformations as suggested for the synthesis of 16S mRNA (see Fig.17). This is because the 16S and 19S mRNAs are spliced from the same primary transcript. That is, differential splicing occurs after the decision to attenuate or not (47). Again, as in our original model the prediction is that the concentration of the agnoprotein, either that of the 16S or 19S or both, is the distinctive factor communicated to the transcribing polymerase in the regulation of termination at the attenuator. According to this suggestion, the amount of agnoprotein in the cell determines, at least in part, the equilibrium between the alternative conformations of the 94 nucleotides at the 5' end of the primary transcript and at the 5' end of the 19S mRNA in the cytoplasm.

Regulation of synthesis of VP_3

Because of the uncertainty about the nature of the VP_3-mRNA (47) it is difficult to speculate whether the synthesis of VP_3 is regulated by a similar feedback mechanism as that which regulates the synthesis of VP_1 and VP_2. The current notion is that VP_3 is synthesized at least in part from a 19S mRNA with a leader that is different from that of the VP_2-mRNA (47). Ghosh et al. (49) suggested that in the 19S species lacking any gap or containing a small gap between residues 444 and 476, base pairing can occur between sequences at and adjacent to the VP_2 AUG and a proximate segment of RNA which is retained in these 19S RNA species (see Fig.21). This base-pairing may make the AUG of the VP_2 translation initiator at residue 480 to 482 inaccessible to ribosomes. .It is interesting to note that in such a situation the AUG of the agnoprotein translation initiator would be sequestered in the 1+2 stem (see Fig.21). Thus as discussed for the regulation of synthesis of VP_1 and VP_2, the stabilization of conformation 1+2 is a prerequisite for the synthesis of VP_3. It is noteworthy that the 19S RNA species lacking any gap or containing a small gap between residues 444 and 476 would encode information for the same 61 amino acid agnoprotein as that encoded in the leader of the 16S mRNA. Since the VP_3 mRNA retains the same sequences involved in the modulation of the secondary structure at the 5' end of the 16S mRNA (see Fig. 17) the production of VP_3 could be regulated by the same feedback mechanism that regulates the synthesis of VP_1. However, an additional base pairing should be formed in order to make the AUG of VP_2 translation initiator inaccessible to ribosomes.

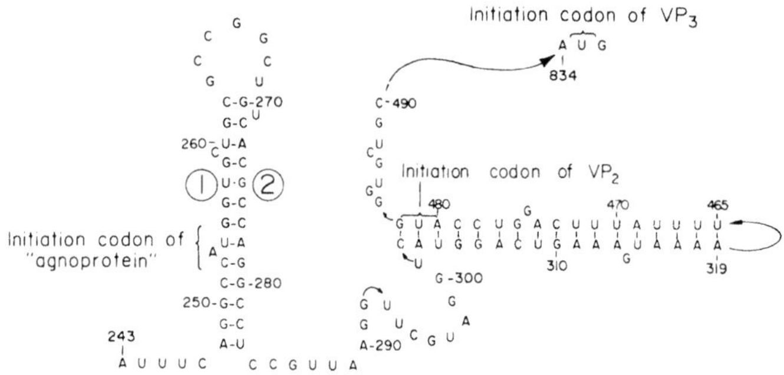

Fig. 21. Schema of a conformation at the 5' end of VP_3 - mRNA. The conformation shows sequestration of initiation codons of agnoprotein and VP_2. The initiation codon of VP_3 is available for ribosome attachment. The base-pairing showing sequestration of initiation codon of VP_2 is from Ghosh et al. (49).

Antiattenuation. We must bear in mind that the models presented here and in our previous communications (37,39), however complex, are still not complete. Thus we can postulate, as in bacteria and their viruses, the existence of a positive control element that suppresses transcription termination, namely a molecule that acts as an antiattenuator leading to the extension of the aborted RNA beyond the termination site. Of the several possible mechanisms suggested previously by which antiattenuation is achieved in procaryotic systems (2-10)we prefer the following two : (i) the antiattenuator interacts with a subunit of the RNA polymerase, enabling it to transcribe through the termination signal; (ii) the anti-attenuator binds to the nascent RNA affecting the secondary structure and preventing transcription termination. This might be accomplished by the movement of the nascent RNA under the action of the antiattenuator into a cellular domain where constraints influence RNA secondary structure formation. Relevant to this proposition is our observation that attenuated RNA is not attached to the nuclear matrix while readthrough transcripts are attached (see Fig.15).

Extension of the model. A speculative extension of the model is that the agnoprotein recognizes its binding sites via an RNA molecule. A possible suggestive candidate is the SV40 associated small RNA (SAS-RNA) (106). SAS-RNA is synthesized from the L(+) DNA strand between nucleotides 2760 and 2825 (62 to 65 nucleotides in length); its synthesis is coupled to the initiation of SV40 late transcription (106). We suggest that the agnoprotein, once it is produced, binds to the SAS-RNA to form a small RNP particle and that this RNP particle recognizes the acceptor sites on the mRNA and at the attenuator region of the primary transcript through base pairing with the SAS-RNA. We have noticed the existence of several homologous sequences between the SAS-RNA and the aborted 94 nucleotide RNA. Thus for example the decanucleotides CCUGGCUGUC present at the 3' end of the SAS-RNA are also found, with the exception of one nucleotide, in the loop of conformation 1+2 (see Fig. 17). An important observation is that the SAS-RNA can also base-pair with sequences in this loop. Thus, the agnoprotein can bind to this loop through the SAS-RNA and stabilize the conformation 1+2, 3+4 of Fig. 17, 1+2, 1* of Fig. 20 and the conformation of Fig. 21. Indeed it has been suggested that a protein may stabilize the stem-and-loop structure by association with DNA or RNA - particularly if there is some specificity of interaction with the loop-containing unpaired nucleotides - and that such an association might be involved in the repression or activation of gene function (107). Other mechanisms for the involvement of the SAS-RNA in the attenuation mechanism are also feasible.

If attenuation is a general mechanism in eucaryotes, then a similar function could be assigned to at least a part of the small RNPs found in the nucleus and cytoplasm of the eucaryotic cell (108). Consistent with this hypothesis are the following observations: (i) low molecular weight RNAs have been found hydrogen-bonded to nuclear and cytoplasmic poly(A)$^+$ RNA from cultured Chinese hamster ovary cells (109). (ii) some of the small RNAs apparently "shuttle" between the nucleus and cytoplasm (110) and (iii) the U_1 and U_2 (D and C) species of small nuclear RNA that are transcribed by polymerase II and derived from transcription units of large size were found to occur in association with ribonucleoprotein particles containing hnRNA (111,112).

Agnoprotein fulfills a role which in the attenuation mechanism in amino acid biosynthetic operons of bacteria is achieved by a successful synthesis of the leader peptide. The similarities between our model and the attenuation model in amino acid biosynthetic operons of bacteria is striking (10,102). If SV40 has inherited the attenuation control mechanism from its procaryotic ancestors then it had to devise mechanisms to overcome the nuclear membrane barrier which uncouples transcription and translation in eucaryotes. It appears that the modification that the eucaryotic virus has made is using the leader protein itself instead of using the translation process of the leader peptide per se (10, 102). In the procaryotic systems successful synthesis of the leader protein and not the final product is the distinctive feature which signals premature termination via stabilization of the 3+4 RNA conformation (10,102). In the eucaryotic system translation has no direct effect on transcription because the two processes occur in separate cellular compartments. Instead, in SV40, the leader protein itself is transported from the cytoplasm ti the nucleus where it might stabilize the 1+2; 3+4 RNA conformation which can lead to transcription termination (see Fig.18). An additional striking similarly between the procaryotic and eucaryotic systems is the location of the pause sites. In both systems it is located at the 1+2 stem-and-loop structure. Fig.22 summarizes the similarities shared by the procaryotic and eucaryotic systems. It should be mentioned that other modifications or supplements that relate to the different natures of the procaryotic and eucaryotic RNA polymerase are conceivable.

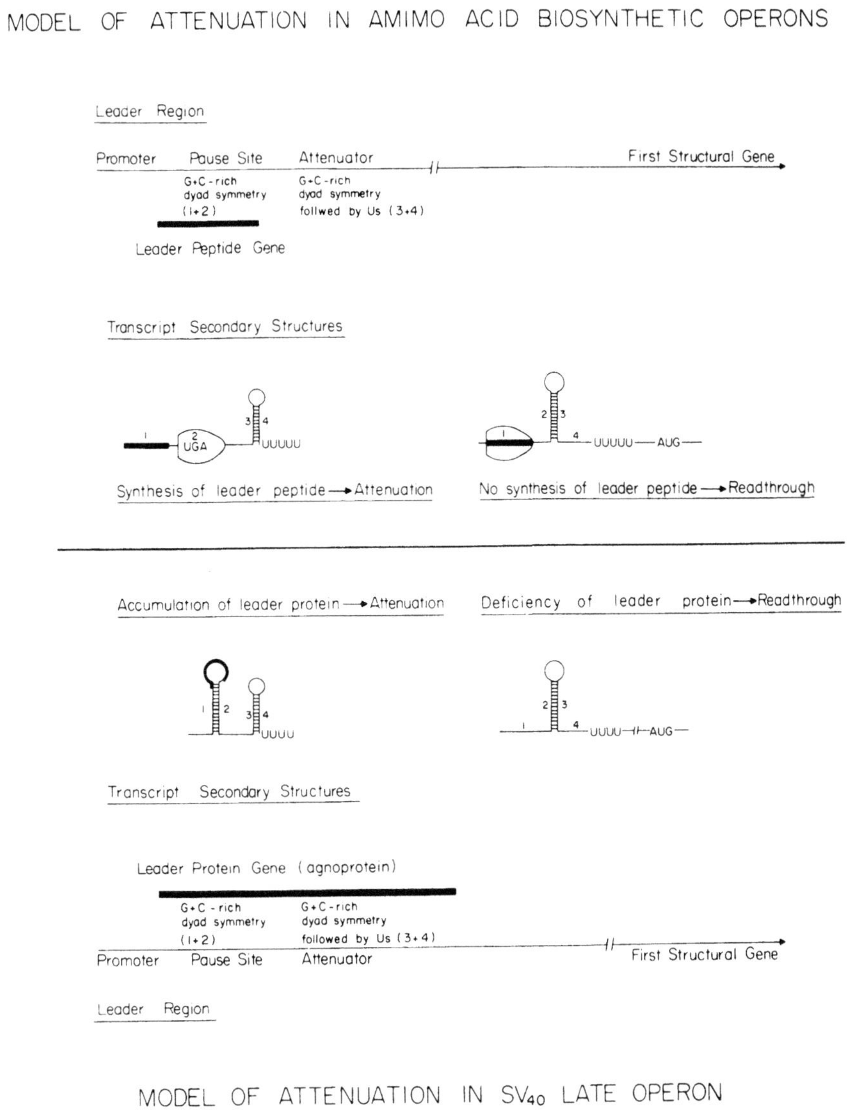

Fig. 22. Similarities shared by the model of attenuation in amino acid biosynthetic operons of bacteria and the model of attenuation in SV40 late operon. The bacterial model was adopted from Kolter and Yanofsky (102). For details see text.

EPILOG

In the long term it should be possible to establish whether attenuation in SV40 and other viruses represents a specialized viral control mechanism or as seems more likely, whether the viruses are mimicking a control mechanism operative in normal eucaryotic cells. The suggested mechanisms described in our model open up approaches to the investigation of attenuation and mRNA modulation as a possible mechanism whereby eucaryotes may regulate transcription in a variety of different circumstances. We hope that testing some of the predictions presented in our model will confirm or rule out particular aspects and in doing so will improve our understanding of how viruses and the eucaryotic cell regulate their gene expression.

ACKNOWLEDGMENTS

We thank Ruchama Leiserowitz for technical assistance. We wish to thank Melvine Stein for expert preparation of the manuscript. This research was supported by USPHS research grant CA 14995 and in part by the Norman V. Wechsler Endowment Fund.

REFERENCES

1. Darnell, J.E. Nature 297:365-371, 1982.
2. Gilbert, W. In: RNA Polymerase (Eds. R. Losick and M. Chamberlin) New York: Cold Spring Harbor Laboratory, 1976, pp. 193-205.
3. Adhya, S. and Gottesman, M. Ann. Rev. Biochem. 47:967-996, 1978.
4. Biro, P.A. and Weissman, S.M. In: Molecular Genetics 3 (Ed. J.H. Taylor) New York Academic Press, 1979, pp. 177-245.
5. Galluppi, G.R. and Richardson, J.P. J. Mol. Biol. 138:513-539, 1980.
6. Gottesman, M.E., Adhya, S. and Das, A. J. Mol. Biol. 140:57-75, 1980.
7. Crawford, I.P. and Stauffer, G.V. Ann. Rev. Biochem. 49:163-195, 1980.
8. Horinouchi, S. and Weisblum, B. Mol. Gen. Genet. 182:341-348, 1982.
9. Platt, T. Cell 24:10-23, 1981.
10. Yanofsky, L. Nature 289:751-758, 1981.
11. Martin, F.H. and Tinoco, L. Nucl. Acids Res. 8:2295-2297, 1980.
12. Celma, M.L., Pan, J. and Weissman, S.M. J. Biol. Chem. 252:9032-9042, 1977.
13. Korn, L.J. and Brown, D. Cell 15:1145-1156, 1978.
14. Bogenhagen, D.F. and Brown, D. Cell 24:200-270, 1981.
15. Brown, D.D. The Harvey Lectures 76 (Academic Press), 1982, pp. 27-44.
16. Jacq, B., Jourdan, R. and Jordan, B.R. J. Mol. Biol. 117:785-795, 1977.
17. Sollner-Webb, B. and Roeder, R. Cell 18:485-499, 1979.
18. Velman, G.M., Klootwijk, S., de Jonge, P., Leer, R. and Planta, R. Nucl. Acids Res. 8:5179-5192, 1980.
19. Din, N., Engberg, J. and Gall, J. Nucl. Acids Res. 10:1503-1513, 1982.
20. Leer, J., Tiryak, D. and Westergaard, O. Proc. Natl. Acad. Sci. U.S.A. 76:5563-5566, 1979.
21. Westergaard, O., Gocke, E. and Leer, J. Nucl. Acids Res. 6:2391-2402, 1979.
22. Kominami, R., Mishima. Y., Urano, Y., Sakai, M. and Muramatsu, M. Nucl. Acids. Res. 10:1963-1979, 1982.
23. Hentschel, C., Irminger, J-C., Bucher, P. and Birnstiel, M. Nature 285:147-151, 1980.
24. Hofer, E., Hofer-Warbinek, R. and Darnell, J.E. Cell 29:887-893, 1982.
25. Fraser, N.W., Nevis, J.R., Ziff, E. and Darnell, J.E. J. Mol. Biol. 129:643-656, 1979.
26. Fraser, N.W. and Hsu, M.-T. Virology 103:514-516, 1980.
27. Birchmeier, C., Grosschedi, R. and Birnstiel, M. Cell 28:739-745, 1982.
28. Stunnenberg, H. and Birnstiel, M. Proc. Natl. Acad. Sci. U.S.A. 79:6201-6204, 1982.
29. Busch, H., Reddy, R., Rothblum, L. and Choi, Y.C. Ann. Rev. Biochem. 51:617-654, 1982.
30. Tamm, I. and Kikuchi, T. Proc. Natl. Acad. Sci. U.S.A. 76: 5750-5754, 1979.
31. Salditt-Georgieff, M., Harpold, M., Chen-Klang, S. and Darnell, J.E. Cell 19:69-78, 1980.
32. Evans, R., Weber, J., Ziff, E. and Darnell, J.E. Nature 278: 367-370, 1979.
33. Tamm, I. and Sehgal, P.B. Effect of Drugs on the Cell Nucleus (Ed. H. Busch) Bristol-Myers Symposium, New York: Academic Press, 1979, pp. 351-274.
34. Laub, O., Bratosin, S., Horowitz, M. and Aloni, Y. Virology 92: 310-322, 1979.

35. Laub, O., Jakobovits, E.B. and Aloni, Y. Proc. Natl. Acad. Sci. U.S.A. 77:3207-3301, 1980.
36. Aloni, Y. In: Protein Biosynthesis in Eucaryotes (Ed. R. Peréz-Bercoff) Plenum Press, 1982, pp. 137-156.
37. Hay, N., Skolnik-David, H. and Aloni, Y. Cell 29:183-193, 1982.
38. Skolnik-David, H., Hay, N. and Aloni, Y. Proc. Natl. Acad. Sci. U.S.A. 79:2743-2747, 1982.
39. Aloni, Y. and Hay, N. Mol. Biol. Rep., In press, 1983.
40. Skolnik-David, H. and Aloni, Y. EMBO J. In press, 1983.
41. Pfeiffer, P., Hay, N., Pruzan, R., Jakobovits, E.B. and Aloni, Y. EMBO J. In press, 1983.
42. Dreyer, C. and Hausen, P. Nucl. Acids Res. 5:3332-3335, 1978.
43. Acheson, N.H. Cell 8:1-12, 1976.
44. Fareed, G. and Davoli, D. Ann. Rev. Biochem. 46:471-552, 1977.
45. Kelly, T.J. and Nathans, D. In: Advances in Virus Research Academic Press Inc. New York, 21:85-173, 1977.
46. Lebowitz, P. and Weissman, S.M. Current Topics in Microbiology and Immunology 87:43-172, 1979.
47. Tooze, J. In: DNA tumor viruses. 2nd ed. Cold Spring Harbor Laboratory 1981, pp. 125-204.
48. Horowitz, M., Laub, O., Bratosin, S. and Aloni, Y. Nature 275: 558-559, 1978.
49. Ghosh, P.K., Reddy, V.B., Swinscoe, J., Lebowtiz, P. and Weissman, S.M. J. Mol. Biol. 126:813-846, 1978.
50. Gidoni, D., Kahana, C., Cannani, D., Groner, Y. Proc. Natl. Acad. Sci. U.S.A. 78:2174-2178, 1981.
51. Aloni, Y. Cold Spring Harbor Sym. Quant. Biol. 39:165-178, 1974.
52. Aloni, Y., Bratosin, S., Dhar, R., Laub, O., Horowitz, M. and Khoury, G. Cold Spring Harbor Sym. Quant. Biol. 43:559-570, 1978.
53. Aloni, Y., Dhar, R., Laub, O., Horowitz, M. and Khoury, C. Proc. Natl. Acad. Sci. U.S.A. 74:3686-3690, 1978.
54. Dhar, R., Subramanian, K.N., Pan, J. and Weissman, S.M. Proc. Natl. Acad.Sci. U.S.A. 74:827-831, 1977.
55. Jay,G., Nomura, S., Anderson, C.W. and Khoury, G. Nature 291: 346-349, 1981.
56. Jackson, V. and Chalkley, R. Proc. Natl. Acad. Sci. U.S.A. 78:6081-6085, 1981.
57. Nomura, S., Khoury, G. and Jay, G. J. Virol. 45:428-433, 1983.
58. Kozak, M. Cell 15:1109-1123, 1978.
59. Kozak, M. In: Protein Biosynthesis in Eukaryotes (Ed. R. Peréz-Bercoff) Plenum Press, 1982, pp. 167-197.
60. Rosenberg, M. and Paterson, B.M. Nature 279:696-701, 1979.
61. Parviakis, G.N., Lockard, P.E., Vanvakopulos, N., Riser, L., Raj Bhandary, U. and Vournakis, J. Cell 19:91-102, 1980.
62. Griffin,B.E. In: DNA Tumor Viruses (Ed. J. Tooze) Cold Spring Harbor Laboratory, 1980, pp. 61-123.
63. Garber, E.A., Seidman, M.M. and Levine, A.J. Virology 107:389-204, 1980.
64. Brady, J.V., Winston, V.D., Conigli, R.A. J. Virol. 27:193-204, 1978.
65. Gariglio, P., Bellard, M., and Chambon, P. Nucl. Acids Res. 9: 2589-2598, 1981.
66. Tweeten, K.A. and Molloy, G. Nucl. Acids Res. 9:3307-3319, 1981.
67. Testa, D., Chanda, P.K. and Banerjee, A.K. Cell 21:261-275, 1980.

68. Reddy, V.B., Thimmappaya, B., Dhar, R., Subramanian, K.N., Zain, B.S., Pan, J., Ghosh, P.K., Celma, M.L. and Weissman, S.M. Science, 200:494-502, 1978.
69. Sanger, F., Brownlee, G.G. and Barrell, B.G. J. Mol. Biol. 13: 373-398, 1965/
70. Saucier, J.M. and Wang, J.C. Nature New Biol. 239:167-170, 1972.
71. Hsieh, T.S. and Wang, J.C. Nucl. Acids Res. 5:3337-3345, 1978.
72. Kingston, R.E. and Chamberlin, M.J. Cell 27:523-531, 1981.
73. Winkler, M.E. and Yanofsky, C. Biochemistry 20:3738-3744, 1981.
74. Farnham, P.J. and Platt, T. Nucl. Acids Res. 9:563-577, 1981.
75. Reisbig, R.R. and Hearst, J.E. Biochemistry 20:1907-1918, 1981.
76. Lycan, D. and Danna, K. J. Virol. 45:264-274, 1983.
77. Andreadis, A., Hsu, Y-P., Kohlhaw, G. and Schimmel, P. Cell 31: 319-325, 1982.
78. Benz, E.W., Wydro, R.M., Nadal-Ginard, B. and Dina, D. Nature 288: 665-669, 1980.
79. Felsenfeld, G. Nature 271:115-122, 1978.
80. Kornberg, R.D.A. Ann Rev. Biochem. 46:931-954, 1977.
81. Chambon, P. Cold Spring Harbor Quant. Biol. 42:1209-1234, 1977.
82. Cremisi, C. Nucl. Acids Res. 9:5949-5964, 1981.
83. Groudine, M., Eisenman, R. and Weintraub, H. Nature 292: 311-317, 1981.
84. Herbormel, P., Saragosti, S., Blangy, D. and Yaniv, M. Cell 25: 651-658, 1981.
85. Keen, M.A., Corces, V., Lowenhaput, K. and Elgin, S.C.R. Proc. Natl. Acad. Sci. U.S.A. 78:143-146, 1981.
86. Scott, W.A. and Wigmore, D.J. Cell 15: 1511-1518, 1978.
87. Stalder, J., Larsen, A., Engle, J.D., Dolan, M., Groudine, M., and Weintraub, H. Cell 20:451-460, 1980.
88. Varshavsky, A.J., Sundin, O. and Bohn, M. Cell 16:453-466, 1979.
89. Mathis, D., Oudet, P. and Chambon, P. Prog. Nucleic Acids Res. Mol. Biol. 24: 1-54, 1980.
90. Weintraub, H. Nucleic Acids Res. 8:4745-4753, 1980.
91. Jakobovits, E.B. and Aloni, Y. Virology 102:107-118, 1980.
92. Jakobovits, E.B., Bratosin, S. and Aloni, Y. Nature 285:263-265, 1980.
93. Saragosti, S., Moyne, G. and Yaniv, M. Cell 20:65-73, 1980.
94. Gariglio, P., Bellard, M. and Chambon, P. Nucl. Acids Res. 9: 2589-2598, 1981.
95. Cannani, D., Kahana, C., Mukamel, A. and Groner, Y. Proc. Natl. Acad. Sci. U.S.A. 76:3078-3082, 1979.
96. Jakobovits, E.B., Saragosti, S., Yaniv, M. and Aloni, Y. Proc. Natl. Acad. Sci. U.S.A. 77:6556-6560, 1980.
97. Beard, P. and Nyffler, K. EMBO J. 1:9-14, 1982.
98. Fey, G. and Hirt, B. Cold Spring Harbor Symp. Quant. Biol. 39:235-241, 1974.
99. Meinke, W., Hall, M.R. and Goldstein, D.A. J. Virol. 15:439-448, 1975.
100. Iverson, L.E., Rose, J.K. Cell 23:477-484, 1981.
101. Remington, J.A. FEBS Letters 100:225-229, 1979.
102. Kolter, R. and Yanofsky, C. Ann. Rev. Genet. 16:113-134, 1982.
103. Piatak, M., Subramanian, K.N., Roy, P. and Weissman,S.M. J. Mol. Biol. 153:589-618, 1981.
104. Mertz, J.E., Murphy, A. and Barkan, A. J. Virol. 45:36-46, 1983.

105. Thimmappaya, B., Reddy, V.B., Dhar, R., Celma, M., Subramanian, K.N., Zain, B.S., Pan, J. and Weissman, S.M. Cold Spring Harbor Symp. Quant. Biol. 42:449-456, 1977.
106. Alwine, J. and Khoury, G. J. Virol. 36:701-708, 1980.
107. Gierer, A. Nature 212:1480-1481, 1966.
108. Ro-Choi, T.S. and Bush, H. In: The cell nucleus (Ed. H. Bush) Academic Press, Vol. 3, 1974, pp. 151-208.
109. Jelinke, W. and Leinwand, L. Cell 15:205-214, 1978.
110. Goldstein, L. and Ko, C. Cell 15:259-]69, 1974.
111. Weinberg, R and Penman, S. J. Mol. Biol. 28:289-304, 1968.
112. Zieve, G. and Penman, S. J. Mol. Biol. 145:501-523, 1976.
113. Laub, O. and Aloni, Y. Virology 75:346-354, 1976.
114. Maniatis, R. and Efstratiadis, A. Meth. Enzymol. 65:299-305, 1980.
115. Smith, S.O. Meth. Enzymol. 65:371-380, 1980.
116. Tinoco, I., Borer, P.N., Dengler, B., Levine, M.D., Uhlenbeck, O., Crothers, D. and Gralia, J. Nature New Biol. 246:40-41, 1973.
117. Shaper, J.H., Pardoll, D.M., Kaufmann, S.A. Barrack, E.R., Vogelstein, B. and Coffey, D. Adv. Anz. Regul. 17:213-248, 1979.
118. Berezney, R. In: The cell nucleus (Ed. H. Busch), 1979, pp. 413-455.
119. Agutter, P.S. and Richardson, J.C. J. Cell Sci. 44:395-435, 1980.
120. Faiferman, J. and Pogo, A.O. Biochemistry 14:3808-3816, 1975.
121. Herman, R., Weymouth, L. and Penman, S. J. Cell Biol. 78:663-674, 1978.
122. Moller, T., Huang, C.Y. and Pogo, A.O. J. Cell Biol. 76:675-691, 1978.
123. Herlan, G., Eckert, W.A., Kaffenberger, W. and Wunderlich, F. Biochemistry 18:1782-1788, 1979.
124. Long, B.H., Huang, C.Y. and Pogo, A.O. Cell 18:1079-1090, 1979.
125. Mariman, E.C.M., Van Eekelen, C.A.G., Reinders, R.J., Berns, A.J.M. and Van Venrooij, W.J. J. Mol. Biol. 154:103-119, 1982.
126. Ross, D., Yen, R.W. and Chae, C.B. Biochemistry, 21:764-771, 1982.
127. Ciezek, E.M., Nordstrom, J.L., Tsai, M.J. and O'Malley, B.W. Biochemistry 21:4945-4953, 1982.
128. Ben-Ze'ev, A., Abulafia, R. and Aloni, Y. EMBO J. 1:1225-1231, 1982.
129. Ben-Ze'ev, A. and Aloni, Y. Virology. In press, 1983.
130. Van Eekelen, C.A.G. and Van Venrooij, W.J. J. Cell Biol. 88:554-563, 1981.
131. Chow, L.T., Broker, T.R., Lewis, J.B. J. Mol. Biol. 134:265-303, 1979.
132. Shaw, A.R. and Ziff, E. Cell 22:905-916, 1980.
133. Nevis, J.R. and Wilson, M.C. Nature 290:113-118, 1981.
134. Akusharvi, G. and Pearson, H. Nature 292:420-426, 1981.
135. Logan, S.J. and Shenk, T. Microbiological Reviews 46:377-383, 1982.
136. Tamm, I. Proc. Natl. Acad. Sci. U.S.A. 74:5011-5015, 1977.
137. Greenblatt, J. Cell 24:8-9, 1981.
138. Astell, C.R., Thomson, M., Merchlinsky, M. and Ward, D.C. Nucl. Acids Res. In press, 1983.
139. Rhode, S.L. and Paradiso, P.R. J. Virol. 45, 173-184, 1983.
140. Valenzuela, P., Quiroga, M., Zaldivar, J., Gray, P., and Rutter, W.J. In: Animal Virus Genetics (Eds. B.N. Fields, R. Jaenisch and C.F. Fox) Academic Press, New York, 1980, pp. 56-70.
141. Tooze, J. In: DNA tumor viruses. 2nd ed. Cold Spring Harbor Laboratory 1981, pp. 915-922.

2

THE USE OF REASSORTANT BUNYAVIRUSES TO DEDUCE THEIR CODING AND PATHOGENIC POTENTIALS

D.H.L. BISHOP, F. FULLER, H. AKASHI, B. BEATY and R.E. SHOPE

I. INTRODUCTION

The Bunyaviridae is a large family of arthropod-borne RNA viruses (1) that has members that are pathogenic for man (e.g. La Crosse, LAC, virus) and, or domestic animals (e.g. Rift Valley fever virus), or are of no apparent impact to man's economy (e.g. snowshoe hare, SSH, virus). The familial characteristics of the member viruses include a genome consisting of 3 negative sense RNA species designated on the basis of size as L (large), M (medium) and S (small). The viruses have an envelope apparently derived from the Golgi membranes of infected cells since the viruses mature by budding into the Golgi cisternae. This latter characteristic is unique to this family of viruses and poses interesting, as yet poorly understood, morphogenetic questions. The outer surface of the virus envelope has projections involving 2 glycoproteins, designated G1 and G2. The viral RNA species are encapsidated by nucleoprotein, N and appear to be associated with small quantities of a large protein (L protein) that may be a transcriptase component. A cartoon of a schematic bunyavirus is given in Fig. 1.

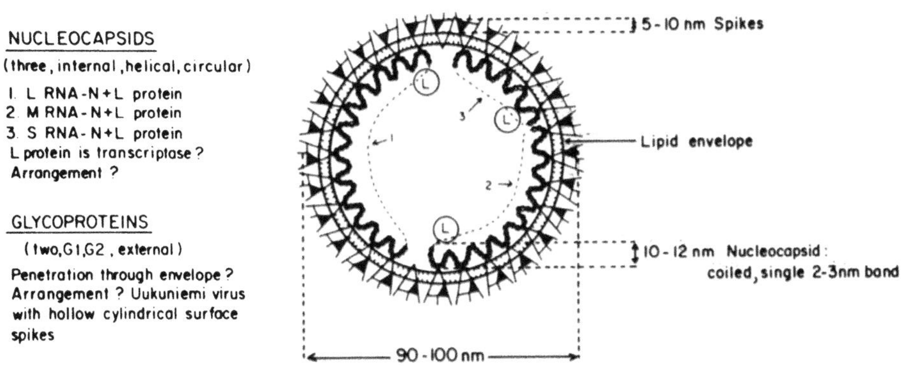

FIGURE 1. A schematic representation of a bunyavirus particle.

Four genera of Bunyaviridae have been defined (Bunyavirus, Phlebovirus, Nairovirus, Uukuvirus) (1). A listing of the accepted and proposed members of the Bunyavirus genus is given in Table 1, updated from an earlier report (2). Based on arthropod isolation studies, most bunyaviruses are believed to be transmitted by mosquitoes. For La Crosse (LAC) bunyavirus transovarial transmission of the virus in the mosquito vector not only allows viruses to be passed from generation to generation, but also provides an overwintering mechanism (2). Most nairoviruses and uukuviruses are transmitted by ticks; phleboviruses are transmitted by sandflies and, or mosquitoes. Some members of the family (e.g., Akabane virus) are transmitted by Culcoides.

Table 1. The Bunyavirus genus (2, 3)

Proposed Serological Classification of Viruses of Family Bunyaviridae, genus Bunyavirus[a].

Anopheles A Group
- Anopheles A
  - CoAr 3624[b]
  - ColAn 57389[b]
  - Las Maloyas[b]
  - Lukuni
  - Trombetas[b]
- Tacaiuma
    - H-32580[b]
    - SPAr 2317[b] (Virgin River[b])
  - CoAr 1071[b] (CoAr 3627[b])

Anopheles B Group
- Anopheles B
  - Boraceia

Bunyamwera Group
- Bunyamwera
  - Germiston
  - Shokwe[b]
  - Batai (Calovo)
  - Ilesha
  - Birao
  - Tensaw
  - Cache Valley (Tlacotalpan)
    - Maguari (CbaAr 426[b])
    - Playas[b]
    - Xingu[b]
  - Northway
  - Santa Rosa
  - Lokern
- Wyeomyia
  - Taiassui
  - Anhembi (BeAr 314206[b], BeAr 328208[b])
  - Sororoca
  - Macaua[b]
- Main Drain
- Guaroa
- Kairi

Bwamba Group
- Bwamba
    - Pongola

C Group
- Caraparu
    - Caraparu (BeH5546[b])
    - Caraparu/Trinidad[b]
    - Ossa
  - Apeu
  - Vinces[b]
  - Bruconha[b]
- Madrid
- Marituba
    - Murutucu
    - Restan
  - Nepuyo (63U11[b])
    - Gumbo Limbo
- Oriboca
    - Itaqui

California Group
- California encephalitis
    - Tahyna (Lumbo[b])
    - Inkoo
    - San Angelo
    - La Crosse (snowshoe hare)
  - Melao
    - Serra do Navio
    - Keystone
    - Jamestown Canyon (South River[b])
    - Jerry Slough
  - trivittatus

Capim Group
- Capim
- Guajara (GU71U350[b])
- BushBush
    - Benfica
    - GU71U344[b]
  - Juan Diaz
- Acara
  - Moriche
- Benevides

Gamboa Group
- Gamboa
  - Pueblo Viejo[b] (75-2621[b])
- Alajuela[b]
  - San Juan (78V2441[b], 75V-2374[b])

Guama Group
- Guama
  - Ananindeua
  - Moju
  - Mahogany Hammock
- Bimiti
- Timboteua
- Catu
- Bertioga
  - Cananeia[b]
  - Guaratuba
  - Itimirim[b]
  - Mirim

Koongol Group
- Koongol
  - Wongal

Minatitlan Group
- Minatitlan
  - Palestina[b]

Olifantsvlei Group
- Olifantsvlei (Bobia)
- Botambi

Patois Group
- Patois
  - Abras[b]
  - Babahoyo[b]
  - Shark River
- Zegla
  - Pahayokee

Tete Group
- Tete
    - Bahig
    - Matruh
    - Tsuruse
- Batama

Simbu Group
- Akabane
  - Yaba-7[b]
- Shamonda
  - Sango
    - Peaton
- Sabo
- Sathuperi
- Shuni
  - Aino (Kaikalur, Samford)
- Simbu
- Thimiri
- Nola
- Manzanilla
    - Ingwavuma
    - Mermet
    - Inini
  - Buttonwillow
- Oropouche
  - Utinga
  - Utive[b]
  - Facey's Paddock[b]

Turlock Group
- Turlock
  - Umbre
  - Lednice
- M'Poko
    - Yaba-1[b]
- Barmah Forest

Unassigned Members
- Jurona
- Kaeng Khoi
- Para[b]

[a]Viruses are classified in three steps indicated by degrees of indentation -- complex, virus, and subtype; viruses in parentheses are varieties.
[b]These viruses are not in the published or working International Catalogue of Arboviruses (Berge, 1975; Karabatsos, 1978).

2. CODING ANALYSES OF CAL GROUP BUNYAVIRUSES

2.1. The formation of recombinant bunyaviruses by RNA segment reassortment

Previous studies have documented the limited ability of California (CAL) serogroup bunyaviruses to form recombinant viruses by RNA segment reassortment (4-6). Thus, from dual virus infections with selected temperature-sensitive (ts) mutants of LAC and snowshoe hare (SSH), or trivitattus (TVT), or Tahyna (TAH), or Lumbo (LUM) viruses, wild-type progeny have been recovered and their reassortant L/M/S RNA genotypes (e.g. SSH/LAC/SSH) characterized by RNA fingerprinting. So far no reassortants have been recovered from crosses involving Bunyamwera and CAL group viruses, or CAL and Group C bunyaviruses, or bunyaviruses and members of other genera of the family (unpublished data). Genotype analyses of natural LAC virus isolates by fingerprinting indicate that although reassortment occurs in nature it is an infrequent event (7).

2.2. The RNA-protein coding assignments of bunyaviruses as deduced from analyses of parent and reassortant CAL group bunyaviruses

Analyses of the genotypes and structural polyproteins of parental and reassortant CAL group viruses have established that the bunyavirus S RNA codes for N protein and a non-structural protein designated NS_S (8,9). It has also been shown that the M RNA codes for the two viral glycoproteins, G1 and G2, and a second non-structural protein, NS_M (9,10). An example of the use of parental and reassortant CAL group bunyaviruses to deduce the coding assignments of the viral N, NS_S and NS_M proteins is given in Fig. 2.

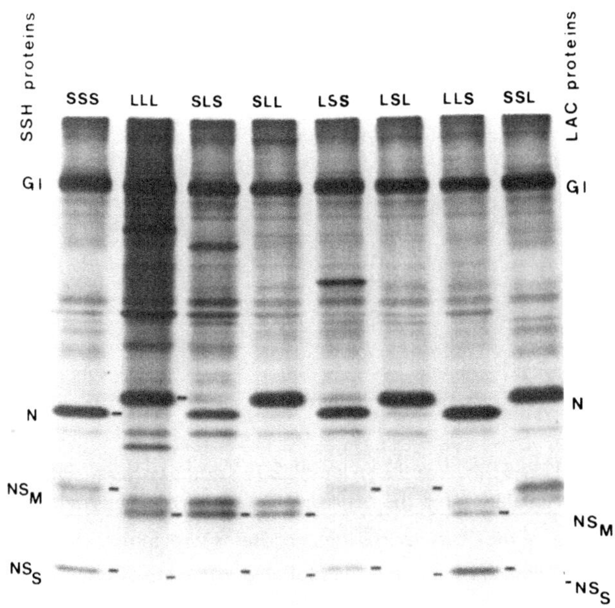

FIGURE 2. The viral proteins induced by LAC (L/L/L), SSH (S/S/S) and six LAC-SSH reassortant viruses. The L/M/S genotypes of the infecting viruses are indicated above the respective lanes of virus induced proteins.

Due to the small quantities of available L protein, it has not been possible to unequivocably prove that the bunyavirus L RNA codes for L protein, however, from their respective sizes, the conclusion appears reasonable. It is not known if the L RNA also codes for non-structural proteins.

2.3. Evidence that the bunyavirus M RNA gene products (the glycoproteins) interact with neutralizing antibodies

It has been shown from analyses of parental and reassortant CAL group viruses that the viral glycoproteins contain the type specific determinants and interact with neutralizing antibodies (11). An example of the data that allowed this conclusion to be drawn is provided in Table 2.

Table 2. Plaque reduction neutralization test reactions of prototype and reassortant CAL group viruses (11).

Virus	Reciprocal of antibody dilution giving 50% reduction in PFU	
	LAC antiserum	TAH antiserum
LAC/LAC/LAC	1280	10
TAH/LAC/TAH	1280	20
TAH/LAC/LAC	640	0
TAH/TAH/TAH	160	1280
LAC/TAH/TAH	320	640
LAC/TAH/LAC	320	640

2.4. Evidence that the bunyavirus M RNA gene products (the glycoproteins) are major determinants of viral virulence for the neurotropic CAL group bunyaviruses

It has been shown that bunyavirus glycoproteins are major determinants of CAL virus virulence (12). Studies of the mortalities of mice following intracerbral or intraperitoneal inoculation of parent, or reassortant CAL group viruses (Table 3), plus analyses of the average survival times after inoculation of lethal doses of the same viruses (Table 4), have established that the principal determinants of virulence are the M RNA gene products. Since the CAL group viruses are neurotropic, it appears that neurotropism is a property determined by the bunyavirus glycoproteins.

Table 3. Intracerebral (ic) and intraperitoneal (ip) virulence in mice of parent and reassortant CAL group viruses (12).

Virus	log LD_{50} per ml. ic inoculation	ip inoculation
LAC/LAC/LAC	7.7	4.4
TAH/LAC/TAH	7.7	4.8
TAH/LAC/LAC	7.7	3.0
TAH/TAH/TAH	9.0	0
LAC/TAH/TAH	8.5	0
LAC/TAH/LAC	7.1	0

Values of 0 represent less than 1.6 log LD_{50} per ml of inoculum.

Table 4. Survival of mice inoculated intracerebrally with lethal doses of parent and reassortant CAL group viruses (12).

Virus	Mean survival $\pm$ SEM 1.0-1.9 log LD_{50}	3.0-3.9 log LD_{50}
LAC/LAC/LAC	5.4 ± 0.6	3.4 ± 0.2
TAH/LAC/TAH	6.0 ± 1.0	3.9 ± 0.3
TAH/LAC/LAC	6.8 ± 0.4	4.0 ± 0.4
TAH/TAH/TAH	2.6 ± 0.2	1.8 ± 0.2
LAC/TAH/TAH	3.6 ± 0.4	2.8 ± 0.4
LAC/TAH/LAC	2.2 ± 0.4	1.6 ± 0.2

2.5. Evidence that the bunyavirus M RNA gene products (the glycoproteins) are major determinants of the success of virus infection and transmission by arthropod vectors.

In the upper midwestern regions of the continental United States LAC and SSH viruses are sympatric in their distribution (2,3). However the two viruses have different veterbrate host preferences (snowshoe hares for SSH virus, chipmunks and tree squirrels for LAC virus), and different preferred arthropod vectors. Most LAC virus isolates have been recovered from *Aedes triseriatus* mosquitoes (2); SSH virus isolates have come from various *Aedes* species e.g., *Aedes canadensis*, *Aedes cinereus*, *Aedes communis* (3). In studies involving measuring the infection rates of colonized *Aedes triseriatus* by LAC, SSH and SSH-LAC reassortants, it has been shown (Table 5) that after oral ingestion both SSH and LAC viruses are equally successful in establishing midgut infections. However subsequent infections of other tissues by SSH virus, or SSH-LAC reassortants with a SSH M RNA species, are less successful than for LAC virus, or SSH-LAC reassortants with a LAC M RNA (13).

Table 5. Infection of *Aedes triseriatus* mosquitoes by SSH, LAC or SSH-LAC reassortants following per os introduction of virus (13).

Virus	Midgut infection		Disseminated infection	
	attempts	Positive	attempts	Positive
LAC/LAC/LAC	45	44 (98%)	44	44 (100%)
SSH/LAC/SSH	40	39 (98%)	39	38 (97%)
SSH/LAC/LAC	32	32 (100%)	32	31 (97%)
SSH/SSH/SSH	26	23 (89%)	23	4 (17%)
SSH/SSH/LAC	40	38 (95%)	38	11 (29%)
LAC/SSH/SSH	13	12 (92%)	12	1 (8%)
LAC/SSH/LAC	21	19 (90%)	19	8 (42%)

The infection attempts that were scored positive were determined from the demonstration of viral antigens in midgut or head squash tissues (disseminated infection).

It has also been shown that the ability of infected Aedes triseriatus mosquitoes to successfully transmit virus to newborn mice is influenced by the origin of the viral M RNA species (14). Again, LAC virus and SSH-LAC reassortants with a LAC M RNA species were found to be more frequently transmitted by these mosquitoes than SSH virus, or SSH-LAC reassortants with a SSH M RNA (Table 6). Without colonized mosquitoes representing the preferred vectors of SSH virus it has not been possible to perform reciprocal experiments to determine if the demonstrable infection and transmission preferences associated with possession of a LAC M RNA are active in alternate mosquito species.

Table 6. Transmission of SSH, LAC and SSH-LAC reassortants by Aedes triseriatus mosquitoes (14).

Virus	No. transmissions	No. attempts	% Transmitted
LAC/LAC/LAC	60	60	100%
SSH/LAC/SSH	35	39	90%
SSH/LAC/LAC	22	23	96%
SSH/SSH/SSH	12	36	33%
SSH/SSH/LAC	11	35	31%
LAC/SSH/SSH	5	14	36%
LAC/SSH/LAC	8	19	42%

Transmissions were documented by the death of newborn mice on which infected (i.e., viral antigen positive) mosquitoes were allowed to feed.

In summary, it has been shown that the M RNA gene products of CAL group bunyaviruses are the principal determinants of virus virulence and neurotopism in model animal systems. Also they are important in determining the success of viral infection and transmission in selected arthropod vectors. Although it has been shown that the M RNA gene products include the 2 viral glycoproteins, G1 and G2, as well as a small non-structural protein, NS_M, it is not known whether the epitopes that interact with neutralizing antibodies are located on the G1 protein or the G2 protein. Nor is it known where the receptor binding sites reside. Further studies using monoclonal antibodies will be required to resolve these issues.

3. GENETIC AND BIOLOGIC PROPERTIES OF THE GROUP C BUNYAVIRUSES

Many of the Group C bunyaviruses (Table 1) have been isolated from man as well as from culicine mosquitoes. These include Apeu, Caraparu (CAR), Itaqui, Madrid, Marituba, Murutucu, Nepuyo, Oriboca, Ossa, and Restan viruses (2). Several cause undifferentiated systemic human febrile illnesses. Unlike the CAL group virus infections these Group C viruses are not associated with human CNS disease. Several of the Group C bunyaviruses have been recovered from rodents and, or marsupials collected in North, Central, or South America, indicating that these small animals plus mosquitoes constitute the primary virus cycle in nature. Human sera surveys have shown that in some regions of the world Group C viruses frequently infect the local human population.

In model animal systems certain Group C viruses are viscerotropic, alternate isolates, or other Group C virus serotypes, or subtypes, are neurotropic. For example, the viscerotropic CAR(v) BeAn 3994 kills 3-day-old baby, or 4-week-old, Swiss outbred mice within 2-3 days following either an ip, or ic, route of inoculation (Tables 7, 8). The neurotropic CAR(n) BeAn 7981 virus isolate kills 3-day-old mice more slowly after inoculation by the ic route (Table 8) and 4-week-old mice ic but at a lower titer than 3-day-old mice (Table 7). The CAR(n) BeAn 7981 virus does not kill 4-week-old mice following ip inoculation (Table 7).

Table 7. Ic and ip titration of viscerotropic and neurotropic CAR virus isolates.

Virus	Titer (log LD_{50} per ml virus stock) Adult mice		3-day-old mice
	ic	ip	ic
CAR (v) BeAn 3994	8.7	$\geqslant$5.5	7.7
CAR (n) BeAn 7981	6.0	$\leqslant$1.5	7.4

Table 8. Average survival time (AST) of 3-day-old mice inoculated ic with viscerotropic or neurotropic CAR virus isolates.

Virus	AST in days as function of dose (log adult mouse ic LD_{50})			
	($\geqslant$5.0)	(2.3-3.0)	(0.3-1.0)	($\leqslant$0.3)
CAR (v) BeAn 3994	2.0	2.4	3.0	-
CAR (n) BeAn 7981	-	3.4	4.3	6.5

In addition to hepatic necrosis that occurs in mice infected with the viscerotropic strain of CAR virus, liver viral infectivity titers are higher than in brain tissues (unpublished data). The neurotropic CAR strain does not induce hepatic necrosis and virus titers are higher in brain tissues than in the liver. Similarly, complement fixing and hemagglutinating antigens of CAR(v) are found in the liver of infected mice, while for mice infected with the CAR(n) strain such antigens are found in the brain.

3.1 Isolation of intertypic reassortant CAR Group C bunyaviruses.

Analyses of the L, M and S RNA fingerprints of the viscerotropic and neurotropic strains of CAR virus indicate that the genotypes of the 2 viruses are very similar. Evidently, though, each RNA species of the 2 viruses can be distinguished (Fig. 3).

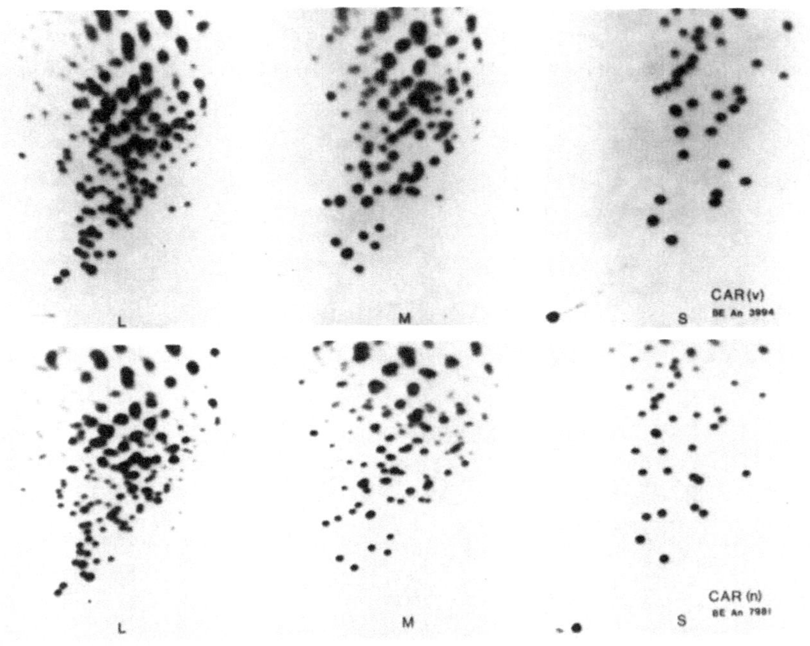

FIGURE 3. Oligonucleotide fingerprints of the L, M and S RNA species of the viscerotropic (CAR(v) BeAn 3994, top panels), and neurotropic (CAR(n) BeAn 7981, lower panels), strains of Caraparu virus.

Analyses of the pathogenic abilities of a number of reassortant CAL group viruses obtained from ts virus coinfections have demonstrated that silent mutations, incidental to the induced ts lesion, are frequently also present in the mutant viruses and can be passed on to "wild-type" reassortant progeny (15). Such silent mutations become apparent when the viruses are assayed in vivo. Thus, it has been shown that "wild-type" LAC/LAC/SSH and LAC/LAC/LAC viruses obtained from particular mutant virus crosses were highly attenuated in vivo by comparison with the parent LAC virus (15). Fewer attenuated reassortants are recovered when dual wild-type parent virus crosses are used to obtain reassortants (unpublished observations). However using wild-type virus crosses more screening is needed to recover particular reassortants.

In order to determine which viral RNA species of the Group C bunyaviruses codes for the gene products responsible for viscerotropism, dual wild-type CAR(v) and CAR(n) virus coinfections were undertaken and progeny virus clones selected at random. The viruses were fingerprinted to determine their genotypes. In addition to the 2 parental type virus, 2 reassortants were obtained from analyses of some 10 progeny virus clones. They had L/M/S RNA genotypes of CARv/CARn/CARn and CARv/CARv/CARn (Fig. 4).

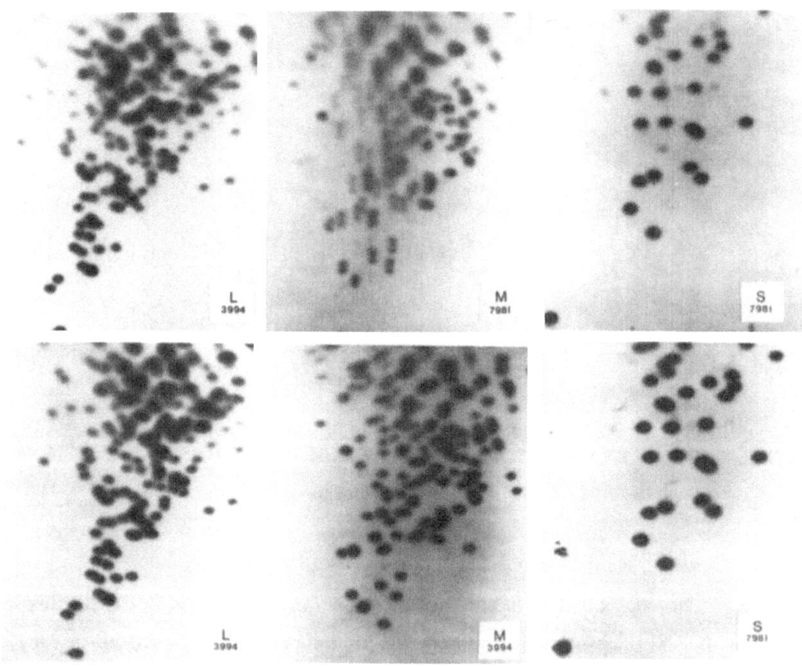

FIGURE 4. L, M and S RNA oligonucleotide fingerprints of the reassortant viruses CARv/CARn/CARn (upper panels) and CARv/CARv/CARn (lower panels).

3.2 Pathogenic properties of intertypic reassortant CAR Group C bunyaviruses.

The pathogenic properties of the 2 intertypic reassortant CAR viruses obtained from dual wild-type CAR(v) and CAR(n) virus infections were determined using baby and 4-week-old outbred Swiss mice. The results obtained are depicted in Tables 9 and 10.

Table 9. Ic and ip titrations of intertypic reassortant CAR viruses.

Virus	Titer (log LD_{50} per ml virus stock)		
	Adult mice		3-day-old mice
	ic	ip	ic
CARv/CARv/CARn	8.7	>4.5	9.1
CARv/CARn/CARn	6.7	<1.5	8.0

Table 10. Average survival time (AST) of 3-day-old mice inoculated ic with intertypic reassortant CAR viruses.

Virus	AST in days as function of dose (log adult mouse ic LD_{50})			
	(>5.0)	(2.3-3.0)	(0.3-1.0)	(<0.3)
CARv/CARv/CARn	2.0	2.2	3.4	-
CARv/CARn/CARn	-	3.8	4.3	5.0

The results of these analyses demonstrated that the CARv/CARv/CARn reassortant resembled the viscerotropic CAR(v) BeAn 3994 parent virus both in terms of the ic and ip titration data and the average survival times of mice receiving different lethal doses of virus (see Tables 8 and 9). Similarly, the CARv/CARn/CARn reassortant resembled the neurotropic CAR(n) BeAn 7981 parent. Since the only genotype difference between the 2 reassortants resides in the origin of the M RNA species, it is reasonable to conclude that the CAR M RNA gene products (presumably the glycoproteins) determine the virulence phenotype. These results need confirmation from analyses of other CAR reassortants, and formal proof that for the Group C viruses that the M RNA species codes for the viral glycoproteins.

4. DISCUSSION

It has been shown for bunyaviruses that the S RNA codes for the $20\text{-}25\times10^3$ dalton nucleocapsid protein, N, and the 10×10^3 dalton non-structural protein, NS_S. The M RNA codes for the 2 viral glycoproteins, the $110\text{-}120\times10^3$ dalton G1 protein and the $30\text{-}40\times10^3$ dalton G2 protein, as well as a 15×10^3 dalton non-structural protein, NS_M. Presumably the L RNA codes for the 180×10^3 dalton L protein that is probably a transcriptase component. Analyses of parent and reassortant CAL group bunyaviruses, and preliminary data with parent and reassortant Group C bunyaviruses, indicate that the M RNA gene products of these viruses determine the virulence characteristics. In what manner the viral glycoproteins interact with cell receptors and target specific cells in particular organs is one of the major unanswered questions of viral pathogenesis.

ACKNOWLEDGEMENTS

These studies were supported by a grant from the USPHS, NIH AI 15400 and AI 19688.

REFERENCES

1. Bishop, D.H.L., Calisher, C., Casals, J., Chumakov, M.P., Gaidamovich, S.Ya., Hannoun, C., Lvov, D.K., Marshall, I.D., Oker-Blom, N., Pettersson, R.F., Porterfield, J.S., Russell, P.K., Shope, R.E. and Westaway, E.G. Intervirol. 14:125-143, 1980.
2. Bishop, D.H.L. and Shope, R. E. In: Comprehensive Virology Vol. 14, (Eds. H. Fraenkel-Conrat and R.R. Wagner), 1979, pp.1-156. Plenum Press, New York.
3. Berge, T.O. (Ed.) International catalogue of arboviruses including certain other viruses of vertebrates. DHEW Publication No. CDC 75-8301, Washington, D.C., 1975.
4. Gentsch, J., Wynne, L.R., Clewley, J.P., Shope, R.E. and Bishop, D.H.L. J. Virol. 24:893-902, 1977.
5. Gentsch, J.R., Robeson, G. and Bishop, D.H.L. J. Virol. 31:707-717, 1979.
6. Bishop, D.H.L. Curr. Top. Microb. and Immunol. 86:1-33, 1979.
7. Klimas, R.A., Thompson, W.H., Calisher, C.H., Clark, G.G., Grimstad, P.R. and Bishop, D.H.L. Amer. J. Epidem. 114:112-131, 1981.
8. Gentsch, J. and Bishop, D.H.L. J. Virol. 28:417-419, 1978.
9. Fuller, F. and Bishop, D.H.L. J. Virol. 41:643-648, 1982.
10. Gentsch, J.R. and Bishop, D.H.L. J. Virol. 30:767-776, 1979.
11. Gentsch, J.R., Rozhon, E.J., Klimas, R.A., El Said, L.H., Shope, R.E. and Bishop, D.H.L. Virology 102:190-204, 1980.
12. Shope, R.E., Rozhon, E.J. and Bishop, D.H.L. Virology 114:273-276, 1981.
13. Beaty, B.J., Miller, B.R., Shope, R.E., Rozhon, E.J. and Bishop, D.H.L. Proc. Nat. Acad. Sci. U.S.A. 79:1295-1297, 1982.
14. Beaty, B.J., Holterman, M., Tabachnick, W., Shope, R.E., Rozhon, E.J. and Bishop, D.H.L. Science 211:1433-1435, 1981
15. Rozhon, E.J., Gensemer, P., Shope, R.E. and Bishop, D.H.L. Virology 111:125-138, 1981.

3

GENETIC DIVERSITY OF BUNYAVIRUSES AND MECHANISMS OF GENETIC VARIATION

R.M. ELLIOTT, J.F. LEES, G.E. WATRET, W. CLARK and C.R. PRINGLE

Medical Research Council Virology Unit, Institute of Virology, Church Street, Glasgow, G11 5JR, Scotland.

1. INTRODUCTION

The family Bunyaviridae currently includes more than two-hundred arthropod-borne viruses which share certain morphological or biochemical characteristics (1). Some of these viruses can infect the central nervous system or cause haemorrhagic fever in man and other animals. The viruses are generally spherical, 80-110 nm in diameter, and contain three helical nucleocapsids. The genome comprises three segments of single-stranded negative sense RNA designated L (large), M (medium) and S (small). For most bunyaviruses the genome encodes at least 4 proteins, including the major virion components which are the nucleocapsid protein N and the two surface glycoproteins G1 and G2. A minor large molecular weight protein designated L is sometimes observed, and this protein is presumed to be the virion transcriptase. The maturation of bunyaviruses occurs at smooth membrane vesicles, predominantly in membranes associated with the Golgi apparatus.

On the basis of serological comparisons and more limited biochemical analyses, four genera of bunyaviruses have been established; Bunyavirus, Nairovirus, Uukuvirus and Phlebovirus. The Bunyavirus and Nairovirus genera are further sub-divided into various serogroups. In this chapter we will describe some of our work on the genetic variation of bunyaviruses. We will consider primarily viruses belonging to the Bunyamwera group of the genus Bunyavirus, but will also describe some features of viruses belonging to the Nairovirus and Uukuvirus genera.

2. PROTEIN SYNTHESIS IN BUNYAVIRUS-INFECTED CELLS

We have compared the patterns of protein synthesis by viruses representing three bunyavirus genera; Bunyamwera (BUN) virus (2) of the Bunyavirus genus, the prototype of the whole family, and two Scottish bunyaviruses - Clo Mor (CM) virus (3) representing the Nairovirus genus, aand St Abb's Head (SAH) virus (4) representing the Uukuvirus genus. These viruses will each replicate in a variety of different mammalian or avian cell lines, but to aid comparison we sought a cell line permissive for all three viruses. Prompted by the observation of Leake *et al*. (5) that XTC-2 cells, an amphibian cell line derived from *Xenopus laevis* (6), permitted the plaque titration of a wide range of arboviruses, we found that these three buyaviruses would replicate in this cell line. The proteins induced in XTC-2 cells by Bunyamwera virus, Clo Mor virus and St Abb's Head virus are shown in Figure 1, and their molecular weights are compared in Table 1. Identification of Bunyamwera virus

Protein	BUN	CM	SAH
L	180*	nd	nd
glyco-protein	115 (30)	90	62-75
N	24	50	25
non-structural	60 50 16 13	45	58 30

* molecular weight x 10^{-3}
nd not determined

TABLE 1. Comparison of the molecular weights of the proteins induced by Bunyamwera, Clo Mor and St Abb's Head viruses in XTC-2 cells.

proteins is facilitated because of the marked inhibition of host cell protein synthesis, in contrast to the lack of inhibition by Clo Mor and St Abb's Head viruses. Clo Mor and St Abb's Head virus specific proteins were identified by immunoprecipitation with specific antisera or by using different radioactive

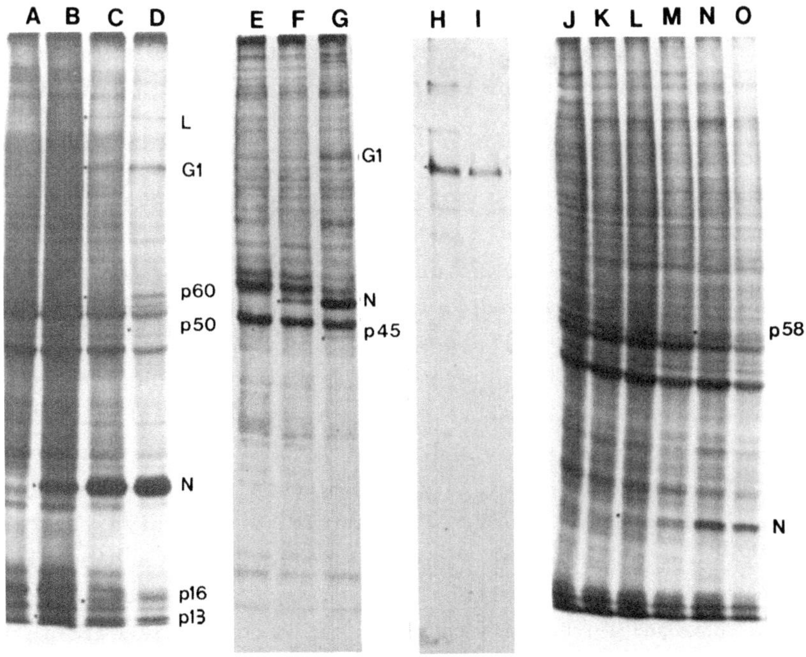

FIGURE 1. Polypeptide profiles of bunyavirus infected XTC-2 cells. Infected cells were pulse-labelled for 1 hour with 35S-methionine (lanes A-G and J-O) or 3H-mannose (lanes H and I) and cell lysates examined by polyacrylamide gel electrophoresis. Lanes B-D = 4, 8 and 16 hours after infection with Bunyamwera virus; lanes F and G = 8 and 16 hours after infection with Clo Mor virus, and lanes H. and I 16 and 24 hours after infection; lanes K - O = 2, 4, 6, 8 and 10 hours after infection with St Abb's Head virus; lanes A, E and J = mock-infected XTC-2 cells.

precursors. The nucleocapsid proteins of Bunyamwera virus and St. Abb's Head virus have similar molecular weights (approx 25000 daltons) but the N protein of Clo Mor virus is about twice as large (48000). The glycoproteins are also distinctive for each virus, as shown in Table 1. (The G2 protein of Bunyamwera virus is expressed more variably and less reproducibly than that of viruses of the California encephalitis serogroup or even other members of the Bunyamwera serogroup). The functions of the minor proteins, which are listed as non-structural in Table 1, are unknown at present. The coding assignments of the proteins and the arrangements of coding sequences within gene segments are not understood for all viruses, and this aspect of the genetic diversity of bunyaviruses is under investigation.

3. MECHANISMS OF GENETIC VARIATION

3.1 Characteristics of homologous and heterologous recombination.

In order to explore the origin of the genetic diversity of the Bunyaviridae, we have concentrated on analysis of mechanisms of genetic variation within a single serogroup of the genus Bunyavirus, namely the Bunyamwera serogroup which includes Bunyamwera virus. Initially three viruses - Batai (BAT) virus , Bunyamwera (BUN) virus and Maguari (MAG) virus - representing the range of serological variation within this serogroup were chosen for detailed study. Temperature-sensitive (ts) mutants of all three viruses were isolated following mutagenesis with 5-fluorouracil. As with all segmented genome viruses recombination in the Bunyaviridae can occur by genetic reassortment of genome sub-units in the course of mixed infection (7). Therefore these ts mutants were classified into groups by recombination experiments, and the homologies of the groups established by heterologous recombination (Table 2).

Bunyamwera serogroup virus	Number of mutants per recombination group				
	I	II	III	(I&II)	Total
BATAI	1	4	0	0	5
BUNYAMWERA	5	3	0	0	8
MAGUARI	12	31	1	1	45
Total	18	38	1	1	58
Gene assignment	L or N	G1/G2	L or N	—	—

TABLE 2. Classification of 5-fluorouracil induced ts mutants of three Bunyamwera serogroup viruses into recombination groups.

The number of recombination groups, therefore, should equal the number of genome sub-units. However, only two recombination groups were identified in three independent genetic analyses of ts mutants of ten different bunyaviruses (8,9,10,11) belonging to both the Bunyamwera and California encephalitis serogroups. Now a mutant of Maguari virus has been described which identifies the missing third group (12). The availability of this mutant provides genetic markers for all three genome sub-units, and unequivocal assignment of recombination group to genome sub-unit can be deduced in principle by determination of the phenotype and genotype of recombinants derived from crosses of different viruses. Certain peculiarities of heterologous recombination, however, which will be discussed later have complicated this procedure. At present it is certain only that the group II mutants represent the G1 polypeptide coding sub-unit (the M-RNA), whereas groups I and III may represent either the N or L polypeptide coding sub-units.

3.2 The extent of sub-unit reassortment.

The apparent ability of different members of the Bunyamwera serogroup to freely exchange genome sub-units suggested that this serogroup probably represents a single gene pool (11). This observation posed the question whether there are any restrictions to recombination between bunyaviruses at this or higher levels of taxonomic divergence? This question has been investigated by two approaches employing on the one hand heterologous crosses of ts mutants of different viruses, followed by screening for non-ts recombinant progeny, and on the other hand crosses of ts and non-ts viruses in combination with the use of specific antisera. The latter procedure requires the use of a ts mutant which will donate a G1 protein coding sub-unit in crosses at restrictive temperature (i.e. a group I or III ts mutant) and a specific antiserum to the non-ts virus. Recombinant progeny will be favoured and parental virus eliminated by screening at the restrictive temperature in the presence of antiserum to the non-ts virus. The former procedure is the more sensitive but depends on the availability of ts mutants of both parental viruses. The latter procedure is the more flexible and universally applicable with any virus against which specific antiserum can be prepared. The recombinant nature of the progeny must be confirmed, however, by genotypic or phenotypic analysis as described below.

It was soon apparent from these experiments that there were barriers to reassortment of sub-units at least in terms of the ease of generation of recombinants in vitro. Figure 2 summarises the results of these experiments so far. Viruses contained within the same box have been shown experimentally to be capable of exchanging genome sub-units. The data for the California encephalitis group viruses comes from the work of Bishop and colleagues (13). Comprehensive attempts to demonstrate recombination between viruses of the Bunyamwera group and viruses of the California encephalitis group were unsuccessful by either approach. A difference in the restrictive temperature appropriate for the mutants of the two serogroups, however, limited the sensitivity of direct screening

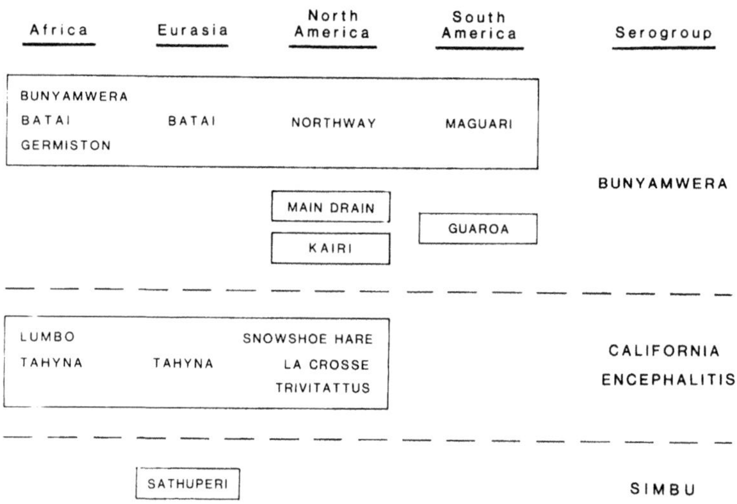

FIGURE 2. A diagrammatic representation of the extent of heterologous recombination among some viruses of the Bunyavirus genus. Viruses within the same rectangle are able to exchange genome sub-units under experimental conditions, but recombination has not been observed between viruses in different boxes under the same conditions. (The data for recombination among viruses of the California encephalitis group are taken in part from the work of D.H.L. Bishop and colleagues).

for non-ts recombinants. Similarly attempts to obtain recombination between Sathuperi virus, a member of the Simbu serogroup, and viruses of either the Bunyamwera or the California encephalitis groups were likewise unsuccessful. Similar restrictions on recombination were observed in both BHK, BSC-1 and Aedes albopictus mosquito cells, although higher recombination frequencies were often recorded in progeny from the latter. These experiments do not exclude the rare occurence of recombination, but they indicate that genome sub-unit exchange is restricted within the Bunyavirus genus. Recombination appears to be possible only between viruses belonging to the same serological grouping, and rare or non-existant between more distantly related members of the genus. In this respect the bunyaviruses are not dissimilar from other segmented genome RNA viruses; for example, recombination among influenza viruses has been demonstrated for viruses belonging to group A or group B, but not between viruses of these two serological types. The bunyaviruses differ from this pattern in that recombination may also be restricted among viruses belonging to the same serogroup. Figure 2 shows that recombination could be demonstrated between five viruses of the Bunyamwera serogroup, but three other viruses - Guaroa virus, Kairi virus and Main Drain virus - appeared to be genetically isolated. Again recombination potential follows serological relationships. The five recombining viruses, though showing a gradation of serological relationships, are more closely related to one another than to the other three viruses (1, 14). These observations together with the apparent uniformity of the terminal nucleotide sequences of the genome sub-units of all members of the same genus (8) suggest that the specificity of recombination may reside at the level of gene products rather than as a result of direct interaction of nucleic acid molecules. Further study of recombination in bunyaviruses may help to elucidate the mechanism of assembly of the correct complement of genome sub-units to produce an infectious virion.

These results are also interesting in that viruses such as Bunyamwera virus and Maguari virus which appear to be totally isolated both geographically and ecologically may yet be capable of exchanging genome sub-units in a laboratory environment. By contrast viruses such as Main Drain virus and Northway virus which naturally co-exixst in the same area remain genetically isolated. It is evident that the bunyaviruses are a group of viruses with potential for sudden variation comparable to the antigenic shift phenomenon associated with influenza viruses. For this reason they merit continual surveillance in view of their disease potential for man and agricultural animals of several bunyaviruses.

3.3 Genotyping by dot-hybridization.

Restrictions in the reassortment of genome sub-units, however, is not confined to interactions between distantly related viruses. For instance Gentsch et al. (15) reported that reassortment almost invariably occured unidirectionally in the California encephalitis group; e.g. in crosses of ts mutants La Crosse virus could act as a donor of M-RNA with snowshoe hare virus as the recipient, but not vice versa. In contrast bidirectional segregation appeared to be the norm in heterologous crosses of viruses of the Bunyamwera serogroup (11). This conclusion was based on analyses of the phenotypes of recombinants, observing the differential electrophoretic mobility of the G1 and N polypeptides in polyacrylamide gel. This type of analysis provides information about the behaviour of two of the three genome sub-units only. The complete genotype of recombinants can be definitively determined by oligonucleotide fingerprinting of each RNA sub-unit (8), but this procedure is not suitable for screening large numbers of recombinants.

Recently we have developed a dot-hybridization assay which provides a simple procedure for genotyping large numbers of recombinants. The method is a modification of a procedure described by Faulkner-Valle et al. (16) for the genotyping of rotavirus recombinants. This type of hybridization assay is possible because we have cloned reverse transcripts of sequences from the L-RNA, M-RNA and S-RNA sub-units of Bunyamwera virus

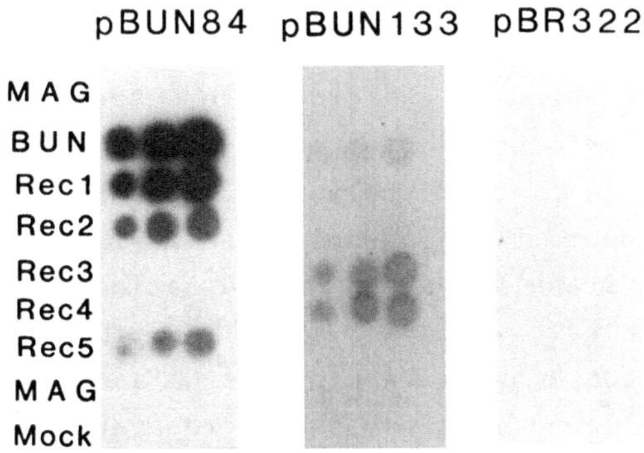

FIGURE 3. Genotyping by dot-hybridization. Cytoplasmic extracts of bunyavirus-infected cells were spotted onto three nitrocellulose filters in 3,6 and 9 ul amounts from left to right on each filter. The filters were hybridised with 32P-labelled plasmid DNA containing sequences from the S-gene (pBUN84) and the M-gene (pBUN133) of Bunyamwera virus, or with plasmid DNA (pBR322). Extracts of cells infected with Bunyamwera virus or recombinants 1,2 and 5 expressing the BUN N protein hybridized with the S-gene probe, whereas extracts of cells infected with Maguari virus or recombinants 3 and 4 expressing the MAG N protein did not hybridize. Likewise the M-gene probe hybridized with extracts of cells infected with Bunyamwera virus or recombinants 3 and 4 expressing the BUN G1 protein, but not with extracts of cells infected with Maguari virus or recombinants 1, 2 and 5 expressing the MAG G1 protein. The pBR322 probe and mock-infected cell extracts did not hybridize.

IxII ts Crosses		Recombinant non-ts virus				
		Genotype by dot hybridization				
Group I parent	Group II parent	L-RNA	M-RNA	S-RNA	No	Total
Bunyamwera	Maguari	MAG	BUN*	MAG*	13	16
Bunyamwera	Batai	BAT	BUN*	BAT*	3	
Maguari	Bunyamwera	BUN	MAG*	BUN*	8	25
Batai	Bunyamwera	BUN	BAT*	BUN*	17	

* independently confirmed by protein analysis

TABLE 3. The genotypes of recombinant clones from crosses of group I and group II mutants of Bunyamwera virus and Maguari virus determined by dot-hybridization.

into plasmids thereby making available large quantities of DNA as specific probes for each sub-unit of the Bunyamwera virus genome. Unfractionated cytoplasmic extracts of infected cells were prepared according to the procedure described by White and Bancroft (17) which dispenses with the need to extract purified mRNA. Up to 96 samples can be applied to a single nitrocellulose sheet. Phosphate-radiolabelled DNA was prepared by nick translation and hybridization carried out at 42 C for about 20 hours. Figure 3 shows the result of a pilot experiment in which Bunyamwera virus M-RNA and S-RNA specific probes were hybridized to cytoplasmic preparations of cells infected with Bunyamwera virus, Maguari virus and recombinants of these two viruses. The pattern of spots accurately reflects the known phenotype of these viruses, the probes hybridizing preferentially with samples expected to contain Bunyamwera virus mRNA. An L-gene probe from Bunyamwera virus behaved similarly (not shown), and all three probes were sufficiently specific to discriminate qualitatively samples containing Bunyamwera virus mRNA from those containing Batai virus mRNA or Maguari virus mRNA.

Table 3 lists the genotypes of some recombinants from several crosses of mutants of group I and group II determined by this procedure. An unexpected feature of these results is that the L and N genes appear to segregate together. No recombinants were obtained with L and N genes derived from different parents. Group II is known to correspond to the G-gene, but assignment of group I is not possible yet from these results. If group I corresponded to the N gene, recombinants would be expected with L-genes derived from either parent. Alternatively if group I corresponded to the L-gene, recombinants would be expected with N-genes derived from either parent. Work is continuing to resolve this problem using the rapid genotyping procedure described above.

3.4 Gene-specific mutation.

Recently a striking instance of gene-specific mutation in bunyaviruses has been observed, which indicates that biologically important variation may be generated by mutation as well as by recombination. It was observed in screening progeny from a recombination experiment that reversion of one of the parental viruses from the ts to the non-ts phenotype was frequently accompanied by a change in mobility of one of the virion polypeptides. Further study revealed that 35 of 36 non-ts revertants from mutant ts 8 (group II) of Maguari virus had G1 polypeptides with atypical electrophoretic mobility. Likewise reversion of two other group II mutants of Maguari virus (ts 14 and ts 45) produced changes in the electrophoretic mobility of G1 at high frequency (0.86 and >0.8 respectively). The frequency of reversion in the ts mutant stock in the absence of temperature selection was not more than 0.001. Figure 4 illustrates G1 electrophoretic mobility changes in several

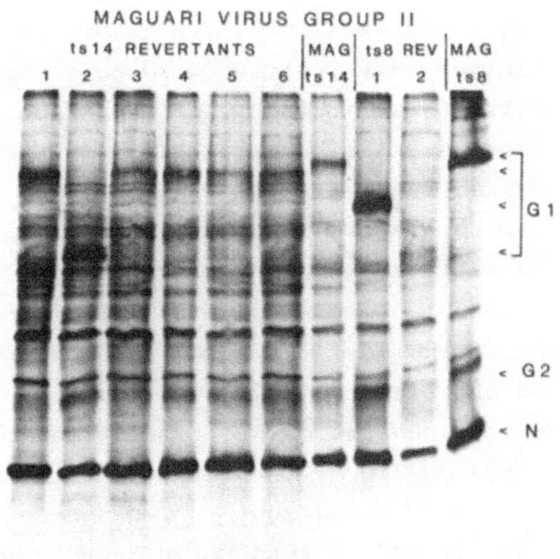

FIGURE 4. Polypeptide profiles of 35S-methionine labelled proteins of Maguari virus group II ts mutants and their non-ts revertants. The G1 polypeptide of the six non-ts revertants of ts 14 and the two non-ts revertants of ts 8 have migrated further than the G1 of their mutant parent.

non-ts revertants of ts 8 and ts 14 of Maguari virus. Contrary to this no changes in virion polypeptide mobility were observed in non-ts revertants of group II mutants of Batai virus (<0,05) or Bunyamwera virus (<0.05), or group I of Maguari virus (<0.05). These results suggest that reversion of ts mutations in the G1 protein-coding gene of Maguari virus occurs mainly by suppression, whereas reversion in the G1 protein gene of Batai virus and Bunyamwera virus occurs by same site mutation. Presumably the G1 protein of Maguari virus is more amenable to conformational change than the G1 protein of the other two viruses. The electrophoretic mobility of G2 protein was not affected by these mutations, and changes in glycosylation were not observed. In one instance neutralisability appeared to be affected.

4. GENETIC MAPPING OF NON-STRUCTURAL PROTEINS

Four non-structural (NS) proteins were observed in Bunyamwera virus infected cells (see Section II). Our cloned cDNAs have been used to investigate the genomic origin of these proteins, and it has been established that NSp13 is encoded by the S-RNA. Recombinant plasmid DNAs were bound to nitrocellulose filters and hybridized to cytoplasmic RNA extracted from Bunyamwera virus-infected cells. Bound RNAs were eluted and translated in vitro in a rabbit reticulocyte system; the result is shown in Figure 5. The N protein and NSp13 were identified as translation products in vitro but no other viral proteins were observed. The reasons for this are unclear though Cash et al. (18) have made similar observations. Two S-gene clones pBUN84 and pBUN85 both hybridized to mRNA which translated to give the N and NSp13 proteins. In contrast, an M-gene probe, pBUN 133, and the plasmid vector pBR322 failed to select any mRNA. This indicates that the S-RNA segment encodes both the N and NSp13 proteins. We are currently attempting to map the other NS proteins by this approach.

5. NUCLEOTIDE SEQUENCE ANALYSIS OF THE BUNYAMWERA VIRUS M-RNA

We have begun nucleotide sequence analyses of our Bunyamwera

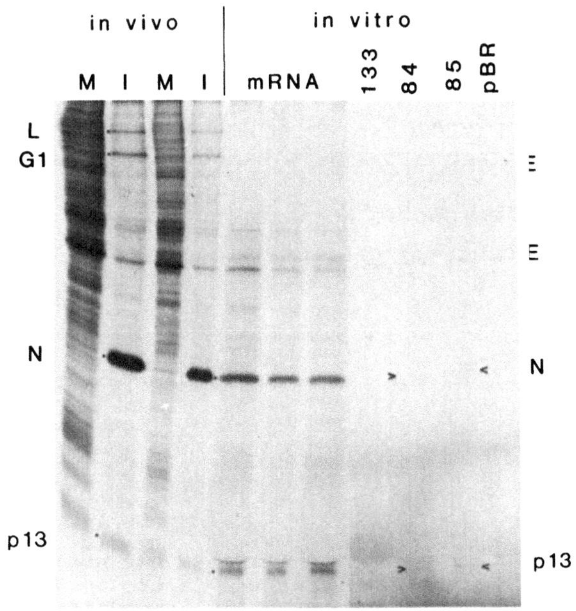

FIGURE 5. Message selection and translation. Filter-bound plasmid DNA was hybridized with RNA extracted from infected cells and the hybridized RNA translated in vitro. Plasmid pBUN84 and pBUN85 selected RNA translated to give N and NSp13 proteins, whereas pBUN 133 and pBR322 failed to hybridize any virus RNA. For comparison, poylpeptide profiles of the translation products of the selected mRNA are presented alongside extracts of radiolabelled Bunyamwera virus infected (I) and mock-infected (M) cells. E = endogenous protein from the in vitro system.

virus cDNA clones, and present a limited comparison of the 3'terminal sequence of the M gene with that of the M gene of La Crosse (LAC) virus obtained by Clerx-Van Haaster et al. (19) by direct RNA sequencing. Figure 6 compares the 3'terminal 160 nucleotides, written in the genomic (-ve strand), and the deduced amino acid sequences from the first translation initiation signal (TAC). The terminal 14 nucleotides are conserved, but there is only scattered homology throughout the remainder of the sequence. The first initiation signals for protein synthesis are found in similar positions in both genes - position 57-59 in Bunyamwera virus and position 62-64 in

```
                                             30
BUN  TCATCACATGATGGCTATGTAGTGTTTGGAAAGTCTCTGTGTAGAAATAAAGGTTCTAC
LAC  TCATCACATGATGGTTCATATCTATTGCAAACTTATAATTTCAAAACTTAGTTTCGGTT
     60                                        90
BUN  TCTTAAGATTATGACGAAAATCGTCAGTGAGTTGACCGACACTCATCGGGTCAATAGTG
LAC  TCTACTAAACATATAACCACGATTAATGTCAACGTCGACGTTCGGGTCACATAGTTTCC
      120
BUN  ATCTACGAAAGTACCACCCGTTGACCAACGTCTTTCCTTAAGG
LAC  ACAAAGGTTCTACCCCGATATCACTTCGTTTTGGGTAGGTTTC
```

```
BUN met arg ile leu ile leu leu leu ala val thr gln leu ala val ser ser pro val ser thr arg cys phe his gly
LAC                 met ile cys ile leu val leu ile thr val ala ala ala ser pro val tyr gln arg cys phe gln asp

BUN gly gln leu val ala glu arg asn ser
LAC gly ala ile val lys gln asn pro ser
```

FIGURE 6. Comparison of the 3'-terminal nucleotide sequences of the M genes of Bunyamwera virus and La Crosse virus. Protein synthesis initiation signals are boxed and termination signals underlined. Below is a comparison of the deduced amino acid sequence originating from the first initiation signal.

La Crosse virus. Other TAC triplets are present in the Bunyamwera virus sequence at positions 122-124 and 130-132; in the La Crosse sequence a TAC triplet is located at position 129-131 but it is closely followed by a termination signal. Thus on this very limited comparison the two genes show a similar architecture.

The deduced amino acid sequences from the first initiation signals have been aligned to maximize hmology between the genes. If the La Crosse virus sequence is displaced by three amino acids, two tripeptide sequences (ser-pro-val and arg-cys-phe) are conserved. The significance of this will not be clear until the complete sequences of both genes have been determined. It should be noted, however, that Clerx-Van Haaster et al. observed that the amino acic sequence preceding ser-pro-val differed when La Crosse virus, a variant of La Crosse virus and snowshoe hare virus were compared, although the amino acids were highly conserved thereafter. The variable region may be a signal sequence similar to that found at the amino terminus of many unprocessed glycoproteins. Furthermore the N-terminal amino acids of the mature protein (ser-pro-val or arg-cys-phe) may be conserved between different bunyaviruses - proof of this suggestion requires amino acid analyses of bunyavirus glycoproteins.

6. OUTLOOK

Some aspects of the genetic diversity and variation of the Bunyaviridae have been desscibed. The availability of cloned cDNAs of the genome sub-units provides tools for the study of genetic processes at the molecular level and ultimately may elucidate the factors controlling this variation.

REFERENCES

1. Bishop, D.H.L., Calisher, C.H., Casals, J., Chumakov, M.P., Gaidamovich, S.Ya., Hannoun, C., Lvov, D.K., Marshal, I.D., Oker-Blom, N.K., Pettersson, R.F., Porterfield, J.S., Russell, P.K., Shope, R.E. and Westaway, E.G. 1980. Bunyaviridae. Intervirology 14; 125-143.
2. Smithburn, K.C., Haddow, A.J. and Mahaffay, A.F. 1946. A neurotrophic virus isolated from Aedes mosquitoes caught in the Semliki forest. Am. J. Trop. Med. 26; 184-208.
3. Main, A.J., Downs, W.G., Shope, R.E. and Wallis, R.C. 1976. Avalon and Clo Mor; two new Sakhalin group viruses from the North Atlantic. J. Med. Ent. 13; 309-315.
4. Nuttal, P.A., Carey, D., Reid, H.W. and Harrap, K.A. 1981. Orbiviruses and bunyaviruses from a seabird colony in Scotland. J. Gen. Virol. 57; 127-137.
5. Leake, C.J., Varma, M.G.R., and Pudney, M. 1977. Cytopathic effect and plaque formation by arboviruses in a continuous cell line (XTC-2) from the toad Xenopus laevis. J. Gen. Virol. 35; 335-339.
6. Pudney, M., Varma, N.G.R. and Leake, C.J. 1973. Establishment of a cell line from the South African clawed toad, Xenopus laevis. Experientia 29; 466-467.
7. Bishop, D.H.L. 1979. Genetic potential of bunyaviruses. Curr. Topics in Microbiol. and Immunol. 86;1-33.
8. Bishop, D.H.L., Clerx, J.P.M., Clerx-Van Haaster, C.M., Robeson, G., Rozhon, E.J., Ushijima, H. and Veerisetty, V. 1981. Molecular and genetic properties of members of the Bunyaviridae. Pp 135-146. In Bishop, D.H.L. and Compans, R,W,(eds), The Replication of Negative Strand Viruses. Elsevier/North Holland. New York and Amsterdam.

9. Ozden, S. and Hannoun, C. 1979. Isolation and preliminary characterisation of temperature-sensitive mutants of Lumbo virus. Virology 84; 210-212.
10. Ozden, S. and Hannoun, C. 1980. Biochemical and genetic characteristics of Germiston virus. Virology 103; 232-234.
11. Iroegbu, C.U. and Pringle, C.R. 1981. Genetic interactions among viruses of the Bunyamwera complex. J. Virol. 37; 383-394.
12. Pringle, C.R. and Iroegbu, C.U. 1982. A mutant identifying a third recombination group in a bunyavirus. J. Virol. 42; 873-879.
13. Bishop, D.H.L. and Gentsch, J.R. 1980. The recombination potential of bunyaviruses. Pp 81-96. In Vesenjak, J. et al.(eds), Arboviruses in Mediterranean Countries. Zbl. Bakt. Suppl. 9. Gustav Fisher Verlag. Stuttgart and New York.
14. Hunt, A.R. and Calisher, C.H. 1979. Relationships of Bunyamwera group virus by neutralisation. Am. J. Trop. Med. Hyg. 28; 740-749.
15. Gentsch, J.R., Rozhon, E.J., Klimas, R.A., El Said, L.H., Shope, R.E. and Bishop, D.H.L. 1980. Limited recombination between bunyaviruses and evidence that bunyavirus M RNA gene products elicit neutralising antibodies. Virology 102; 190-204.
16. Faulkner-Valle, G.P., Lewis, J., Pedley, S. and McCrae, M.A. 1983. Isolation and characterisation of ts mutatns of a bovine rotavirus. In Bishop, D.H.L. and Compans, R.W., Double-stranded RNA viruses. (In press).
17. White, B.A. and Bancroft, F.C. 1982. Cytoplasmic dot hybridization. Simple analysis of relative mRNA levels in multiple small cell or tissue samples. J. Biol. Chem. 257; 8569-8572.
18. Cash, P., Vezza, A.C., Gentsch, J.R. and Bishop, D.H.L. 1979. Genome complexities of the three mRNA species of snowshoe hare bunyavirus and in vitro translation of S mRNA to viral N polypeptide. J. Virol. 31; 685-694.
19. Clerx-Van Haaster, C.M., Akashi, H., Auperin, D.D. and Bishop, D.H.L. 1982. Nucleotide sequence analyses and predicted coding of bunyavirus genome RNA species. J. Virol. 41; 119-128.

4

VARIATIONS IN RIFT VALLEY FEVER VIRUS

R. BARZILAY, D. BEN-NATHAN and A. KEYSARY
Israel Institute for Biological Research, P.O.Box 19, Ness-Ziona 70450, Israel

INTRODUCTION

By 1977, Rift Valley Fever (RVF) has covered most of the African continent, and reached the Mediterranean basin. The recent notification that Zinga virus (hitherto ungrouped arbovirus) is antigenically identical to RVF (RVFV) expands the RVF-covered Africa to include West Africa and Madagascar.

As a member of the Bunyaviridae family, RVFV is endowed with a tripartite segmented genome. This feature indicates a potential for high frequency genetic variation that takes place through exchange and reassortment of genomic segments. There are two indications that such variations do indeed occur in nature. The RNA from six different RVFV isolates were subjected to fingerprint analysis. All but two, which were identical, displayed unique and distinctive fingerprint pattern. The two identical patterns were derived from the two Egyptian strains isolated during the 1977-78 outbreak (1). The other variation concerns the pathogenicity of RVFV strains. All known pre-1975 cases of RVF in humans lacked haemorrhagic manifestations. RVF cases since 1975, however, included significant number of the haemorrhagic fever syndrome (2). In contrast, two naturally occurring viruses, Lunyo and Zinga (considered to be RVFV variants) display extensive attenuation, and none is known to cause an epizootic.

In our laboratory, we had the chance to observe and record two additional items of evidence indicating the high potential of variation inherent to RVFV. All the observations to be described were obtained with the Smithburn neurotropic RVFV strain (RVFV-N). This is an attenuated strain used extensively for vaccination and is not known to induce encephalitis upon non-neural inoculation in mice (3). This strain (RVFV-N) was obtained from Dr. McIntosh, S.A.R.

Changes in Pathogenicity.

In order to obtain plaque-purified clones of RVFV-N, we tried to adapt the virus to grow in BHK cells. The various passages in BHK were monitored for virus titer by intercerebral (ic) inoculation of suckling mice (SM). Some dams of the inoculated broods became moribund and died about 2 weeks after inoculating the SM. Table 1 presents the summarized mortality data, and the distribution of the observed mortality along the time axis.

Table 1. Mortality in mothers of RVFV-injected broods

A. Overall mortality rate:

Total number of injected broods	130
Total of dead dams	55
% mortality	42

B. Distribution of recorded deaths (days p.i.):

Day (pi)	1-6	7	8	9	10	11	12	13	14
Dead	0	3	1	8	16	11	10	5	1
% mortality	0	5	2	15	29	20	18	9	2

Peak mortality was reached at day 10-11 post infection (Table 1B). Since ic inoculation of SM with RVFV-N caused death within 2-3 days, this time table suggested a reinfection cycle. We were unable to correlate the observed mortality with virus titer or passage level in BHK cells. However, in several cases, signs of cannibalization preceded mortality in dams. This observation, again, corroborated the suggestion of reinfection.

The first set of experiments was designed to investigate possible modes of reinfection. Suckling mice from two groups of families were infected ic with a uniform dose of RVFV-N. In the first group, all SM in each brood were inoculated. In the second group, only one half (4 of 8) of the SM in each brood were infected, and the other half served as "sentinels". The results (Table 2) indicated that (a) there were cases of reinfection; (b) all reinfections were of the ascending transmission type; (c) there was not a single case of horizontal transmission.

Table 2. Secondary infections with RVFV-N

Brood * injection	No. of families	Mortality in: broods	dams
8/8	10	8/8 x 10	2/10
4/8	11	4/8 x 11	3/11

* 1.2 x 10^5 pfu in 0.03 ml

The demonstrated ascending transmission together with the observed cannibalization implicated the naso-pharyngeal cavities as the port of entry. This notion was investigated in the next experiment, designed to test (a) whether adult mice can be productively infected with RVFV-N via the intranasal (in) route, and (b) whether descending i.n. transmission can be demonstrated. Dams were infected with RVFV-N by the i.c. or the i.n. routes. Mortality was followed up in the infected dams and in their broods. The intranasal route allowed productive infection with RVFV-N (Table 3).

Table 3. Intranasal infections with RVFV-N

No. of families	Animal* inoculated	Route	Mortality dams	broods
8	dams	i.c.	8	11/64
8	dams	i.n.	8	40/64

* virus dose is the same as in Table 2.

Two additional conclusions were drawn: (a) descending transmission occurred from dams to progeny; (b) the rate of transmission was about 4 times higher when the dams were inoculated by the intranasal route.

Initially, these findings could have comprised a demonstration that the RVFV-N strain changed its characteristics in a way that enabled it to infect animals via a non-neural route. On the other hand, at least the ascending transmission clearly involved the possible neural route of the olfactory bulbs, and thus did not qualify as a "non-neural" infection. Also, although not studied in depth, the descending transmission probably occurred through virus-contaminated milk, and again, the olfactory bulbs could have been involved. One feature, however, was significantly prominent: RVFV-N was able to disseminate through the recipients' bodies, at levels sufficient to reinfect the progeny. Moreover, the dissemination level was significantly higher when the primary infection occurred via the intranasal route. Continuation of this line of reasoning called for the hypothesis that the progeny of RVFV-N after intranasal inoculation was somehow different from the parental population. Indeed, a blood sample from a moribund dam (infected due to ascending transmission) was assayed for virus, and was found to contain $10^{3.5}$ $MICCD_{50}$ per ml. A stock of this virus was prepared in brains of SM. The stock was titrated in three systems: i.c. in SM, i.c. in adult mice (AM) and intraperitoneally (i.p.) in AM. The results presented in Table 4 show that virus titer by the i.c. route was the same in SM and in AM. Extensive mortality, however, was observed in the i.p. inoculated adult mice. Therefore, progeny RVFV-N population obtained after ascending transmission was able to productively infect by an extraneural route. End-point determination (Reed & Muench) showed that the titer by the i.p. route was only about 10-fold lower than by the i.c. route. To our knowledge, this is the first demonstration that RVFV-N yields variants which are pathogenic by the extraneural route. These findings may have been due to pre-existing virus variants that were somehow enriched in numbers after ascending transmission. Similar experiments were performed with clones of RVFV-N (obtained by 3 cycles of plaque purification). One cycle of ascending transmission increased the virulence of the virus upon i.p. inoculation, by a factor of 100.

Table 4. Titration of MB-stock from a viremic dam

Dilution	SM/i.c.	AM/i.c.	AM/i.p.
-2	-	-	8/8
-3	-	-	6/8
-4	-	-	6/8
-5	8/8	8/8	7/8
-6	8/8	8/8	6/8
-7	8/8	8/8	6/8
-8	6/8	8/8	7/8
-9	4/9	5.8	5.8
-10	0/8	0/8	0/8

Antigenic changes

Changes in the pathogenicity of the virus comprised one type of variation observed in the RVFV-N strain. A second type of variation that we have encountered was expressed by antigenic modification. As we shall show, RVFV-N is antigenically distinct from the wild-type, pantropic RVFV, and that subpopulations of RVFV-N may exhibit different antigenic behaviour.

In order to simplify future serosurveys, an ELISA system was developed in our laboratory. The sensitivity of this system was compared to plaque reduction neutralization (PRN), hemagglutination-inhibition (HI) and reverse passive HI (RPHI). These methods were simultaneously applied to several serum samples of bovine origin. The samples were kindly provided to us by the Veterinary Services, from a collection representing various stages of active RVF-immunization in cattle. The RVFV-specific antibody levels of each sample, obtained by the 4 methods is listed in Table 5.

Table 5. RVF-antibody titers in bovine sera assayed by various techniques.

Serum Number	80% PRN	HI	RPHI	ELISA
1. 3012	20	20	< 10	< 20
2. 3013	< 10	10	< 10	< 20
3. 3014	< 10	20	< 10	< 20
4. 3015	80	40	20	20
5. 3016	< 10	20	< 10	< 20
6. 3027	< 10	10	< 10	< 20
7. 3028	80	40	< 10	< 20
8. 3030	80	80	< 10	< 20
9. 3038	1280	+320	+60	200
10. 3040	640	320	+80	200
11. 3051	640	320	+40	80
12. 3055	320	160	160	40
13. 3112	< 10	< 10	< 10	< 20
14. 3113	< 10	< 10	< 10	< 20
15. 3114	< 10	< 10	< 10	< 20
16. 3115	< 10	< 10	< 10	< 20
17. 3116	< 10	< 10	< 10	< 20
18. 3118	< 10	< 10	< 10	< 20
19. 3119	< 10	< 10	< 10	< 20

+; End-point differs in 1 dilution step between two separate determinations, and only the lower dilution factor is recorded.

Table 6 compares the ability of the 4 methods to score "positive" samples at various serum dilution levels.

Table 6. Number of sera scored as positive (out of 19) by different methods at various titer levels

Level	80% PRN	HI	RPHI	ELISA
≥10	8	12	5	n.d.
≥20	8	10	5	6
≥40	7	7	4	4
≥80	7	5	2	3
≥160	4	4	1	2

Contrary to what could have been expected, fewer examples were scored as positive by ELISA than by PRN or HI. Our particular version of ELISA employed was extremely heterogenous. The bovine antibodies, elicited against an inactivated South American wild type virus were asked to compete with Lapine antibodies, elicited against RVFV-N, for binding to an inactivated Entebbe strain. This raised the possibility that the observed "inferiority" of the ELISA was due to some non-identity of the antigens involved. Indeed, a scan through the published RVFV literature yielded a clear indication that at least the Entebbe RVFV strain was antigenically non-identical to RVFV-N (4). This notion was borne out when the same rabbit IgG-enzyme conjugate was assayed for binding to various dilutions of live RVFV-N or to inactivated Entebbe antigen. As shown in Fig.1, more antibodies were bound in the homologous reaction than in the heterologous one. The next experiment was a competition assay, in which antibodies elicited by the wild type antigen (P-Ab) or by the RVFV-N strains (N-Ab) competed with the anti-RVFV-N conjugate (N-Ab*) for binding to (a) the wild-type antigen (P-Ag) or (b) to RVFV-N (N-Ag). Fig. 2 shows that both P-Ab and N-Ab competed with N-Ab* equally well when the target sites were P-Ag. When the target was the N-Ag, however, the two Ab classes were clearly distinguishable. The N-Ab competed completely with the conjugate, whereas the P-Ab were only partially so. Therefore, we surmised, there were sites in the N-Ag that were recognizable only by the N-Ab but not by the P-Ab, and thus probably absent in the P-antigen. We are unable yet to do the complementary experiments, in which the same non-labeled Ab will compete with a P-type conjugate,since this reagent is not available to us at this time.

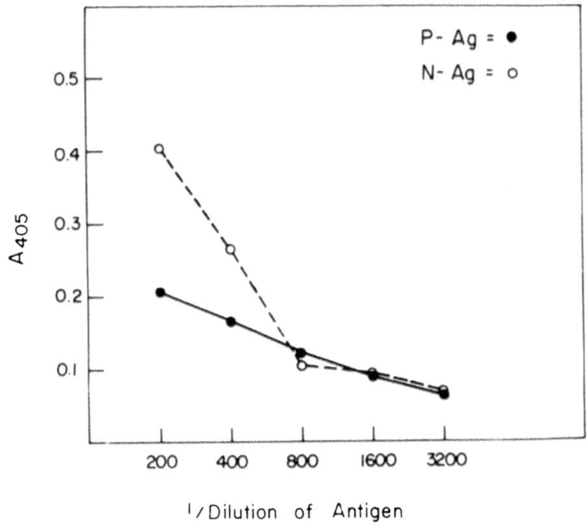

FIGURE 1. Binding of anti RVFV-N IgG-enzyme conjugate (N-Ab*) to various concentrations of the pantropic (P-Ag; ●) or the neurotropic (N-Ag; o) antigens.

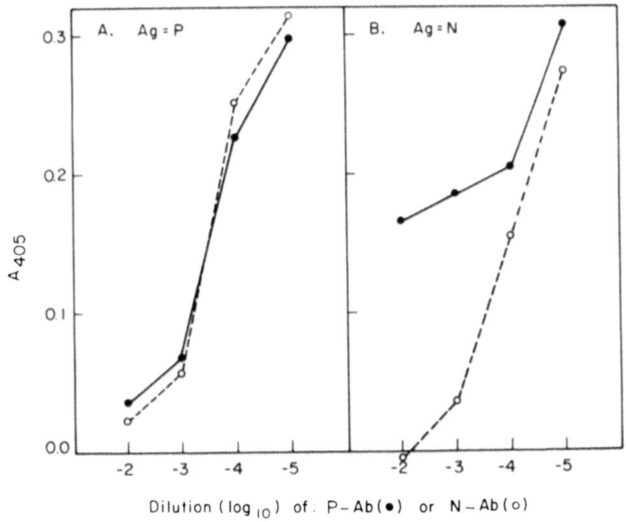

FIGURE 2. Competition of pantropic RVFV-directed antibodies (P-Ab; ●) or RVFV-N-directed antibodies (N-Ab; o) against anti RVFV-N IgG for binding to the pantropic (A) or to RVFV-N (B) antigens.

The ELISA system was thus shown to be sufficiently sensitive so as to differentiate between two strains of RVFV. Next, this system was asked to differentiate sub-population of a single strain. An IgG-enzyme conjugate was prepared from the serum of a rabbit hyperimmunized with plaque purified (once) RVFV-N. This conjugate was assayed for direct binding to various clones of RVFV-N, obtained by 3-fold plaque purification. The assay also included two lots of P-Ag (I-78; I-79). As can be seen (Fig.3), different slopes were obtained. Only one clone, 3005, exhibited similarity to the L-clone. Clones 3010 and 3015 yielded steeper slopes. Preliminary statistical analysis of several assays indicate that the differences are significant.

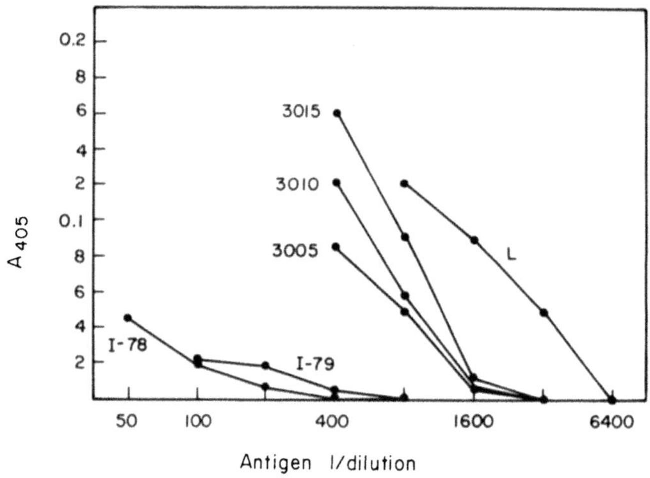

FIGURE 3. Binding of N-Ab* to various RVFV antigens. I-78 and I-79 are two lots of pantropic RVFV antigen preparations. 3005, 3010, 3015 and L are cloned sub-populations of RVFV-N.

We feel that our results may indicate that antigenic differences can be demonstrated in sub-populations of RVFV. Whether these differences are qualitative or only quantitative has not yet been studied.

REFERENCES

1. Cash, P., Robeson, G., Erlich, B.J. and Bishop, D.H.L. Contr.Epidem. Biostatist. 3:1-20, 1981.
2. Smithburn, K.C. Br.J.Exp.Path. 30:1-16, 1949.
3. Meegan, J.M., Watten, R.H. and Laughlin, L.W. Contr.Epidem.Biostatist. 3:114-123, 1981.
4. Binn, L.N., Randall, R., Harrison, C.J., Gibbs, C.J. Jr. and Aulisio, C.G. Am.J.Trop.Med.Hyg. 12:236-238, 1963.

5

PATHOGENIC ASPECTS OF MURINE CORONAVIRUS INFECTION IN RATS

V. TER MEULEN, H. WEGE and R. WATANABE
Institute of Virology, University of Würzburg
Versbacherstr. 7, 8700 Würzburg, Germany

Coronaviruses are a group of agents which are widespread in nature and associated with a great variety of acute, subacute and chronic disease processes of clinical and economic importance. In mice and rats subacute and chronic CNS diseases have been observed as a result of a persistent viral infection. In rats, the disease course is of remitting and relapsing nature accompanied by demyelination. Virological and immunological studies suggest that the persistent virus infection in brain tissue induces an autoimmune response to brain antigens which contribute to the disease process.

6

PERSISTENT PARAMYXOVIRUS INFECTIONS IN VITRO AND IN VIVO

ERLING NORRBY, HOOSHMAND SHESHBERADARAN, CLAES ÖRVELL and KRISTER KRISTENSSON.
Department of Virology, Karolinska Institute, School of Medicine (E.N., H.S.)
Department of Virology, National Bacteriological Laboratory (C.Ö.) Stockholm
and Department of Pathology, Huddinge Hospital, Huddinge, Sweden (K.K.)

Epidemic disease based on a chain of virus transmission between acutely infected individuals is of a relatively recent origin. Prior to the existence of societies including in excess of tens of thousands of people a continued virus perseverance required persistence in individuals of the infectious agent in the absence of obvious signs of disease. Thus not many millenia ago viruses only occurred in the form of persistent infections. This talent for persistence has remained with most different kinds of viruses till today. Persistence of viruses of different nature infer various forms of virus-cell interactions. DNA viruses or retroviruses may sequester into forms of DNA which persist in an episomal state or integrated into host cell DNA. RNA viruses lacking a reverse transcriptase only can persist under conditions when the expression of the genome is quantitatively or qualitatively limited so that the lytic destruction of cells is suppressed. In addition these viruses in spite of their continued replication should not reveal their presence in cells by any introduction of changes of antigens in the cytoplasmic membrane. If such changes occurred a normally functioning immune surveillance system could identify and possibly remove infected cells. The persistence of a lytically replicating virus under conditions of a yielding immunity represents a special situation which will not be discussed in this article. However in some of the conditions to be presented the virus initiates an infection in the central nervous system of an immature individual. This may increase the possibilities for conversion of an acute into a persistent infection due to the absence of a vigorous immune response or due to special host cell conditions in the incompletely developed brain.

Persistent infections with different paramyxoviruses - measles, mumps and parainfluenza 1 (Sendai) - will be discussed. Among these only measles virus has previously been documented to be capable of establishing a persistent infection in man, subacute sclerosing panencephalitis (SSPE) (1).

However, recently a case of chronic encephalomyelitis in man caused by mumps virus was described (2). In the present studies the character of virus-cell interactions in in vitro and in vivo systems was identified by determining the presence of 5 different structural components nucleoprotein - NP -, polymerase - P -, matrix - M -, fusion factor - F - and hemagglutinin or hemagglutinin-neuraminidase - H or HN - by use of sets of monoclonal antibodies in immune fluorescence analyses.

Synthesis of measles virus structural components by laboratory strains in persistently infected cell cultures

Cell cultures persistently infected with measles virus under varying conditions have been established in different laboratories. It should be noted that the conditions for establishment of persistent infections in vitro markedly differs from those characterizing the in vivo systems. Thus in the former situation the immune reactions do not exert any pressure on the system and further the persistent infection can be maintained by continued nondisturbed division of cells.

One persistently infected cell line which produced small amounts of infectious virus but considerable quantities of noninfectious material was established with Edmonston virus in HeLa(Lu 106) cells in this laboratory (3). Characterization of the synthesis in these cells of five structural components was made by use of monoclonal antibodies (4). The immune fluorescence analysis revealed that all cells synthesized NP and P proteins but that detectable amounts of H, F and M components were found in approximately 50, 10 and 30% of cells, respectively. The occurrence of F component synthesis in only a small percentage of cells could explain why the persistently infected cell culture was not deranged by cell fusion. Further studies of 3 persistently infected cell lines by the same approach, however, showed that a restricted synthesis of F protein was not of importance in all situations of virus persistence in vitro (5). Thus one line of persistently infected, established cells of human prostate origin, MAS-SSPE cells, produced demonstrable quantities of F component in virtually all cells. In spite of this no cell fusion occurred. This was found to be due to that the MAS cells had an exceptional resistence to measles virusinduced cell fusion. In another persistently infected cell line established in Vero cells by infection with a hamster neurotropic strain, HNT-Vero cells, no infectious virus was produced and the maintenance of the persistent infection involved a variable fraction of infected cells, generally only a few per cent. The infected cells synthesized readily detectable quantities of NP, P and H components but only small quantities of M component. No F component was demonstrable. The epitope characteristics of the NP and P components identified by a set of monoclonal antibodies showed variations contrasting with the previously identified stable NP and P component epitope pattern identified in studies of 9 lytically replicating strains of measles virus (6). Thus it appears that under conditions when an effective synthesis of complete infectious virus is not required more dynamic variation in both qualitative and quantitive appearance of structural components are found than in lytic infections.

Synthesis of measles virus structural components by defective SSPE virus strains in cell cultures

The synthesis of 5 different structural components of measles virus in Vero cells infected with 3 strains of defective SSPE virus strains, Biken, IP-3 and DR, and for comparison the lytic laboratory strain Edmonston, were studied (7). The SSPE strains showed varying degree of defectiveness. The Biken and to a lesser extent IP-3 strains synthesized small quantities of infectious virus and infected cells showed some capacity for hemadsoption. In contrast the DR strain was strictly cellassociated. No detectable infectious virus was produced and infected cells did not show any hemadsorption. Immune fluorescence analysis with monoclonal antibodies showed that all 4 virus strains produced large amounts of NP and P antigen. The Edmonston strain also produced readily detectable amounts of envelope components. In contrast the synthesis of envelope components, especially the M component was markedly reduced in cells infected with the Biken and IP-3 strains ,and in cells infected with the DR strain none of the 3 envelope components H, F and M were detectable.

Epitope characteristics of the matrix protein of lytically replicating SSPE and other measles virus strains

Epitope characterization of 9 different strains of measles virus by use of sets of monoclonal antibodies against 5 structural components has been performed (6). The NP, P and F proteins appear stable, but there were minor variations in the epitope pattern of the H component and pronounced variations in the M component. Studies of strains with the same designation obtained from different laboratories and further of 3 strains passaged 30 consecutive times at low multiplicity have shown that the epitope pattern of the M component of different strains has a high stability (Norrby, to be published).

Although it is not clear to what extent lytic SSPE virus isolates retain characteristics of the original defective virus causing disease it was considered of interest to determine the epitope characteristics of the M protein of such strains. One complication in this study was the fact that a LEC virus strain presumed to represent an SSPE virus isolate was used to generate the monoclonal antibodies employed in these studies (11). Since the LEC strain used had been passaged many times both in other and in this laboratory after its release from the Wistar institute, Pa, a low passage LEC virus isolate was obtained from Dr. H. Koprowski at this institute for comparative purposes. The results obtained in radioimmune precipitation assays with 9 different monoclonal antibodies identifying 6 different epitopes and 14 different strains of measles virus including 6 lytically replicating SSPE strains are summarized in Table 1. Five SSPE strains, which all were obtained from Dr. J. Sever, National Institutes of Health, Bethesda, Md, out of the 6 studied showed an identical M epitope pattern. The same M epitope pattern was also shown by some fresh isolates of virus from cases of regular measles. The early passage LEC SSPE virus showed an M epitope pattern different both from the other SSPE strains and also the LEC strain used for generation of the hybridomas producing monoclonal antibodies. Thus the significance of the finding of similar M protein characteristics in 5 out of 6 lytically replicating SSPE virus strains can not as yet be conclusively evaluated. More SSPE strains from other laboratories have to be studied.

Hybridoma		LEC-KI, Woodfolk Schwartz, Moraten	LEC-WI	MVO, MVP, Hallé McClellan, Dean Zistev, Mantooth	Hu 2	Edmonston DP
Group no	Clone no					
1	10EF10	+	+	+	+	+
2	16BB2*	+	+	+	+	+
3	19AG10	+	-	+	+	+
4	19CG6 19GF6	+	+	+(weak)	+	-
5	19DC5	+	+	-	+	-
6	19DF10 19HC5 19HF6	+	-	+	-	+

* Differs from 10EF10 in precipitation of breakdown products and canine distemper virus M protein

Table 1. The reactivity of 9 monoclonal antibodies against 6 different epitopes on the M protein with 14 different strains of measles virus in radioimmune precipitation assays. See refs 3, 4, 6 for methodology. Halle-, McClellan, Dean, Zistev, Mantooth and LEC (KI-Karolinska institute, WI-Wistar Institute) represent lytic SSPE strains.

Synthesis of measles virus structural components and genome material in acute and persistent infections in the hamster brain

Three weeks old hamsters were inoculated intracerebrally with a neurotropic strain of measles virus as described by Byington and Johnson (8). This animal model allows an analysis of the change of an acute into a persistent infection in hamster central nervous system. The synthesis of viral structural components was analysed by immune fluorescence using specific rabbit hyperimmune sera against whole measles virus and its NP and M components (9) and also by use of monoclonal antibodies (Norrby and Johnson: unpublished). In addition the presence of virus genomic material was determined by in situ hybridization (10).

During the first 12 days after infection the five different structural components were demonstrable in hamster brain tissue, but at later stages there was a pronounced reduction of synthesis of envelope components. At day 30 after infection H, F and M components were no longer detectable. Concomitantly with this disappearance of envelope components the animals developed an immune response against the virus. This was interpreted to signify that possibly the mounting of the antiviral immune response caused a selection of a defective cell-associated virus variant which no longer synthesized any envelope components.

Comparison of replication and gene expression also revealed pronounced differences between the acute and persistent phase of measles virus infection in the central nervous system of hamsters. The average number of copies of viral RNA per infected cells was about 200 to 600 times higher in acutely as compared to persistently infected cells. These results indicate that in measles virus persistence there is both a change in the qualitative expression of the virus genome in that the synthesis of envelope components is restricted and also a quantitative reduction in the overall turnover of viral genomes.

Characterization of persistent infections with Sendai (parainfluenza 1) and mumps virus in mouse brain

Sendai virus was inoculated intracerebrally into newborn, 12 and 21 days old mice (12). Maximum titer of infectious virus in brain tissue was reached 1-2 days after infection and the virus had disappeared after 3 days in older mice but after 6 days in animals inoculated immediately after birth. The latter animals showed a slower appearance of an immune response and in contrast to the situation in older mice the virus infection persisted. Immune fluorescence analysis with monoclonal antibodies against the NP, P, M, HN and F components of Sendai virus (13) showed that there was a change in the virus-cell interaction in the acute and persistent phase of the infection. During the acute, productive phase all virus structural components were synthesized but during virus persistence NP, P and M but not HN and F components were produced.

Partly similar findings were made in studies of mice infected as newborns with mumps virus (14). Virus strains of high and low neurovirulence, Kilham and RW, respectively, were employed and for comparison newborn hamsters were inoculated with the Kilham virus. The hamsters developed an acute encephalitis and died on days 8 to 9 after inoculation. Large concentrations of infectious virus was present in the brain of these animals. In contrast mice did not develop any signs of disease and no infectious virus could be identified in their brain. However the appearance of ar immune response indicated that a certain replication of virus had occurred. Analysis by monoclonal antibodies with specifities analogous to those described for measles and Sendai virus (Örvell to be published) showed that a persistent infection became established in these animals (Fig 1) but that no synthesis of the envelope components HN, F and M was demonstrable. Thus in this

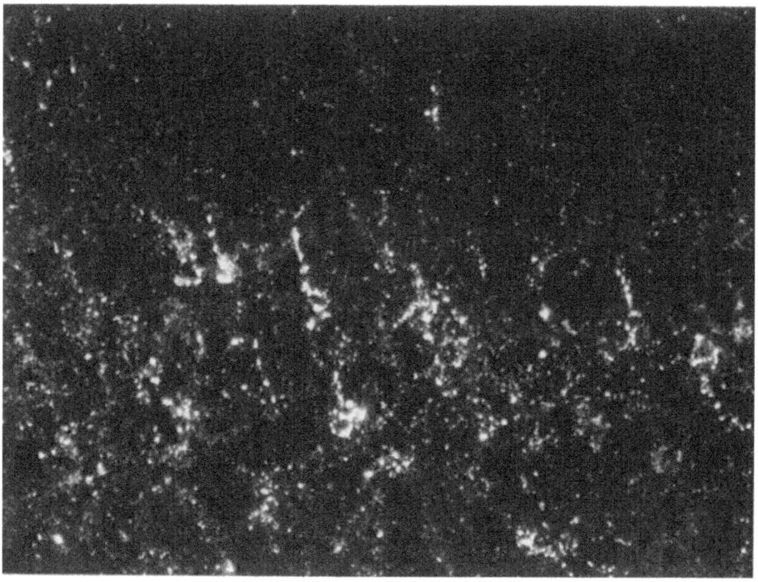

FIGURE 1. Brain from a mouse 9 days after infection with the Kilham strain showing numerous neurons with mumps virus antigen. Staining with monoclonal antibodies against the NP antigen. Magnification x 50.

this system the virus either goes through a very limited and short-lasting phase of complete replication before establishment of a persistent infection or it immediately establishes a defective form of virus-cell interaction. In the latter situation host-cell restriction would appear to be a pathogenetic factor of major importance.

The condition of measles virus persistence in the central nervous system of patients with SSPE

The current state of knowledge regarding measles virus infection in the brain of SSPE patients was recently reviewed (15). The virus causing the disease has a defective nature. Serological studies and a direct characterization of virus specific polypeptides in brain tissue indicate a major defect in the synthesis of the M protein. It is also possible that the synthesis of other envelope components, in particular H, may be reduced. Although a local synthesis of HI antibodies in the central nervous system and an increase in HI serum titers is seen in SSPE patients, this increase in H compared to that of antibodies against other components (except M) is less pronounced (16). Further, the use of gene-specific cDNA clones for hybridization with 3 SSPE brain materials indicate a reduced expression of synthesis of both M and H-specific messenger RNA (Rozenblatt; personal communication). The use of genomic cDNA in in situ hybridization studies (10) showed that like in brain material from hamsters with a persistent measles virus infection, brain material from patients with SSPE harbors a measles virus infection with more than 100 times reduced gene expression compared to that of an acute infection. It remains to characterize SSPE brain material by use of monoclonal measles antibodies to directly determine the nature of virus-cell interaction in terms of synthesis of different structural components.

Epilogue

Persistent paramyxovirus infections in the central nervous system studied by direct analysis of infected tissues or in cell culture systems are characterized by a reduced expression of one or more of the 3 envelope components H (or HN), F and M. As a consequence the virus does not mature normally and the surface of the infected cells may in some cases remain unaltered, whereby their identification by immune surveillance mechanisms is prevented. It should be emphasized that paramyxoviruses, also when not subjected to the selective pressure by the immune system, have a tendency to restrict the synthesis of envelope components when establishing persisttent infections. Thus such a change may represent an inherent weakness of the mechanisms for replication of paramyxoviruses, but it remains to be determined if the reduced genome expression acts on the level of transcription or translation. Further it remains to be elucidated to what extent special features of cells in the central nervous system may determine the change in paramyxovirus replication, which leads to viral persistence.

REFERENCES

1. Norrby, E., Haase, A.T., Johnson, K.P. and Örvell, C. In: Medical Virology (Eds.L. de la Maza and E.M. Peterson).Elsevier Science Publishing Co. Inc. New York, 1982, pp. 217-238.
2. Vaheri, A., Julkunen, I. and Koshiniemi, M-L. Lancet II:685-688, 1982.
3. Norrby, E. Arch. Ges. Virusforsch 20:215-224, 1967.
4. Norrby, E., Chen, S.N. Togashi, T., Sheshberadaran, H. and Johnson, K.P. Arch. Virol. 71:1-11, 1982.
5. Sheshberadaran, H., Norrby, E. and Rammohan, K.W.: Submitted for publication, 1983.
6. Sheshberadaran, H., Chen, S.N. and Norrby, E.: Virology. In press. 1983.
7. Johnson, K.P., Norrby, E., Swoveland, P. and Carrigan, D.R. Arch Virol. 73:255-262, 1982
8. Byington, D.P. and Johnson, K.P. J. Inf. Dis. 126:18-26, 1972.
9. Johnson, K.P., Norrby, E., Swoveland, P. and Carrigan, D.R. J. Inf. Dis. 144:161-169, 1981.
10. Haase, A.T., Swoveland, P., Stowring, L., Ventura, P., Johnson, K.P., Norrby, E. and Gibbs, C.J., Jr. J. Inf. Dis. 144:154-160, 1981.
11. Togashi, T., Örvell, C., Vartdal, F. and Norrby, E. Arch. Virol. 67: 149-157, 1981.
12. Kristensson, K., Örvell, C., Leestma, J. and Norrby, E. J. Inf. Dis. 147:297-301, 1983.
13. Örvell, C. and Grandien, M. J. Immunol. 129:2779-2787, 1982.
14. Kristensson, K., Örvell, C., Malm, G. and Norrby, E. Submitted for publication. 1983.
15. Wechsler, S.L. and Meissner, H.C. Progr. Med. Viral. 28:65-95, 1982.
16. Norrby, E., Enders-Ruckle, G. and ter Meulen, V. J. Inf. Dis. 132: 262-269, 1975.

7

FACTORS INVOLVED IN MEASLES VIRUS PERSISTENCE.

M.J. CARTER, P.N. BARRETT, M.M. WILLCOCKS, K. KOSCHEL
and V. ter MEULEN
Institut für Virologie und Immunbiologie, Universität Würzburg, D-8700 Würzburg.

SUMMARY

Recent reports have emphasised two aspects of measles persistence. Firstly, an apparent lack of immune response towards the matrix protein is associated with an apparent defect in the production of this polypeptide, and secondly antibody can act at the cell membrane to affect the synthesis of intracellular polypeptides and might thus account for the observed lack of matrix protein. We have investigated possible reasons underlying inability to produce M protein and now report that this may be accomplished at the level of mRNA turnover and function. In addition we have examined the mechanism of antibody-induced antigenic modulation and report that this process apparently involves alterations in the synthesis of virus-specific RNAs. These findings may be of significance in the analysis of all cell lines in which antibody has been used to produce cell curing.

INTRODUCTION

Measles virus is known to be associated with the fatal human CNS disease SSPE. It is thought that the virus establishes a persistent infection in the host during acute measles which gives rise to the disease SSPE some years later. SSPE can only be diagnosed after clinical onset and, since the frequency of disease is low (approximately one case per million acute measles), little is known about the mechanism by which persistence is established and maintained in the years between acute measles and SSPE. For this reason, much importance has been placed on the study of persistent measles virus infections <u>in vitro</u> as a model for this event <u>in vivo</u>. It is therefore necessary to summarise the key features of the disease in order to assess how successful this modelling may be and to describe the systems currently in use, before discussing some data obtained more recently. The evidence on which these conclusions are based has been extensively reviewed (1,2,3).

Features of SSPE

The SSPE patient is typically in a state of hyperimmunisation towards measles virus. Antibodies are present in both serum and CSF and can be demonstrated by radioimmune precipitation of infected cell lysates. These antibodies react with all measles virus structural proteins but fail to recognize matrix protein. The absence of matrix protein directed antibodies in immunoglobulins produced by lymphocytes invading CNS tissue has become a pathognomonic feature of SSPE. Direct searches for M protein in infected brains have so far failed to detect this molecule, which is a major structural component of the virus. Histopathological examinations have shown that all patients possess inclusion bodies in the cells of the CNS, and these have been identified as measles virus nucleocapsids by immune fluorescence and electron microscopy. However, the typical measles associated cytopathic effect, giant cell formation, is absent; and no virus particles either infectious or non-infectious have ever been detected. Virus budding has never been observed and it has therefore been concluded that SSPE is associated with a defect in the maturation process.

The patients' lack of immune response to matrix protein and the absence of this molecule in infected brain has provided a plausible rationale for the mechanism underlying such a maturation defect. M is thought to act as the final trigger in virus assembly, bringing together viral glycoprotein inserted in the plasma membrane with intracellular nucleocapsid structures prior to particle formation. The association of M with the cell membrane leads to local thickening and the formation of an "inner leaflet". This morphological feature has also not been detected in SSPE patient brain tissue.

Since virus maturation is apparently defective, direct isolation of virus from infected tissue has not proved possible. However, co-cultivation of diseased tissue with cell lines susceptible to virus infection has occasionally resulted in the rescue of infectious budding virus, therefore this defect may be to some extent host specific. These

Figure 1

The SSPE patient immune response.

Measles virus (Edmonston) proteins were immunoprecipitated from infected Vero cell lysates using: Rb, rabbit anti-measles Edmonston hyperimmune serum; CSF sample from an SSPE patient and serum from the same patient. Radiolabelling of protein and immunoprecipitation were performed using ^{35}S-methionine exactly as described by Stephenson and ter Meulen (15). Proteins were separated on 10 % polyacrylamide/SDS gels (16). Measles virus proteins comprise: the large protein (L), the haemagglutinin (H), the phosphoprotein (P), the nucleocapsid protein (N) and its major degradation product (NC), the fusion protein, both uncleaved (F_0) and large cleavage product (F_1), and the matrix protein (M). A small polypeptide (x) is also recognized which is related to the P protein. In our work we seldom observed appreciable amounts of P or F proteins.

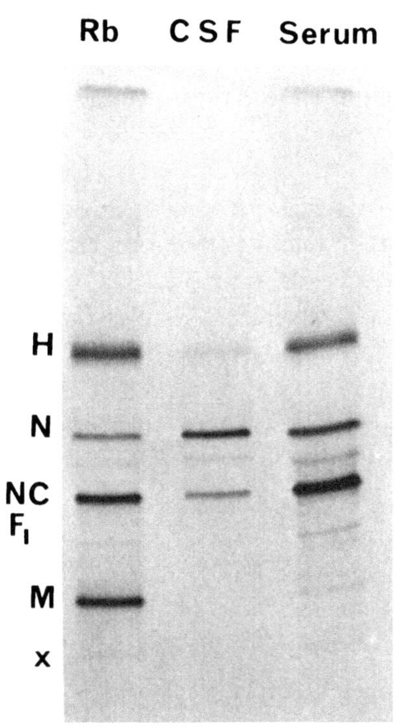

viruses, referred to as SSPE viruses, are very similar to measles viruses, cultures infected with them display all of the features of a lytic measles virus infection and the essential features of clinical SSPE are therefore not preserved. However, these viruses have been used to study the establishment and maintenance of persistent infections in vitro.

SSPE is more accurately modelled by SSPE cell lines. Frequently, the co-cultivation procedure described above, fails to yield a budding virus. Instead a cell associated cytopathic agent is obtained which can be maintained in tissue culture. These SSPE-cell lines preserve some essential features of SSPE in that intra-cellular nucleocapsids are present but virus is not released and M protein is apparently not produced. However, extracellular virus membrane antigens are expressed and cpe is evident.

It is known that the increase in anti-measles antibody titres is present some years before the onset of SSPE. It is possible therefore that this response is involved in the maintenance of the persistent infection. It has recently been shown that measles-specific antibody is able to modulate the expression of virus proteins inside the cell as well as on the plasma membrane. This antibody induced antigenic modulation (AIAM) could therefore be responsible for the inhibition of M protein synthesis.

We have used persistently infected cell lines, to investigate the defects underlying the failure in M protein production, and the mechanism of AIAM. We report here that two SSPE cell lines possess different lesions in M protein manufacture, one transcriptional and one translational, and that AIAM involved an effect on intracellular RNA synthesis. During the AIAM process in vitro some cell functions disrupted by the virus can be restored and this could model the symptomless years separating acute measles and SSPE.

FAILURE TO PRODUCE MATRIX PROTEIN IN SSPE CELL LINES.

The persistent infections used in this work were derived in vitro either by infection with a lytic virus or by co-cultivation with

Figure 2
Immune precipitation of N-1 cell, measles-specific polypeptides. Track 1, uninfected cell lysate; track 2, measles virus Edmonston-infected Vero cell lysate; tracks 3-7, N-1 cell lysates. Precipitated with: tracks 1, 2, and 7, hyperimmune serum; track 3, control ascites fluid; tracks 4-6, monoclonal antibodies specific for matrix protein.

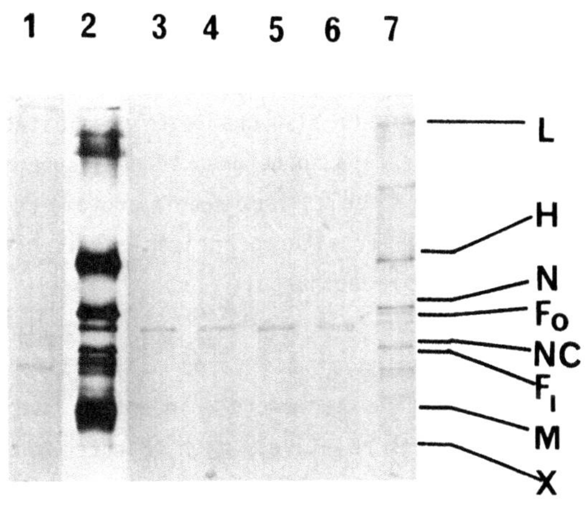

diseased tissue. The CLu106 is an Edmonston measles virus persistently infected Lu106 cell line established by Norrby (4). C6PI is a rat glioma cell line persistently infected with the SSPE virus Lec (5). The SSPE cell lines N-1 and MF were obtained by co-cultivation experiments, the MF in Würzburg (6),and the N-1 in Japan (7).

We have examined the two SSPE cell lines and neither produced infectious virus during ordinary cell culture procedures or cell fusion and co-cultivation experiments. In neither case could we detect production of M protein by radioimmune precipitation using anti-SSPE and anti-measles virus antiserum raised in rabbits. Furthermore, four monoclonal antibodies directed against M of measles virus Edmonston also failed to reveal this protein in immune precipitation experiments (Figure 2). Immune fluorescence or radioimmunoassay procedures were likewise negative. Serum and CSF samples are still available from the patient whose tissue was used in the production of the MF cell line. These samples were also unable to precipitate matrix protein from cell lysates. This data therefore strongly supports the conclusion that the late passage MF cells lack matrix protein completely. Even if M protein had been greatly altered antigenically, the homologous serum and CSF should still recognise it.

Analysis of cell proteins by immune precipitation was complicated since expression of all virus polypeptides was low. The majority of cells in each culture were uninfected. Consequently non-specific precipitation was relatively high (Figure 2, track 1), and the situation was further confused by breakdown products of the nucleocapsid protein, some of which migrated very close to the matrix protein. This difficulty could be substantially overcome by the technique of in vitro translation which leads to a great decrease in background (8). Since co-cultivation has occasionally resulted in the rescue of infectious virus, host factors may be involved in suppression of virus protein synthesis. Therefore, in vitro translation experiments can also be used to examine the possibility of host control (9). Putative regulatory factors are absent in these experiments. Proteins formed in vitro were immunoprecipitated using anti-SSPE virus antiserum raised in rabbits and pre-adsorbed with uninfected Vero cell antigens. Messenger RNA was extracted from uninfected Vero cells, N-1 or CLu106

Figure 3

Translation of N-1 and CLu cell mRNA.

mRNA was extracted from uninfected Vero cells, CLu106 cells and N-1 cells as described by Barrett et al. (17) and fractionated on poly-U sepharose columns (18). Poly-A containing fractions were translated in a rabbit reticulocyte lysate system (19). With the modifications suggested by Siddell et al. (20).

Panel A. Immunoprecipitation of translation products from Vero and CLu cells. Track h, rabbit hyperimmune serum preadsorbed with uninfected cell antigens; tracks 1-4 monoclonal antibodies specific for matrix protein. Panel B. Products of the in vitro translation of mRNAs from immunoprecipitated as above.

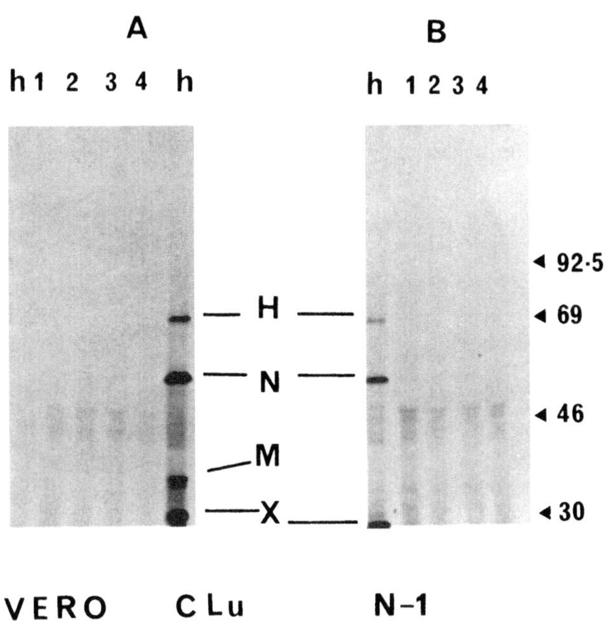

Figure 4

Southern blot analysis of mRNA extracted from SSPE cell lines. RNA samples were denatured with formamide and formaldehyde, and analysed on 1.5 % agarose gels containing 15 % formaldehyde and buffered with morpholinopropane sulphonic acid (Mops). Gels were blotted onto nitrocellulose filters using the method of Southern (21) and RNA was fixed by baking at 80°C for 2 hours. Immobilised RNA was then hybridised to ^{32}P-labelled, cloned DNA copies of measles virus sequences contained in the mRNAs for M, NC or H proteins. Clones used were a gift from Dr. S. Rozenblatt, Weizmann Institute, Israel (23,24).

In each panel: N, mRNA from N-1 cells; C, mRNA from CLu cells; Cd, mRNA from CLu cells after deadenylation; MF, mRNA from MF cells.
Panel A. mRNA blot hybridized to ^{32}P-labelled DNA copies of nucleocapsid mRNA sequences.
Panel B. mRNA blot hybridized to ^{32}P-labelled DNA copies of matrix protein mRNA.
Panel C. Long exposure of panel B.
The positions of influenza WSN kinase-labelled RNA segments 4-8, electrophoresed and blotted with the RNA samples are indicated on panels A and B. The intermediate-sized, M-specific RNA maintained in MF cells is marked on panel C.

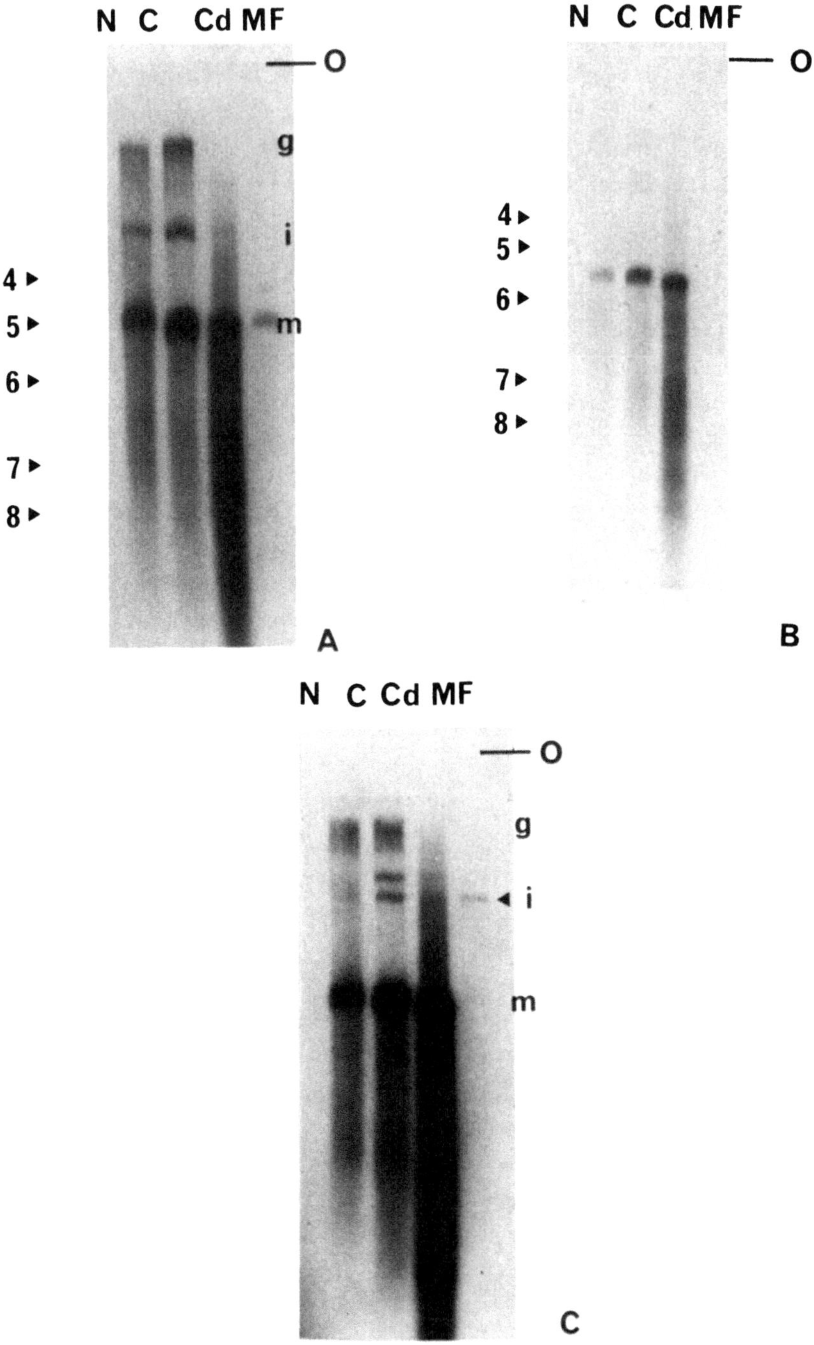
N C Cd MF
O
g
i
m
4
5
6
7
8
A
N C Cd MF
O
4
5
6
7
8
B
N C Cd MF
O
g
i
m
C

cells and translated (Figure 3).

It was apparent that matrix protein was not produced by the N-1 cell line. This was true whether ^{35}S-cysteine or ^{35}S-methionine were used as the label. Furthermore no differences were observed in this region of the gel between the translation products of RNA extracted from N-1 and uninfected cells in the absence of immunoprecipitation. The MF cell line was also negative in this assay. In contrast, translation of RNA extracted from the CLu106 cell line produced much matrix protein. This cell line also released small amounts of infectious virus and matrix protein was easily demonstrable in cell lysates.

This experiment enabled us to completely exclude the possibility that M protein is very unstable in the N-1 cell cytoplasm and rapidly degraded (10). Such processing is absent in in vitro translation systems.

This result suggested that mRNA for matrix protein was either defective in the SSPE cell lines or absent. A small peptide product, or one which differed completely in antigenicity might escape detection. We therefore subjected the mRNA samples to blot analysis using ^{32}P labelled, cloned DNA copies of the nucleocapsid or matrix message. The nucleocapsid clone acted as an internal positive control (Figure 4). In all analyses we have conducted, we have observed three size classes of RNA, the largest (4.5×10^6 mol. wt.; 11) comprises genome, the intermediate species ($1.2\text{-}1.7 \times 10^6$) are of positive sense (K. Baczko, personal communication) and may represent dimeric messages (Billeter et al., submitted for publication) similar to those obtained in VSV infections (12). The smallest size class ($0.5\text{-}0.9 \times 10^6$) is the most abundant and increases in mobility following treatment with oligo-$dT_{(12-18)}$ and RNase-H. It seems therefore to possess a poly-A tail some 60-90 residues long and presumably represents the messenger RNA. The intermediate size class was not obviously altered in mobility by this treatment (Figure 4C, tracks C and Cd) and was found mainly in the unpolyadenylated RNA fraction. However, it could contain internal poly-A which is prevented from attaching to the poly-U column due to secondary structure constraints.

All cell lines examined were positive for nucleocapsid message (Figure 4A). A similar analysis was therefore conducted using cloned

DNA copies of matrix protein mRNA sequences (Figure 4B). It was apparent that whilst CLu 106 and N-1 cell RNA contained mRNA-sized molecules which possessed matrix-specific sequences, MF cell RNA did not. This was true even on prolonged exposure (Figure 4C). However, matrix protein-specific information was still contained in the intermediate size class RNA. In all cell lines examined quantities of mRNA exceed those of intermediate-sized RNA. Consequently, the failure to detect mRNA while intermediate-sized RNA was clearly present, argues for a true imbalance in intracellular RNA species. This could be produced by an effect on mRNA transcription or degradation.

Interestingly, N-1 cells contain a mRNA which is apparently unable to give rise to a normal matrix protein product. The nature of this defect is unknown, but preliminary data suggests this mRNA, unlike that for nucleocapsid, does not sediment with polysomes. However, a very early termination signal might prevent appreciable ribosome binding to the molecule and thus also lead to this result. The small peptide product formed by such a process would probably not be detected.

These data therefore suggest that failure to produce M protein is accomplished in two different SSPE cell lines by two different mechanisms. In MF cells the cause may lie in the production or degradation of mRNA, and in N-1 cells, mRNA formed cannot be used to form normal protein. Work in our laboratory is currently being directed towards the identification of one or both of these mechanisms operating in human brain.

ANTIBODY INDUCED ANTIGENIC MODULATION

The possible significance of antibody in the modulation of virus protein expression has been mentioned above. During virus persistence it is likely that viral glycoproteins will be inserted into the cell membrane. The CNS is a functionally complex tissue, intercellular communication is extensive and to this end modified plasma membranes have developed bearing the complex apparatus required for neurotransmission. It is likely that haphazard insertion of virus proteins could disrupt such a system and we have used a cell line derived from rat CNS tissue, and persistently infected with SSPE virus Lec, to investigate

Figure 5 A

Removal of virus antigen from the C6PI cell surface. Antibody-induced antigenic modulation was performed using human antiserum obtained from SSPE patients, heat inactivated at 56°C for 30 minutes and filter sterilized. The serum was then diluted in DMEM containing 10 % heat inactivated FCS to a final haemagglutination titre of 50 U/ml. The parent C6 and the C6 cells persistently infected with SSPE virus Lec (C6PI) were grown in this antiserum-containing medium as monolayers in plastic tissue culture flasks and passaged every three days by standard treatment with a 0.05 % trypsin-EDTA solution. Fresh antiserum-containing medium was then added. At each passage C6PI cells undergoing antigenic modulation (○) and control C6PI cells (●) were incubated for 1 hour under SSPE serum. The amount of human antibody bound was measured using ^{125}I-labelled anti-human IgG. This technique reflects the removal of virus antigen and corresponding decrease in human antibody bound.

Panel a) Reduction in antibody binding during AIAM.

Panel b) Restoration of antibody binding capacity after removal of antiserum. Cells removed from antiserum were incubated with human serum in parallel with C6PI control cells for 1 hour before bound antibody was measured as above.

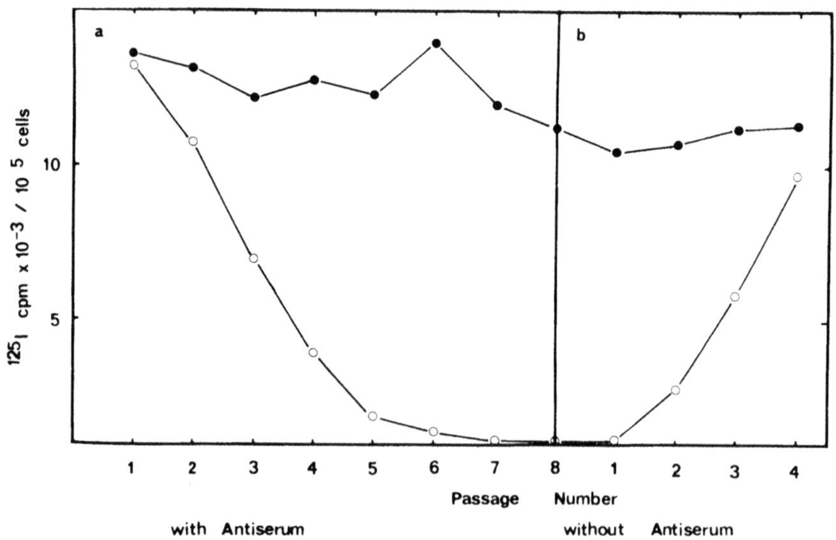

Figure 5B

Determination of isoproterenol stimulated cAMP synthesis.

Cells were stimulated with isoproterenol at each passage and extracts were prepared (13). Levels of cAMP in the extracts were measured by the method of Gilman (22). Results are expressed as percentage of uninfected control C6 cell levels.

● C6PI maintained without SSPE serum

○ C6PI maintained in SSPE serum.

Panel a. Restoration of isoproterenol stimulation to control cell levels during AIAM.

Panel b. Reversal of isoproterenol stimulation enhancement on removal of antiserum.

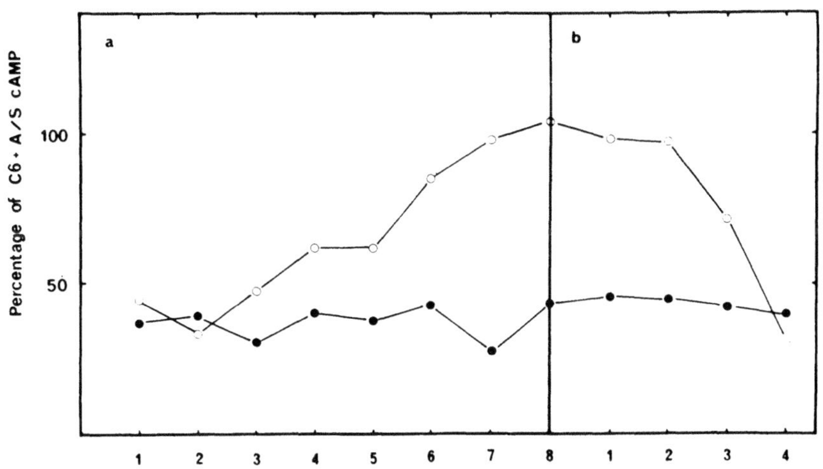

this phenomenon.

It is known that uninfected C6 cells can be stimulated by isoproterenol, which mimics the action of catecholamine hormones and acts at the ß-adrenergic receptors located on the cell membrane causing an increase in intracellular cAMP. This ß-receptor/adenylate cyclase response is greatly inhibited by virus proteins present in the C6PI cell plasma membrane. However, when C6P1 cells were cultured under human serum possessing anti-measles virus activity, virus membrane antigens were gradually removed. This removal of cell surface antigen, reflected in decreased binding of the human antibody, could be quantitated and the effect is shown in Figure 5A. The process was fairly rapid, and corresponded with restoration of isoproterenol sensitivity (Figure 5B) (13). Internal virus antigen persisted but this too was affected by antiserum treatment and could no longer be detected after 7-8 passages in medium containing antiserum. However, the cells were not cured, removal of antibody lead to the rapid spread of virus antigen through the culture until all cells were once more expressing virus antigen (Figure 6). A process such as this could explain the apparently normal CNS function observed in the years between acute measles and SSPE whilst an effect on the M protein could result in the establishment of persistence. Indeed these events may be favoured in the CNS where complement is not abundant and antibody-induced cell lysis is correspondingly inefficient (14).

We have used the C6P1 system in an attempt to elucidate the mechanisms underlying AIAM. Messenger RNA extracted from these cells revealed a similar pattern of virus protein expression when translated in vitro. Virus-specific proteins were not detected in the translation products of mRNA extracted from cells under antiserum either with or without immune precipitation (Figure 7A, 7B). In both C6PI and antiserum-removed C6PI cells, virus products were easily demonstrated. The small band migrating below M protein (h) is also precipitated from uninfected cells and presumably represents a host protein contaminent of the original virus preparation used to raise the antiserum (8). Direct analysis of RNA extracted from these cells revealed two distinct situations. RNA extracted from cells within the first seven passages revealed a specific decrease in the mRNA. Intermediate and

Figure 6

Antibody induced antigenic modulation.

Fixed cells were processed for immunofluorescence at the passages indicated using rabbit anti-measles antiserum and fluorescein-conjugated goat anti-rabbit antibody.

A C6PI control cells

B C6PI after 10 passages in antiserum-containing medium

C C6PI passage 3 after removal of antiserum

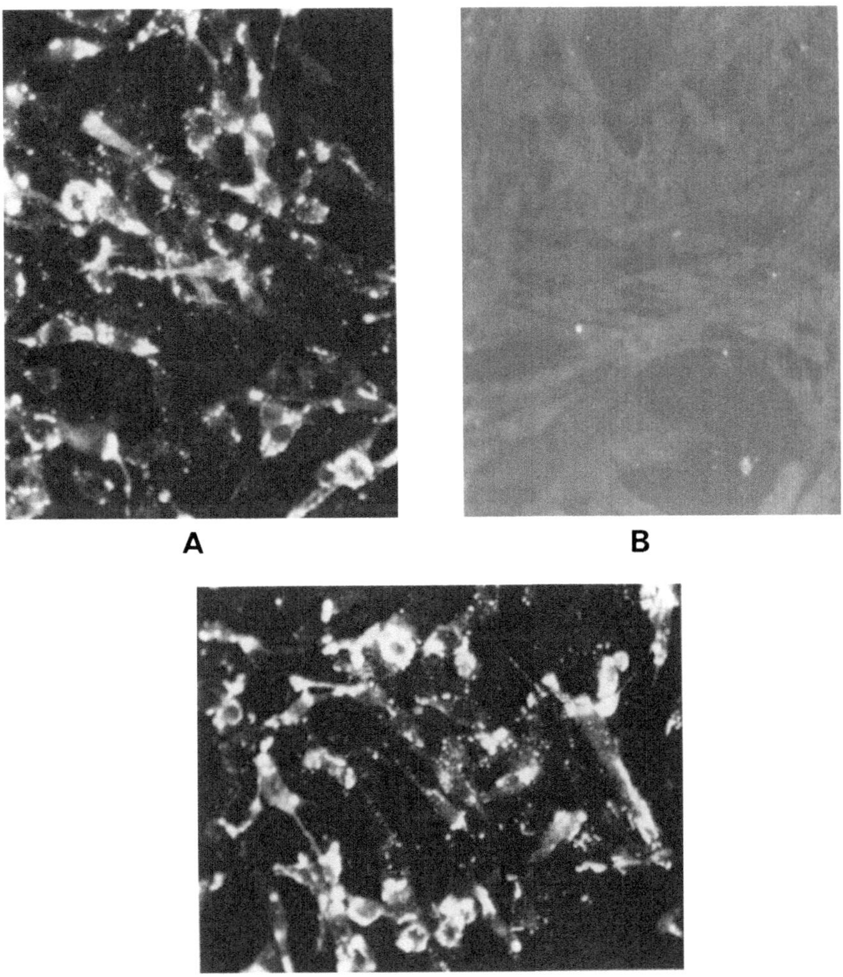

A B

C

Figure 7

In vitro translation of mRNA extracted from C6PI cells undergoing antibody-induced antigenic modulation.

In both panels: Track 1, mRNA isolated from C6PI cells; track 2, mRNA isolated from C6PI cells after 10 passages in medium containing SSPE serum; track 3, mRNA extracted from cells, 12 passages after removal of antiserum.

Panel A, products of the reticulocyte lysate translation system without immunoprecipitation.

Panel B, products of _in vitro_ translation precipitated with rabbit anti measles antiserum which had not been preadsorbed with uninfected cell antigens.

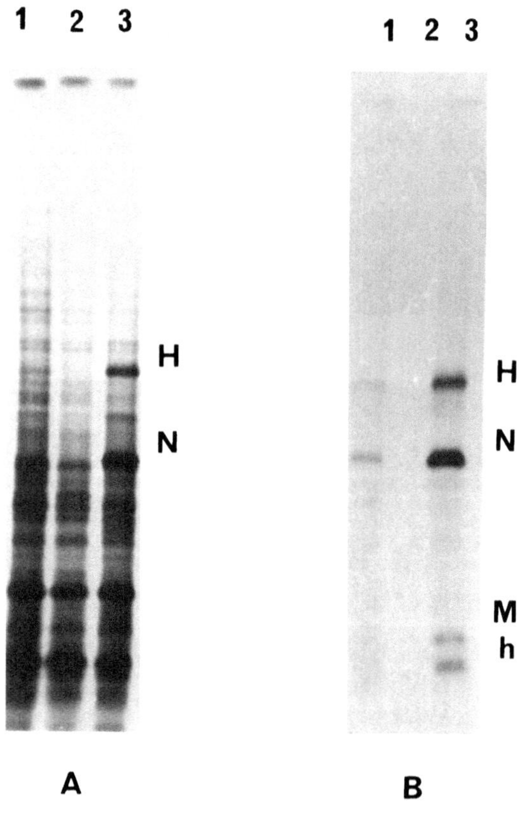

genome size classes were relatively unaffected (Figure 8). This was true whether the M, H or N-specific clones were used. RNA extracted later (passage 20) showed a general decrease in all virus-specific RNA size classes.

During this work it was necessary to consider the possibility that incubation under antiserum led to the curing of the vast majority of cells, but that infection could later spread from a few remaining infected cells, when the antiserum was removed. We therefore cloned antiserum treated cells which no longer expressed virus antigen in the continued presence of antiserum and in this way obtained 20 subclones of the C6P1 cell line. When antiserum was removed from the subclones, virus reactivation was observed in only two. This would tend to support the "curing hypothesis" outlined above. However, on continued passage, another clone developed virus antigen, and on examination, every clone has proved to be harbouring virus-specific RNA. Genome could no longer be detected in these cells, but the intermediate size and mRNA size classes were demonstrable. It would seem therefore that the lowered amounts of virus RNA observed after many passages under antiserum is due to a low level of virus expression in each cell, and not due to expression within a limited subpopulation of infected cells. Furthermore, this situation is apparently stable, although spontaneous activation of virus can occur for as yet unknown reasons. During the growth of these clones none has lost viral information which suggests some apparatus may exist for partioning the information at cell division.

DISCUSSION

The antigenic modulation described above, provides a convenient model, by which measles virus persistence could be established. In this model, antibody brings about a decrease in the expression of all virus polypeptides, and not merely in M as has been previously suggested. The evidence above hints at a sequence of events.

Firstly, antibody acting at the plasma membrane, could strip off external virus antigen, and then, over a longer timescale, bring about a decrease in all virus mRNAs in the cytoplasm. It cannot be concluded that this is accomplished by a direct inhibition of the viral trans-

Figure 8

Analysis of RNA extracted from C6PI cells undergoing AIAM.

RNA (10 μg/track) extracted from: Track 1, C6PI cells; track 2, uninfected C6 cells; track 3, C6PI cells after 7 passages in SSPE serum; track 4, uninfected Vero cells; track 5, C6PI cells, 12 passages after removal of antiserum. Southern blot analysis hybridized to cloned, ^{32}P-labelled copies of nucleocapsid mRNA sequences.

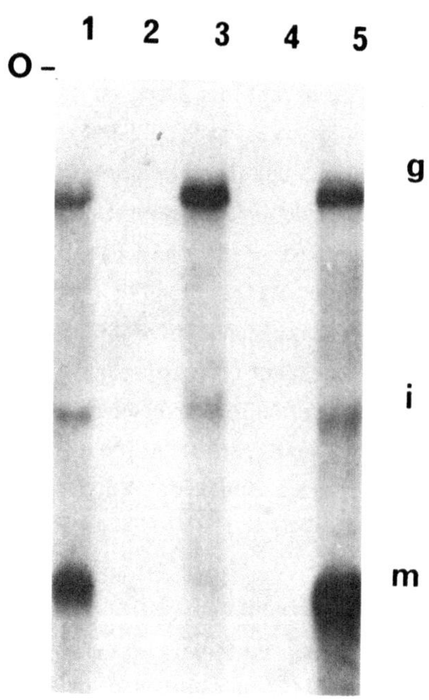

cription process rather than an effect on the translation of proteins required for mRNA formation or stability. In any event the net result would be decrease in mRNA levels. Protein synthesis must therefore fall and the disappearance of virus antigen could thus be explained. As proteins disappear, synthesis of all virus RNAs must be inhibited, total levels of virus-specific RNA would then fall. The situation at high passage under antibody could thus be created. At even higher passage levels, by which time direct action of antibody is no longer required (indeed antigen is no longer present in the membrane), the amount of genome has fallen to undetectable levels. Since the process of transcription is essentially one of amplification, the continued detection of intermediate and message-size RNA can be explained.

Since virus information is retained by the C6PI subclones it can be deduced that this situation is apparently stable at cell division. This need not be so *in vivo*, where CNS cell division does not occur. Finally, during AIAM we can detect mRNA, and fail to detect protein either in cells or *in vitro*. This situation is reminiscent of that in the N-1 cells but differs because in the modulated C6 infection, the effect is general and applies to all mRNAs. Amounts of these apparently "non-translated" mRNAs are much decreased. This is not so in N-1 cells where matrix protein mRNA is readily detectable in near normal amounts. It seems therefore that the situation in N-1 cells represents a true translational defect. Since blotting is more sensitive than *in vitro* translation or immunoprecipitation from cell lysates, it is likely that a low level translation and replication must occur in C6P1 tissue cultures allowing a "tick over" of virus information, and its replication for cell division. We are currently searching with more sensitive techniques for such low-level expression, and conducting investigations into the mechanism by which such infections can be reactivated.

ACKNOWLEDGEMENT.

This work was supported by the Deutsche Forschungsgemeinschaft and Volkswagen-Stiftung. We thank Professor E. Wecker and Dr. S. Siddell for helpful discussion, Helga Kriesinger for typing the manuscript and Annette Rummel and Helga Sennefelder for technical assistance. We also thank Drs. Doi and Homma for providing the N-1 cell line.

REFERENCES

1. Stephenson, J.R. and ter Meulen, V. In: Aspects of Slow and Persistent Virus Infection. 2. New Perspectives in Clinical Microbiology (Ed. D.A.J. Tyrrell) Martinus Nijhoff Publishers, The Hague, Boston, London, 1979.
2. ter Meulen, V. and Carter, M.J. In: Virus Persistence (Eds. B.W.J. Mahy, A.C. Minson and G.K. Darby), S.G.M. Symposia 33, C.U.P., 1982, 97-132.
3. Wechsler, S.L. and Meissner, H.C. Prog. Med. Virol. 28: 65-95, 1982.
4. Norrby, E. Arch. Virol. 20 215-224, 1967.
5. Halbach, M. and Koschel, K. J. Gen. Virol. 42: 615-619, 1979.
6. Kratzsch, V., Hall, W.W., Nagashima, K. and ter Meulen, V. J. Med. Virol. 1: 139-154, 1977.
7. Doi, Y., Sanpe, T., Nakajima, M., Okawa, S., Katoh, T., Itoh, H., Sato, T., Oguchi, K., Kumanishi, T. and Tsubaki, T. Japan. J. Med. Sci. Biol. 25: 321-333, 1972.
8. Carter, M.J. and ter Meulen, V. Advances in Brain Research, Vol. 59, in press.
9. Stephenson, J.R., Siddell, S.G. and ter Meulen, V. J. Gen. Virol. 57: 191-197, 1981.
10. Roux, L. and Waldvogel, F.A. Cell 28: 293-302, 1982.
11. Baczko, K., Billeter, M.A. and ter Meulen, V. J. Gen. Virol. in press.
12. Herman, R.C., Schubert, M., Keene, J.D. and Lazzarini, R.A. Proc. Natl. Acad. Sci. USA. 77: 4662-4665, 1980.
13. Barrett, P.N. and Koschel, K. Virology, in press.
14. Oldstone, M.B.A. and Fujinami, R.S. In: Virus Persistence (Eds. B.W.J. Mahy, A.C. Minson and G.K. Darby) SGM Symposia 33, C.U.P. 1982, 185-202.
15. Stephenson, J.R. and ter Meulen, V. Proc. Natl. Acad. Sci. USA 76: 6601-6605, 1978.
16. Laemmli, U.K. Nature (Lond.). 227: 680-685, 1971.
17. Barrett, T., Wolstenholme, A.J. and Mahy, B.W.J., Virology 98: 211-225, 1979.
18. Lindberg, U. and Persson, T. Methods in Enzymology, 34: 496-499, 1974.
19. Pelham, H.R.B. and Jackson, R.J. Eur. J. Biochem. 67: 247-256, 1976.
20. Siddell, S.G., Wege, H., Barthel, A. and ter Meulen, V. J. Virol. 33: 10-17, 1980.
21. Southern, E.M. J. Mol. Biol. 98: 503-517, 1975.
22. Gilman, A.G., Proc. Natl. Acad. Sci. USA. 67: 305-312, 1970.
23. Gorezki, M. and Rozenblatt, S. Proc. Natl. Acad. Sci. USA 77: 3686-3690, 1980.
24. Rozenblatt, S., Gesang, C., Lavie, V. and Neuman, F.S. J. Virol. 42: 790-797, 1982.

8

MECHANISM OF ADENO-ASSOCIATED VIRUS LATENT INFECTION

KENNETH I. BERNS
Department of Immunology and Medical Microbiology, University of Florida, College of Medicine, Gainesville, Florida 32610, U.S.A.

SUMMARY

Adeno-associated virus (AAV) is a defective parvovirus with a linear, single-stranded DNA genome. In the absence of a helper virus coinfection AAV DNA is uncoated in the nucleus and can integrate into the cellular DNA. The integrated DNA exists as a head-to-tail tandem repeat and the junctions with cellular DNA specifically involve the terminal repetition of the virion genome. The latent viral genome is efficiently rescued by superinfection of the latently infected cells by a helper virus. Recently, the AAV genome has been cloned intact into pBR322. Upon transfection of the recombinant plasmid into human cells in the presence of a concomitant adenovirus helper infection the AAV genome is rescued, replicated, and infectious virus is produced. This system has allowed detailed study of the rescue process. To date the primary conclusions derived are that the terminal repetition is essential to the rescue process and that the symmetrical nature of the terminal repetition affords remarkable capabilities of self repair in the subsequent process of replication after the rescue is effected.

INTRODUCTION

The adeno-associated viruses (AAV) belong to the family Parvoviridae, which are small DNA viruses (1). The virion has icosahedral symmetry with a diameter of 18-26 nm and is composed of three coat proteins that encapsidate a linear, single polydeoxynucleotide chain genome with a molecular weight of 1.5-1.8 x 10^6 daltons. In addition to the absolutely defective AAV (genus dependovirus), the family contains two genera of autonomous viruses; the parvoviruses, which infect many vertebrate species, and the densoviruses, which are insect viruses.

AAV is absolutely dependent upon coinfection with a helper virus for productive multiplication (2-4). Although in cell culture either adenovirus or herpes simplex virus may serve as a complete helper (2-5), it seems likely that adenovirus serves as the primary helper in nature. Isolation of AAV from patients has only been achieved in the presence of a concomitant adenovirus infection (6) and, indeed, AAV was originally isolated as a contaminant of what were thought to be purified adenovirus stocks (2-4).

As an absolutely defective virus AAV has a special problem in terms of the biological continuity of its genome in the absence of a helper virus coinfection. Under these conditions the AAV virion can penetrate to the cell nucleus where the genome is uncoated, but there is no detectable virus-specific DNA, RNA, or protein synthesis (7). In a biological sense the virus has solved the problem by being able to establish a latent infection in the absence of the helper virus. A stable relationship with the host cell genome is formed from which the AAV genome can be efficiently rescued by superinfection of the latently infected cell with a suitable helper virus at a later time.

AAV latent infection was discovered by Hoggan _et al_. (8) as a consequence of a U.S. government program to screen primary cell cultures intended for vaccine production for the presence of cryptic viruses. Up to 20% of primary African green monkey kidney cell lots were found to produce AAV after infection with adenovirus, although all the cell lots were initially negative for AAV antigens and infectivity. Additionally, 1-2% of the lots of human embryonic kidney cells tested were also latently infected with AAV. Thus, in monkeys, and possibly in people (80% of U.S. adults are seropositive for AAV), AAV latent infections seemed to be a fairly common phenomenon.

It proved to be readily possible to establish AAV latent infections in a continuous line of human cells (Detroit 6) in cell culture by infection with AAV alone at a high moi (250 tissue culture infections doses$_{50}$/cell)(9). Such cultures remained positive for 100 passages and, when cloned at the 39th passage, 30% of the clones were positive for AAV rescue by adenovirus superinfection. Several of the clones were tested for the number of copies of AAV DNA present by liquid DNA:DNA hybridization and on average there were 3-5 copies of the AAV genome per diploid amount of cell DNA (9). Independently derived clones of latently infected human KB cells were tested to see whether the AAV DNA present was covalently linked to high molecular weight DNA in the cells. Total cellular DNA was denatured and reannealed briefly so that only highly reiterated sequences found in chromosomal DNA could anneal to form large networks that were then precipitated in high salt. The AAV sequences were found to coprecipitate with the chromosomal DNA and it was concluded that the AAV DNA was indeed integrated into chromosomal DNA (10).

In this paper the results of recent work on the molecular basis of AAV latent infection are presented.

RESULTS

The existence of human cell clones latently infected by AAV has offered a unique opportunity to study the arrangement of the viral DNA sequences at the molecular level in cells derived from the normal host. In this particular instance the defectiveness of the virus has provided a significant operational advantage. These studies have involved Southern blot (11) analyses of total cellular DNA after restriction endonuclease digestion and electrophoresis. The following results and conclusions have been obtained (12). 1) The AAV genome is integrated into cellular DNA. The apparent molecular weight of the DNA species containing AAV sequences was reduced by digestion with restriction enzymes that do not cleave virion DNA to a size that exceeded 20 x 10^6 daltons. 2) The integrated copies of the AAV genome were arranged as a head-to-tail tandem repeat. Sequences from opposite ends of the genome were found to be contiguous in the integrated DNA. In some cases the terminal sequences from both ends were separated by small stretches of apparent cell DNA. In the clones examined there were 3-5 copies of the genome per diploid amount of cell DNA. Within the resolution afforded by the agarose gel electrophoresis only one species of fragment containing AAV sequences was detected. These data, together with the evidence for a tandem array, suggested that all of the AAV sequences in a given clone were likely to be integrated at a single site. 3) The terminal sequences of the virion genome were at the junctions between viral and cellular DNAs. With one exception all internal sequences from integrated DNA migrated identically with the internal restriction fragments from virion DNA. The single exception involved the fusion of two adjacent internal fragments. All terminal sequences were either fused to sequences from the opposite end of the virion genome or were associated with nonviral DNA. 4) In different clones that were independently derived the AAV sequences were integrated at different sites in the cellular DNA. The species containing terminal viral sequences differed in gel mobility after digestion of DNA from different clones (13). Internal sequences remained associated with fragments of consistent size. 5) With continued passage of the cells the virion genome was excised at a slow rate. In total cell DNA from passages 8-9 (early) all of the AAV sequences were associated with high molecular weight cell DNA. However, in total cell DNA from passage 118 (late) approximately 10-20% of the AAV DNA migrated identically with virion DNA in the double-stranded form. Furthermore, the arrangement of the sequences in the low molecular weight DNA was identical to that in virion DNA within the limits of resolution afforded by restriction analysis. Whether the free form of AAV DNA could replicate or simply represented a continual slow excision of integrated copies of the genome is not known. Multiple analyses of the integrated sequences by restriction enzyme digestion showed only one difference between early and late passage cells. Sma I cuts only within the terminal repetition in AAV DNA. Two of the four species seen after digestion of the integrated sequences were altered markedly in the late passage cells, suggesting that the terminal repetition had been genetically active during continued passage of the cells (12). 6) An adenovirus mutant that served as a helper for AAV DNA replication failed to rescue latent AAV (14). Recent studies have demonstrated that the early adenovirus genes Ia, II, and IV are required for AAV replication (15,16). Early region Ib is absolutely required for adenovirus transformation (17). A mutant in region Ib (hr 6) that helped AAV DNA replication in an exogenous coinfection failed to rescue the integrated form of AAV. Thus it would appear that a specific adenovirus function is required for rescue of latent AAV, in addition to those functions required to help AAV multiplication.

The studies of the clones of latently infected cells strongly suggested that the terminal sequences of the AAV virion genome played an important role in the integration of the viral DNA and apparently remained genetically active once in the integrated form. AAV DNA has an inverted terminal repetition of 145 bases (18, Fig. 1). The terminal 125 bases are a palindromic sequence; bases 1-41 are complementary to 85-125. There are two shorter palindromes in the intervening sequences, 42-62 and 64-84. Thus, when folded over on itself to maximize the number of potential base pairs, this terminal sequence forms a T-shaped structure in which only 7 of the 125 bases are not paired. Of the unpaired bases 6 are needed to allow the two internal palindromes to fold over and the 7th

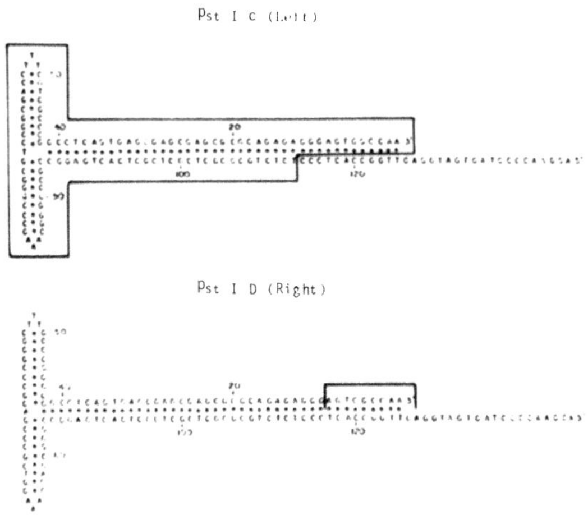

FIGURE 1. The sequence of the inverted terminal repetition in AAV2 DNA has been folded to form the structure that maximizes the possible self base pairing. The boxed sequences were deleted in pSM609. In pSM1205 bases 47-57 in the right terminal repetition of pSM609 were also deleted.

base separates the two internal palindromes. The highly organized terminal sequence in AAV DNA has been demonstrated to play a critical role in the replication of the viral DNA (19-24) and appears also to be important in the ability of the virus to establish latent infection.

Studies of the significance of the sequence organization of the terminal repetition have been facilitated by construction in vitro of a recombinant plasmid in which the double-stranded form of the AAV genome has been inserted intact into the bacterial plasmid pBR322 (24). When this plasmid was transfected into human cells that had been coinfected with adenovirus, the AAV genome was rescued, replicated, and infectious virus was produced. It thus appeared that the plasmid represented a completely defined model for the rescue of AAV from the integrated state since the complete sequences of both pBR322 (25) and AAV (26) DNAs are known. The first conclusion was that the tandem arrangement of viral sequences noted in the cloned cells in culture was not necessary for rescue of the viral genome. The only viral sequence present more than once was the terminal repetition.

Even the terminal repetition did not have to be present in more than one copy. The recombinant plasmid was formed by tailing the 3' termini of the double-stranded form of AAV virion DNA with poly dC and of pBR322 cut at the Pst I site with poly dG. After annealing the unligated DNA was transfected into E. coli and sealing occurred within the bacterial cell. In this process many clones were produced which had deletions at one or both ends of the AAV genome. The clone pSM609 had deletions at both termini; the 7 terminal nucleotides were missing on the left side and the 111 terminal nucleotides were missing on the right (Fig. 1). When this recombinant DNA clone was transfected into human cells coinfected with adenovirus, the DNA was again rescued, replicated, and infectious virus were produced. Much more interestingly, when the DNA from the virions produced was analyzed, it was found that the original wild type sequence had been restored (27). The fact that the terminal 125 bases in AAV DNA are both palindromic and repeated at both ends means that effectively the sequences within this region are present in four copies. Thus, the highly organized structure is capable of an extraordinary degree of self repair. One possible model for the repair of pSM609 is the following. A cut in the plasmid at the 3' end missing 7 bases would allow the end to hairpin and serve as a primer for DNA synthesis (because of the self complementary palindromic sequence). Under these conditions the first 7 bases inserted would repair the deletion. Repair of the more extensive 111 base deletion would be somewhat more complex, but a displaced single strand of AAV DNA (displaced by the growing strand primed at the other end of the insert in the plasmid) could form a circle stabilized by hydrogen bonding between the inverted terminal repeats. At the deleted end 32 bases remain, enough to stabilize the circle. In this structure the deleted end would be the 3' end which could serve as a primer for a repair type of DNA synthesis using the intact 5' terminal repeat sequence as a template (Fig. 2). The circular structure formed has been hypothesized to be an intermediate in AAV DNA replication (28).

In contrast to the type of repair possible for the extensive deletions present in pSM609 above, removal of an additional 11 base symmetrical sequence from the relatively intact end of pSM609 (Fig. 1) means that these bases are no longer present at all in the AAV insert in pBR322. Although the DNA from this additionally deleted recombinant plasmid could still be rescued, it was not replicated to form virion DNA and so no infectious virions were formed. On a larger scale this pattern was followed with other deletions studied. Deletions of less than 115 bases on either

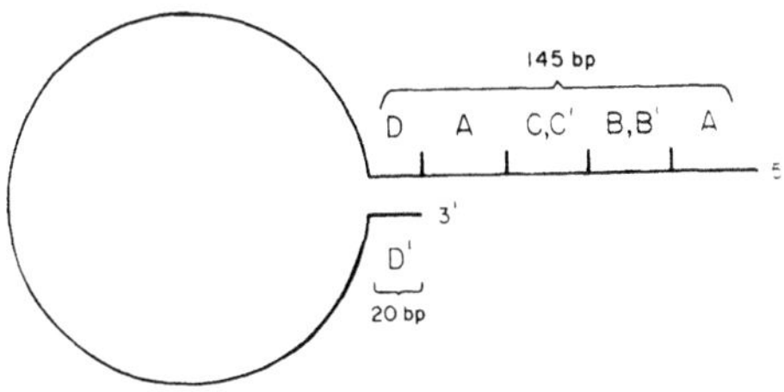

FIGURE 2. The hypothetical single-stranded circle formed by the AAV2 sequence in pSM609 would be stabilized by base pairing between the repaired right inverted terminal repetition and the 32 bases remaining in the left terminal repetition. This structure could serve first as a template-primer for repair of the left terminal repetition and then as an intermediate in normal DNA replication.

end were rescued and replicated. If both ends had deletions, rescue did not take place, nor did it occur if a deletion extended beyond the terminal repetition (27). By an extension of these studies it should prove possible to develop a detailed knowledge of the specific structural and/or sequence requirements for effecting rescue of the AAV genome from the integrated state.

The actual mechanism for rescue is not known. Conceivably the AAV insert could first be excised and then replicated, or replication could begin while the DNA is still inserted and free copies could be generated as a consequence of replication. At present the latter seems more likely. It has been possible to follow the fate of the parental DNA after transfection. To date free forms of pBR322 DNA which would result from primary excision of the AAV insert have never been detected. This tentative conclusion must, of course, be tempered by the possibility that the insert is rescued from only a small fraction of the input molecules and thus is below the level of detection possible by Southern blotting.

DISCUSSION

The study of AAV latent infection has reached a stage at which the detailed molecular biology is capable of elucidation. The defective nature of the virus has permitted the establishment of latent infections in cells of the normal host in culture. This is in distinct contrast to other nuclear DNA viruses and permits in depth studies of the state of the viral DNA in cells from which the virus can be readily rescued. Studies with clones of latently infected cells are under way to determine the cellular sequences flanking the integrated viral genome (R. Bohenzky and K. Berns, unpublished data). The existence of recombinant clones of AAV DNA in the plasmid pBR322, a completely defined system, from which the viral genome can be rescued after transfection of human cells permits definition of the minimum molecular requirements for rescue, in addition to general studies of the molecular biology of viral replication. Also underway are studies at the molecular level of the requirements for integration of the viral DNA into other DNAs in eukaryotic cells. AAV can recombine with SV40 after either coinfection or cotransfection of monkey cells. The requirements for and the products of recombination are being characterized (Z. Grossman, E. Winocour, and K. Berns, unpublished data).

The successful rescue of clones with deletions at both ends of the AAV genome (e.g., pSM609) raises an interesting question with regard to the specificity of the process. From studies of independently derived clones of latently infected human cells no specificity of flanking sequences has been detected. On the other hand, the presence of the terminal repetition at junctions with cellular DNA and the inability to rescue the viral genome from recombinant plasmids with significant deletions (>50 base pairs) at each end imply that this sequence or structure does play a critical role in the process of rescue. These facts plus the ability to rescue pSM609 suggest two possibilities; either there is sequence specificity in the terminal repetition, but the signal sequence is located internally, or the critical feature is the self complementarity of the overall sequence, enough of which must be retained to allow potential formation of a cruciform structure. Although the latter possibility is especially intriguing in light of its similarity to putative recombination intermediates, resolution of this problem awaits experiments in which part or all of the AAV terminal repetition is replaced with substitute sequences that retain the ability to form the cruciform structure.

The biological consequences of AAV latent infection are potentially of equal interest to the molecular findings, but may be much more difficult to define. AAV infects many vertebrate species, including man (85%

of adults in the U.S. are seropositive), but no disease has ever been associated with infection. Infectious AAV has only been isolated from people suffering concurrent adenovirus infections (6) and the diseases associated with such infections are attributable to the helper virus co-infection. Indeed the possibility exists that AAV coinfection results in the amelioration of disease in such infections. AAV inhibits the replication of adenovirus (29) and the oncogenicity of both adenovirus and herpes simplex virus in cell culture and in experimental animals (30-34). It is possible that latent infection _in vivo_ may be protective as well. Indeed, this has been suggested in the case of AAV latent infection in chickens where adenoviruses are pathogens of serious economic consequence (35). The extent, as well as the consequence, of AAV latent infection in humans in uncertain. One suggestive study of the possible consequences measured the frequency of antibodies to AAV in patients with cancer of the cervix or prostate (thought to be associated with herpes simplex virus type II) in comparison to matched controls. In the control group 83% of the population had antibodies to AAV whereas only 14% of the cancer patients had detectable levels of such antibodies. In both groups 80-90% of the patients had antibodies to herpes simplex viruses (36). The underlying assumption in this study was that latent AAV infection required a prior AAV infection which would have produced a measurable antibody response. In chickens the possibility has been raised that the AAV genome may be transmitted via the germ line (35); a phenomenon that has not been demonstrated for any other DNA animal virus. The rather low frequency (1-2%) of latent infection of the original lots of human embryonic kidney cells tested by Hoggan _et al_. (8) would imply that germ line transmission is unlikely to be a common event in humans.

Because of its defectiveness and lack of readily detectable pathogenicity, AAV is among the more unobtrusive human viruses. Yet its small size, unusual biology, and prevalence make it an interesting subject of study. Because its biological interactions with both adenovirus and herpes viruses suggest a commonality in certain steps vital to virus replication, many of the data obtained are likely to be of quite general interest.

ACKNOWLEDGEMENT

This work has been supported by a grant from the U.S. Public Health Service, AI 16326.

REFERENCES

1. Bachmann, P.A., Hoggan, M.D., Kurstak, E., Melnick, J.L., Pereira, H.G., Tattersall, P. and Vago, G. Intervirol. 11:248-254, 1979.
2. Atchison, R.W., Casto, B.C. and Hammon, W.McD. Science 149:754-756. 1965.
3. Hoggan, M.D., Blacklow, N.R. and Rowe, W.P. Proc.Natl.Acad.Sci. USA 55:1457-1471, 1966.
4. Parks, W.P., Melnick, J.L., Rongey, R. and Mayor, H.D. J. Virol. 1: 171-179, 1967.
5. Buller, R.M.L., Janik, J.E., Sebring, E.D. and Rose, J.A. J. Virol. 40:241-247, 1981.
6. Blacklow, N.R., Hoggan, M.D. and Rowe, W.P. Proc.Natl.Acad.Sci. USA 58:1410-1415, 1967.
7. Rose, J.A. and Koczot, F. J. Virol. 10:1-8, 1972.
8. Hoggan, M.D., Thomas, G.F., Thomas, F.B. and Johnson, F.B. In: Proceedings of the Fourth Lepetit Colloquium, North Holland Publishing Co., Amsterdam, 1972, pp. 243-249.
9. Berns, K.I., Pinkerton, T.C., Thomas, G.F. and Hoggan, M.D. Virology 68:556-560, 1975.
10. Handa, H., Shiroki, K. and Shimojo, H. Virology 82:84-92, 1977.
11. Southern, E.M. J. Mol. Biol. 98:503-518, 1975.
12. Cheung, A., Hogan, M.D., Hauswirth, W.W. and Berns, K.I. J. Virol. 33:739-748, 1980.
13. Berns, K.I., Cheung, A.K-M., Ostrove, J.M. and Lewis, M. In: Virus Persistence (Eds. B.W.J. Mahy, A.C. Minson, and G.K. Darby) Cambridge University Press, 1982, pp. 247-279.
14. Ostrove, J. and Berns, K.I. Virology 104:502-505, 1980.
15. Richardson, W.D., Carter, B.J. and Westphal, H. Proc.Natl.Acad.Sci. USA 77:931-935, 1980.
16. Janik, J.E., Houston, M.M. and Rose, J.A. Proc.Natl.Acad.Sci. USA 78:1925-1929, 1981.
17. Graham, F.L., Harrison, T. and Williams, J. Virology 86:10-21, 1978.
18. Lusby, E.W., Fife, K.H. and Berns, K.I. J.Virol. 34:402-409, 1980.
19. Straus, S.E., Sebring, E.D. and Rose, J.A. Proc.Natl.Acad.Sci. USA 73:742-746, 1976.
20. Hauswirth, W.W. and Berns, K.I. Virology 78:488-493, 1977.
21. Hauswirth, W.W. and Berns, K.I. Virology 93:57-68, 1979.
22. Berns, K.I. and Hauswirth, W.W. Adv.Virus.Res. 25:407-449, 1979.
23. Spear, I.S., Fife, K.H., Hauswirth, W.W., Jones, C.J. and Berns, K.I. J. Virol. 24:627-634.
24. Samulski, R.J., Berns, K.I., Tan, M. and Muzyczka, N. Proc.Natl.Acad. Sci. USA 79:2077-2081, 1982.
25. Sutcliffe, J.G. Cold Spring Harbor Symp. Quant.Biol. 43:77-90, 1978.
26. Srivastava, A., Lusby, E.E. and Berns, K.I. J. Virol.45:555-564, 1983.
27. Samulski,R.J., Srivastava, A., Berns, K.I. and Muzyczka, N. Cell, in press, 1983.
28. Berns, K.I. and Hauswirth, W.W. In: Organization and Replication of Viral DNA (Ed. A. Kaplan), Chemical Rubber Company, 1982, pp. 3-35.
29. Parks, W.P., Casazza, A.M., Alcott, J. and Melnick, J.L. J. Exp. Med. 127:91-108, 1968.
30. Kirchstein, R.L., Smith, K.O. and Peters, E.A. Proc.Soc.Exp. Biol. Med. 128:670-673, 1968.
31. Mayor, H.D., Houlditch, G.S. and Mumford, D.M. Nature (London) New Biol. 241:44-46, 1973.

32. de la Maza, L.M. and Carter, B.J. J.Nat.Cancer Inst. 67:1323-1326, 1981.
33. Ostrove, J.M., Duckworth, D.H. and Berns, K.I. Virology 113:521-533, 1981.
34. Blacklow, N.R., Cukor, G., Kibrick, S. and Quinman, G. In: Replication of Mammalian Parvoviruses (Ed. D.C. Ward and P. Tattersall) Cold Spring Harbor Laboratories, Cold Spring Harbor, N.Y. 1978, pp.87-98.
35. Dawson, G.J., Yates, V.J., Chang, P.W. and Oprandy, J.J. Nature 298: 580-582, 1982.
36. Mayor, H.D., Drake, S., Stahmann, J. and Mumford, D.M. Am.J.Obstet. Gynecol. 126:100-105, 1976.

9

HOST RANGE TRANSITION OF LYMPHOTROPIC MINUTE VIRUS OF MICE DURING PERSISTENT INFECTION IN L-CELLS

DINA RON, ESTHER GUETTA and JACOV TAL
Department of Biology, Ben-Gurion University of the Negev, Beer Sheva, Israel

SUMMARY

Infection of A9 strain of mouse L-cells with the lymphotropic strain of minute virus of mice (MVM(i)), resulted in initial arrest of viral replication and a subsequent establishment of a short term, 150 day long persistent infection. The virus produced by the carrier culture differed from MVM(i) in its ability to replicate lytically in A9 cells and the parallel loss of infectivity to T-lymphocytes. The newly acquired host range was a stable property of the progeny virus, suggesting that it is genetically determined. Restriction endonuclease analysis failed to detect differences between the mutant viral DNA and that of MVM(i), indicating that the host range determining region in the viral genome is small.

INTRODUCTION

Minute virus of mice (MVM) is an autonomous parvovirus containing a linear, single stranded (ss) DNA genome, some 5000 nucleotides in length (1). In susceptible cells, MVM establishes a lytic infection which results in the appearance of cytopathic effects (CPE) and cell death within 48-72 hours. Following penetration of the infected cell nucleus by the parental ssDNA two separate types of DNA synthesis are observed: conversion of the viral DNA to a double stranded (ds) replicative form (RF) DNA, and DNA amplification synthesis (1). The assembly of viral particles is done in the cell nucleus.

Two strains of MVM are known to date: the prototype MVM(p) and the immunosuppressive virus MVM(i) (2). Although they are serologically identical (3) and have very similar DNA sequences according to restriction endonuclease mapping criteria (2), they differ in their host range properties: MVM(p) replicates lytically in L-cells and MVM(i) grows in T-lymphocyte cell lines (4). The reciprocal infections, i.e., MVM(p) in T-cells and MVM(i) in L-cells are both non-productive, resulting in limited cell lysis and virus production but not in the destruction of the infected cultures (4). In both cases the blocks to viral replication have been shown to be intracellular ones. Hence, these growth properties, together with their limited genetic capacity make these viruses an attractive model system for the study of the genetic determination of viral host range.

Our aim in this study was, originally, to gain an understanding of the mechanisms by which MVM(i) replication in A9 cells (L-cell derivative line) was restricted. Further experiments with this system revealed a more complex virus-cell relationship: after an initial period in which MVM(i) replication was blocked, a persistent infection emerged. We show here that the virus shed by the persistently infected culture had undergone host range transition, and present evidence that this transition is genetically determined.

MATERIALS AND METHODS

Cells

A9, a derivative of mouse L-cells (7), EL4, a mouse lymphoma cell line (8) and hyb 1/11, a hybrid line between ouabain resistant A9 and EL4 cell lines were obtained from P. Tattersall. The mouse testicular cell lines TM3 and TM4 (9) were provided by J.P. Mather. All cells were grown in Dulbecco Minimal Essential Medium (Gibco) supplemented with 5% fetal calf serum (Seralab). HAT solution was added to the hyb 1/11 growth medium (10).

Viruses

MVM(p) and MVM(i) were obtained from P. Tattersall.

Virus titrations

Hemagglutination and plaque assays were done according to previously described procedures (11). Infectivity of all 3 MVM strains was determined by plaque assays using hyb 1/11 as indicator cells.

Immunofluorescence straining

The indirect method was used (12). Fluorescein conjugated goat anti-rabbit IgG was purchased from Miles Yeda.

Dispersed cell assay

The method described by Lavi and Etkin was used (13). In short, infected or uninfected cells (between 10^5 and 10^6) were trapped on 25 mm, 0.2 μ nitrocellulose filters (Schleicher and Schull) by slow filtration. The filters were air dried, denatured by alkali, neutralized, baked in an oven at 80°C for 4 hrs and hybridized to double stranded viral DNA which was labelled with ^{32}P by nick translation. The viral DNA, cloned in pBR 322, was obtained from D.C. Ward.

RESULTS

Establishment of a persistent infection of MVM(i) in A9 cells

The infection of A9 cells with the lymphotropic strain MVM(i) does not initially affect their growth rate and viability. The single stranded viral DNA is converted to a double stranded RF, and is efficiently amplified, but very small amounts of progeny ssDNA are synthesized (Ron and Tal, manuscript in preparation). When the infected culture (designated $A9_{MVM(i)}$) was maintained in monolayer culture for a period exceeding 10 days, extensive cell lysis began and variable amounts of virus were detected by hemagglutination and plaque assays, but complete destruction of the culture did not occur. A persistent infection was thus established, in which virus production and cell destruction were maintained alongside with cell proliferation. The percentages of cells involved in virus production at different stages during the entire 150 day long process were determined by immunofluorescence. The results (table 1) show that the peak of cell lysis was between 40 and 60 days post infection. This period was also characterized by high levels of virus production and by a marked decrease in the overall growth rate of the culture (data not shown). After this "crisis" period, the culture resumed normal growth rate but low level virus production was maintained for an additional 90 days. Eventually, the cells were spontaneously cured from detectable infectious virus as well as from viral proteins and viral DNA.

TABLE 1. Percentages of A9 cells involved in virus production during the persistent infection.

Days post infection (dpi)	% Fluorescent nuclei
5	0.4
9	0.3
16	10.0
40	16.0
56	17.0
85	4.0
91	3.0
150	<0.1

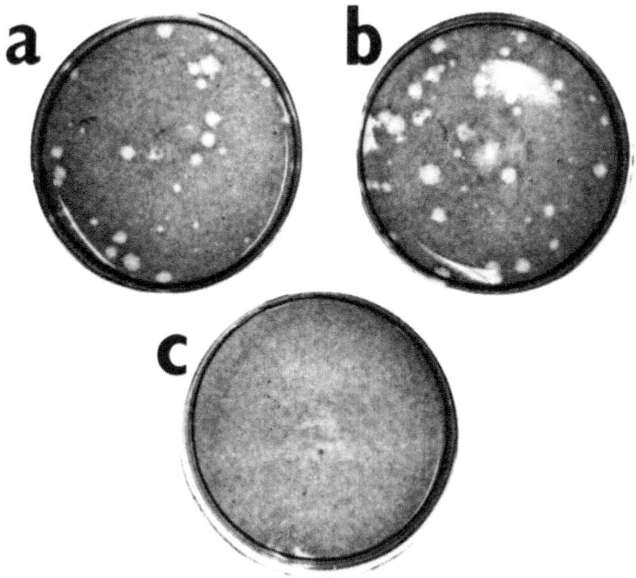

FIGURE 1. Plaque formation in A9 cells by MVM(i) and $MVM(i)_{A9}$. $5x10^5$ cells in monolayer culture were infected with 40 pfu of $MVM(i)_{A9}$ or with $4x10^5$ pfu of MVM(i) and overlayed with agar for direct plaque assay. (a) $MVM(i)_{A9}$ from passage 8. (b) $MVM(i)_{A9}$ from passage 15. (c) MVM(i).

Evidence for host range transition of MVM(i)

To study the growth characteristics of the persistent infection progeny virus (designated $MVM(i)_{A9}$), we prepared a plaque purified stock of virus from passage 15 (100 dpi), and compared its ability to replicate in fresh A9 cells to that of MVM(i). Figure 1 shows that unlike MVM(i), $MVM(i)_{A9}$ has acquired the ability to produce plaques in A9 cell monolayers. The same MVM(i) stock did produce plaques in the permissive hyb 1/11 cells, indicating that its inability to grow in A9 cells was due to its restriction in these cells and not to a possible loss of infectivity.

The change in viral host range was also evident from visualizing viral capsid proteins in situ: A9 cells were infected with MVM(p), MVM(i), and $MVM(i)_{A9}$, and 20 hours later subjected to immunofluorescence staining. The results, some of which are shown in Figure 2, demonstrate the absence of fluorescence upon infection with MVM(i) (Fig. 2b). However, infection of

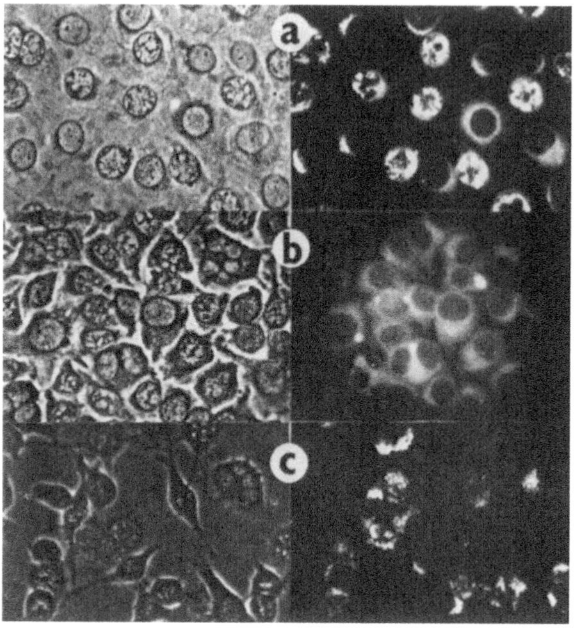

FIGURE 2. Indirect immunofluorescence assay of MVM(p), MVM(i) and $MVM(i)_{A9}$ infected A9 cells.

The cells were infected with 10 pfu/cell of each virus. The left hand columns are phase contrast micrographs of representative fields and the right hand columns are immunofluorescence of the same fields. (a) MVM(p). (b) MVM(i). (c) $MVM(i)_{A9}$.

these cells with either MVM(p) or $MVM(i)_{A9}$ (Fig. 2a and c, respectively) resulted in similar percentages of fluorescent nuclei (around 20% at 10 pfu/cell). To further compare the ability of EL4 and A9 cells to support $MVM(i)_{A9}$ replication, two parameters were chosen: virus production, which was assayed by hemagglutination, and viral DNA synthesis by the dispersed cell assay. The results of this analysis (Table 2) clearly show that $MVM(i)_{A9}$ had lost its ability to grow or to replicate its DNA in EL4 cells, and had adapted to grow in A9 cells with efficiency comparable to that of MVM(p). Hence, the $MVM(i)_{A9}$ host range is an altered, rather than an expanded one.

It could be argued that $MVM(i)_{A9}$ was not a host range mutant of MVM(i) but rather MVM(p) which was present as a minor contaminant in the original

TABLE 2. Growth properties of MVM(p), MVM(i) and $MVM(i)_{A9}$

	Viral DNA synthesis			Virus production		
	MVM(p)	MVM(i)	$MVM(i)_{A9}$	MVM(p)	MVM(i)	$MVM(i)_{A9}$
		(CPM)			(HAU/cell x 10^5)	
EL4	1,160	21,490	1,040	N.T.(-)	2,200(+)	0.9(-)
A9	70,340	5,390	50,260	2,000(+)	1.2(-)	1,600(+)

Cells were infected at a moi of 20 pfu/cell. 24 hours later the relative viral DNA content of $8x10^5$ cells was determined by the dispersed cell assay. Hemagglutination assay was performed at 4 dpi. The (+) and (-) symbols indicate the presence or absence of characteristic CPE, visualized microscopically.

MVM(i) stock and was enriched when passaged in A9 cells. To test this possibility, DNA from $MVM(i)_{A9}$ virus was extracted, converted to a double stranded form in vitro by DNA polymerase I and was labelled with ^{32}P by the inclusion of α-^{32}P-dATP in the reaction. MVM(i) and MVM(p) DNAs were similarly treated and the 3 DNAs were subjected to cleavage with 7 restriction endonucleases (AluI, MboI, HinFI, PstI, HhaI, FnuDII and Hind III) which had been previously shown to cleave MVM(i) and MVM(p) differently in at least one of their sites (2). The cleavage products were separated by agarose gel electrophoresis and the results, some of which are shown in Figure 3, clearly

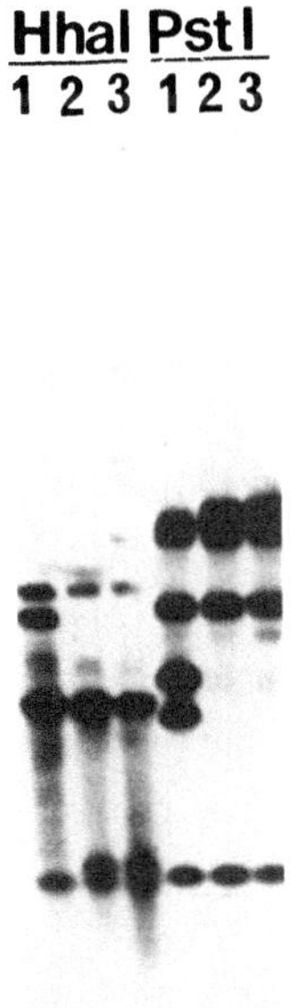

FIGURE 3. Comparisons of MVM(p), MVM(i) and $MVM(i)_{A9}$ DNAs by restriction and nuclease cleavage.

See text for details. (1) MVM(p). (2) MVM(i). (3) $MVM(i)_{A9}$

indicate that $MVM(i)_{A9}$ was a host range mutant of MVM(i), and not MVM(p). The virus was designated hr 301. It should be noted that none of the enzymes used in this analysis (altogether accounting for about 80 separate cleavage sites) revealed any differences between the DNAs of MVM(i) and hr 301, suggesting that if such a difference exists, it must be very small.

Hr 301 replication in EL4 cells is blocked between RF DNA synthesis and amplification

To localize the block to its replication in EL4 cells, EL4 and A9 cells were infected with hr 301 virus at 10 pfu/cell. 4 and 24 hours post infection (hpi) the cells were extracted by the Hirt procedure, the ss and ds viral DNAs were separated by agarose gel electrophoresis, blotted onto nitrocellulose membrane filters and hybridized to viral DNA probe. The results (not shown) revealed the following: (i) the block was not due to a membranal barrier in EL4 cells, since at 4 hpi the parental, ssDNA contents of both EL4 and A9 cells were quantitatively identical. (ii) Complete conversion of parental DNA to ds RF DNA was observed at 24 hpi, in both A9 and EL4 cells. However, there was little or no amplification of viral RF DNA in EL4 cells. (iii) Progeny ssDNA was not made in EL4 cells. These results localize the block to hr 301 replication between RF DNA synthesis and amplification. It should be noted that MVM(p) replication in EL4 cells is blocked at the same stage (Ron and Tal, manuscript in preparation).

Phenotypic stability of hr 301

In the absence of a direct proof that the host range transition of MVM(i) resulted from a mutation in the viral DNA, the stability of the acquired host range was used as an indication to support genetic change. To test this, NB324K, an SV40 transformed human cell line was used (5). Being permissive to both MVM(i) (4) and hr 301, it enabled the replication of the mutant virus without subjecting it to selective pressures. The cells were infected with hr 301 and the progeny virus was used to reinfect fresh NB cells. After 9 additional passages the host range of the progeny virus was identical to that of hr 301, i.e., it grew well in A9 cells but not in EL4 cells, suggesting that the host range transition is genetically controlled. Furthermore, repeated attempts to establish a lytic infection of hr 301 in EL4 cells have failed even when high moi (over 100 pfu/cell) was used.

To gain better understanding of the relationship between the host range

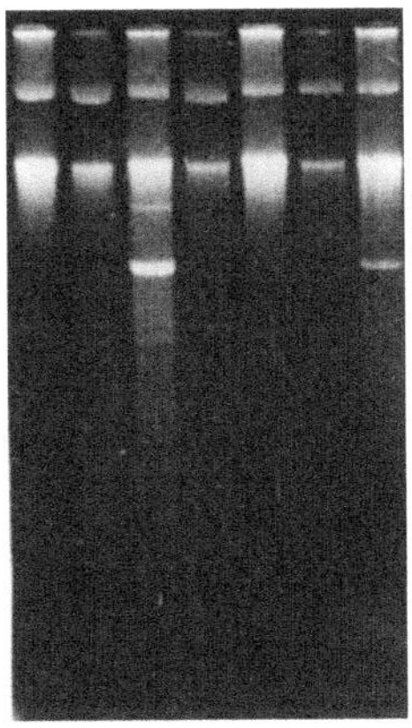

FIGURE 4. Intracellular viral DNA forms in infected TM3 cells.

The cells were infected with 10 pfu/cell each of MVM(i), MVM(p) and hr 301. Viral DNAs were extracted from the cells by the Hirt procedure, separated by agarose gel electrophoresis and stained with ethidium bromide. RF1 and RF2 are monomeric and dimeric forms of viral RF DNA. ss, single stranded viral DNA (barely visible in the photographic reproduction). (1) Mock infected. (2) MVM(p), 4 hpi. (3) MVM(p), 24 hpi. (4) MVM(i), 4 hpi. (5) MVM(i), 24 hpi. (6) hr 301, 4 hpi. (7) hr 301, 24 hpi.

properties of MVM(p), MVM(i) and hr 301, several cell lines were screened for susceptibility to the 3 viruses. From this survey, one general conclusion emerged. Every cell line tested that was permissive to MVM(p) also supported hr 301 replication. This correlation was true not only for mouse fibroblastic cell lines but also for TM4, a mouse cell line of epithelial origin (9) and for ATla, a line of rat cells which had been transformed by avian erythroblastosis virus (6). Both these cell lines were permissive to MVM(p) and hr 301 but restrictive to MVM(i). Furthermore, this generality seems to apply not only to virus production, but also to partial expression of the viral functions in restrictive cells. One such example is in TM3, another mouse testicular cell of epithelial origin, believed to originate from Leydig cells (9). Although these cells are non-permissive to either MVM(p), MVM(i) or hr 301, the restrictive stage is not identical for all 3 viruses. A gel electrophoretic analysis of the viral DNAs at 4 and 24 hpi (Figure 4) shows that MVM(p) DNA (lanes 2 & 3) and hr 301 DNA (lanes 6 & 7), but not MVM(i) DNA (lanes 4 & 5) were converted to double stranded replicative forms. Further experiments showed that MVM(i) penetrates TM3 cells efficiently, but persists in single stranded form in the nucleus (Guetta and Tal, in preparation).

DISCUSSION

There is an extensive literature dealing with evolution of virus populations during the course of persistent infections. It is generally found that cytolytic viral infections are converted to more temperate host-virus interactions, which allow the virus to reside within the culture without destroying it. This coexistence is made possible by one or more of four mechanisms recognized to date: (i) the generation of viral mutants, generally conditional mutants such as temperature sensitive (ts), which alter the growth pattern of the virus (14,15,16); (ii) appearance of defective interfering (DI) particles among the progeny virus population. Such particles, which contain shortened viral genomes within normal protein capsids, interfere with the replication of the wild-type viruses (17,18); (iii) production of interferon by the infected cells. Interferon has been shown responsible for the maintenance of persistent infections in several systems (19,20); and (iv) integration of the viral DNA into the host genome (21,22). While integration is found among certain DNA viruses and retroviruses, ts and DI mutant particles have been implicated mainly in persistent infections by RNA viruses. The reasons for this is most probably the relatively high occurrence of errors in replication of the RNA genome, due to the absence of RNA proof reading enzymes, which renders these genomes more prone to the induction of mutants (23).

The results presented here show that upon infection of A9 cells with MVM(i), a persistent infection resulted, in the course of which a host range mutant virus, designated hr 301, was produced. The events leading to the establishment of this persistent infection are not clear. During the initial 10 days, small amounts of virus were detected by hemagglutination, and preliminary results indicate that most, if not all of this virus was wild type MVM(i). It seems, therefore, that the series of events leading to persistent infection includes: (i) formation of "restricted" MVM(i) RF DNA; (ii) its partial amplification and (iii) introduction of mutation, or mutations, into the restricted RF molecules which can overcome the block and subsequent establishment of a lytic infection in a fraction of the infected culture. According to this scheme, the persistent infection is established in A9 cells by hr 301 virus and not by MVM(i). However, it is not clear why, once a mutant virus appeared, complete destruction of the culture did not occur. The nature of the "restricted" RF DNA, too, is unclear. Since DNA methylation has been shown to interfere with gene expression in various systems (24,27,28) we investigated the possibility that MVM(i) DNA was methylated in A9 cells, but no methylations in viral Hpa II and Hha I sites were detected (Segev, Ron and Tal, unpublished results).

We have recently found that none of the known mechanisms used to establish or maintain persistent infections (i.e., interferon, DI particles, ts mutants and integration) was involved in this persistent infection (Ron, Tattersall and Tal, in preparation). It is, therefore, possible that hr 301 is not a uniform virus throughout the 150 day process, but that rather it undergoes sequential changes, which enable it to establish productive infection in only a fraction of the cells at each stage. Alternatively, or perhaps in addition to changes in the viral population, the cell population may be changing and gradually selecting out those which resist viral infection. That the carrier state is maintained by a sequence of changes that render a transient fraction of the culture susceptible to lytic infection is supported by our ability to cure the culture with anti-MVM serum at any stage in the persistent infection. Sequential changes in the cell population are supported by the spontaneous termination of the persistent infection and the elimination of any traces of virus. Similar spontaneous curing has been previously observed in L-cells persistently infected with VSV (25), and coevolution of virus and cells during in vitro persistent infection was demonstrated by Ahmed et al. in reovirus infected L-cells (26).

Finally, it is interesting to point out that the change in MVM(i) host range was contained within the limits of the two allotropic states known to us. It seems at present that MVM can exist in either one of the two host range phenotypes, defined by the ability to grow in discriminating cell lines such as L-cells and T-lymphocytes.

ACKNOWLEDGMENTS

We thank Peter Tattersall and David Ward for valuable discussions and materials, Dominique Stehelin for ATla cells, Jenifer Mather for TM3 and TM4 cells, Alan Mehler for critical reading of this manuscript and Nitza Peleg for technical assistance. This work was supported by grants from the Israel Cancer Society and the U.S.-Israel Binational Science Foundation (BSF).

REFERENCES

1. Bournguignon, G.J., Tattersall, P. and Ward, D.C. J. Virol. 20: 29-306, 1976.
2. McMaster, G.K., Beard, P., Ingres, H.D. and Hirt, B. J. Virol. 38: 317-326, 1981.
3. Bonnard, G.D., Manders, E.K., Campbell, D.A. Jr., Herberman, R.B. and Collins, M.J. J. Exp. Med. 143: 187-205, 1976.
4. Spalholz, B.A., Bratton, J., Ward, D.C. and Tattersall, P. J. Cell Biochem., in press.
5. Stein, H.M. and Enders, J.F. Proc. Soc. Exptl. Biol. Med. 109: 495-499, 1962.
6. Quade, K. Virology 98: 461-465, 1979.
7. Littlefield, J.W. Nature 203: 1142-1144, 1964.
8. Shevach, E.J., Stabe, J.D. and Green, I. J. Immunol. 108: 1146-1151, 1972.

9. Mather, J.P. and Sato, G.H. Exp. Cell Res. 124: 215-221, 1979.
10. Littlefield, J.W. Science 145: 709-710, 1963.
11. Tattersall, P. J. Virol. 10: 586-590, 1972.
12. Tattersall, P. In: Replication of Mammalian Parvoviruses (Eds. D.C. Ward and P. Tattersall) Cold Spring Harbor Laboratory, Cold Spring Harbor, N.Y., 1978, pp. 131-149.
13. Lavi, S. and Etkin, S., Carcinogenesis 2: 417-423, 1981.
14. Youngner, J.S., Dubovi, E.J., Quagliana, D.O., Kelly, M. and Preble, O.T. J. Virol. 19: 90-101, 1976.
15. Youngner, J.S. and Preble, O.T. Comprehensive Virology 16: 73-135, 1980.
16. Yoshida, T., Hamaguchi, M., Naruse, H. and Nagai, H. Virology 120: 329-339, 1982.
17. Holland, J.J., Kennedy, S. Ian T., Semler, B.L., Jones, C.L., Roux, L. and Grabau, E.A. Comprehensive Virology 16: 137-192, 1980.
18. Huang, A.S. and Baltimore, D. Nature 226: 325-327, 1970.
19. Ramseur, J.M. and Friedman, R.M. J. Gen. Virol. 37: 523-533, 1977.
20. Mathews, J.H. and Vorndam, A.V. J. Gen. Virol. 61: 177-186, 1982.
21. Weinberg, R.A. Ann. Rev. Biochem. 9: 197-226, 1980.
22. Handa, H., Shiroki, K. and Shimojo, H. Virology 82: 84-92, 1977.
23. Holland, J.J., Spindler, K., Horodyski, F., Grabau, E., Nichol, S. and Vandepol, S. Science 215: 1577-1585, 1982.
24. Youssoufian, H., Hammer, S.M., Hirsch, M.S. and Mulder, C. Proc. Natl. Acad. Sci. USA 79: 2207-2210, 1982.
25. Holland, J.J., Semler, B.L., Jones, C., Perrault, J., Reid, L. and Roux, L. In: ICN-UCLA Symp. Molec. Cell. Biol., Vol. 11 (Eds: J.G. Stevens, G.T. Todaro and C.F. Fox) Academic Press, 1978, pp. 57-73.
26. Ahmed, R., Canning, W.M., Kauffman, R.S., Sharpe, A.H., Hallum, J.V. and Fields, B.N. Cell 25: 325-332, 1981.
27. Stewart, C.L., Stuhlmann, H., Jähner, D. and Jaenisch, R. Proc. Natl. Acad. Sci. USA 79: 4098-4102, 1982.
28. Kruczek, I. and Doerfler, W. The EMBO Journal 1: 409-414, 1982.

10

TRANSPLANTATION OF MOUSE GENES INTO RAT CELLS CHRONICALLY INFECTED BY MOLONEY MURINE LEUKEMIA VIRUS (M_O-MuLV)

RACHEL OREN[1] and YAEL TEITZ[2]

[1]Israel Institute for Biological Research, Ness-Ziona, Israel and
[2]Sackler School of Medicine, Tel-Aviv University, Ramat-Aviv, Israel

SUMMARY

Three rat cell lines (RE, ES and RT21) chronically infected with Moloney Murine Leukemia Virus (M_O-MuLV) became sensitive to infection with EMC virus and to the antiviral effect of mouse interferon. These two properties are specific to mouse cells only. EMC virus multiplies in M_O-MuLV infected rat cells to a titer of 10^8 pfu and destroys the infected cells. Mouse interferon induces an antiviral state in these cells when challenged with EMC or VSV viruses, and causes inhibition of 97 per cent of M_O-MuLV release. These properties are observed only in rat cells chronically infected with M_O-LV obtained from mouse, but not in those infected with Moloney Leukemia Virus obtained from rat (M_O-RLV).

INTRODUCTION

Retroviruses are eukaryotic RNA viruses which during their multiplication cycle produce intermedial proviruses. The provirus hypothesis suggests that retroviruses evolve in nature by recombination of replication-competent retrovirus with a portion of the cell genome involved in normal transfer of genetic information in and between cells (1, 2). These viruses incorporate their genomes into the host cell chromosomes during one phase of their growth cycle, a property that enables them to become "transducers" of cellular genes (1).

Retroviruses are viruses that carry genes that are without essential replicative or structural function. The role, origin, purpose and even number of the viral non-essential genes that retroviruses may carry from one host cell to the other are still under investigation. Recently the oncogenes carried by retroviruses which code for transforming proteins have been studied very intensively in many laboratories (2, 3).

In the present study, we have been able to demonstrate the transplantation of mouse genes into three rat cell lines, chronically infected with Moloney Murine Leukemia Virus (M_0-MuLV) (4). This virus replicates in the rat cells without affecting their normal growth. Rat cell lines infected chronically with M_0-MuLV express characters typical to mouse cells, such as sensitivity to infection with EMC virus and an antiviral response to mouse interferon.

Treatment of cells with interferon causes them to become unable of supporting replication of a broad range of viruses. Interferon inhibits viral multiplication most effectively in cells of species in which it has been produced. Several pathways have been suggested which are operative in interferon-treated virus-infected cells. The biochemical basis of inhibition of replication of EMC or VSV virus seems to be an inhibition of viral RNA or protein synthesis, or both. Treatment with interferon on retrovirus chronically-infected cells causes inhibition of virus release. Retrovirus production is blocked at the plasma membrane level where the virions assemble and from where they bud (5, 6, 7, 8).

RESULTS

Sensitivity to infection with EMC virus.

Different cell lines were assayed for their sensitivity to infection with Encephalomyocarditis virus (EMC). Mouse cells, but not rat cells, are sensitive to this infection. Thus, cell lines originated from mouse non-infected or infected with Moloney Murine Leukemia Virus (M_0-MuLV) or Moloney Rat Leukemia Virus (M_0-RLV) (9, 10) were sensitive to EMC infection. Cytopathic effect was observed after 24 - 28 hrs and virus titers usually reached levels of 10^8 pfu/ml. Cells originated from rat, non-infected or infected with M_0-RLV were resistant to EMC infection. However, when the same rat cells were chronically infected with M_0-MuLV, they became sensitive to EMC infection, yielded high titers of virus and showed cytopathic effect similar to that observed in cells originated from mouse. These results are summarized in Table I.

Table 1. Sensitivity of different cell lines to infection with EMC virus

Origin of cells	Cell line	Sensitivity to infection	
		Cytopathic effect	Titers (pfu/ml)
Rat embryo	RE	-	-
	RE M_0-RLV	-	-
	RE M_0-MuLV	+	10^8
Rat thymus	ES	-	-
	ES M_0-RLV	-	-
	ES M_0-MuLV	+	10^7
Rat thymus	RT21	-	-
	RT21 M_0-RLV	-	-
	RT21 M_0-MuLV	+	10^8
Mouse embryo	3T3 M_0-RLV	+	10^8
	3T3 M_0-MuLV	+	10^8
Mouse	L 929	+	10^8

Confluent cultures on 4 cm plastic plates from the different cell lines were infected with EMC virus in various titers from 10^1-10^7 pfu/ml. After 24 - 28 hrs, the cultures were assayed for virus yield and cytopathic effect (CPE).

Virus yield was assayed on mouse cell line L 929 and expressed as pfu per 1 ml.

(-) no CPE
(+) CPE

Antiviral activity induced by interferon treatment.

Because of the species specificity of interferon, mouse interferon induces an antiviral state in mouse cells but not in rat cells. Protection induced by interferon treatment to viral infection was tested in the different rat cell lines. Table 2 presents the experimental results. Rat cell lines, non-infected and chronically infected by M_0-MuLV or M_0-RLV were treated with various doses of mouse or rat interferon. After 24 hrs cell response to infection with 10^4 pfu of VSV was examined. As can be seen in Table 2, only the rat cell line RE chronically infected with M_0-MuLV established protection against VSV by previous treatment with mouse interferon, and, of course, the mouse cell line 3T3 chronically infected with M_0MuLV which served as a positive control in the experiment. In rat cell lines, non-infected or chronically infected with M_0-RLV, only rat interferon induces antiviral activity. Similar antiviral response after treatment with mouse interferon was shown in the M_0-MuLV chronically-infected rat cells when infected with EMC virus. (Results are not presented).

Table 2. Sensitivity of different rat cell lines to antiviral effect on mouse interferon

Origin of cells	Cell line	Interferon unit/ml Mouse -			Rat -		
		$5x10^{-4}$	$5x10^{-3}$	$5x10^{-2}$	10^{-4}	$5x10^{-3}$	$5x10^{-2}$
Rat embryo	RE	+	+	+	-	-	+
	RE M_0-RLV	+	+	+	-	-	+
	RE M_0-MuLV	-	-	+	-	-	+
Rat thymus	ES	+	+	+	-	-	+
	ES M_0-RLV	n.d.	n.d.	n.d.	-	-	+
	ES M_0-MuLV	n.d.	n.d.	n.d.	-	+	+
Rat thymus	RT21 M_0-RLV	+	+	+	-	+	+
Mouse	3T3 M_0-MuLV	-	-	+	+	+	+

Cell lines were treated with various amounts of mouse or rat interferon for 24 hrs. Protection induced by interferon was assayed by detection of CPE in these cells following infection with 10^4 pfu of VSV per culture.

(-) no CPE
(+) CPE
(n.d.) not done

Inhibition of Moloney leukemia virus release by interferon.

The effect of Mouse interferon on virus production was examined in the mouse cell line 3T3 chronically infected with Moloney leukemia virus. Virus release was followed for 18 hrs after interferon treatment. Virus yield in the extracellular medium was measured by the amount of ^{3}H-uridine in pulse-labeled virus particles and by reverse transcriptase activity. As shown in Figure 1, the accumulation of virus particles in the culture media as measured by radioactivity and/or reverse transcriptase activity was similar. In the untreated cells, virus accumulated at a linear rate for the first 6 hrs and reached maximal level after 18 hrs. Virus release from the interferon treated cells was almost completely inhibited during the first 9 hrs, after which it rose slightly. At 18 hrs following interferon treatment, virus reverse transcriptase was inhibited by 83 per cent and the release of ^{3}-H-uridine-labeled virus by 85 per cent.

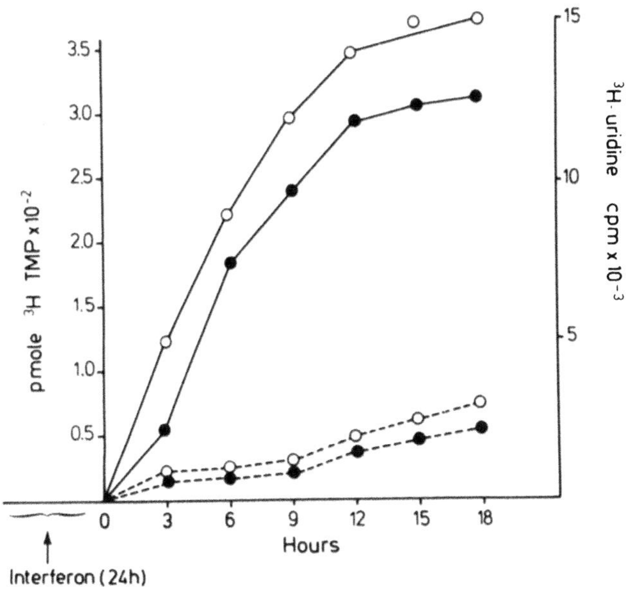

FIGURE 1.

Reverse-transcriptase activity: control (•——•)
interferon treated (•- - - - -•)

^{3}H-uridine labeled particles : control (o——o)
interferon treated (o- - - - - -o)

Release of Moloney leukemia virus after interferon treatment. 3T3/M_0-RLV confluent cultures in 75 cm plastic bottles were treated for 24 hrs with 20 u/ml mouse interferon. Interferon was removed and half of the cultures were labeled with 30 uCi/ml ^{3}H-uridine for 30 minutes. Cells were washed with PBS containing 10^{-3}M uridine and reincubated in fresh medium. At indicated intervals reverse-transcriptase activity in unlabeled cultures was assayed in 0.1 ml of supernatant fluids (6). The amount of ^{3}H-uridine labeled virus was determined in 2 ml samples of the supernatant fluids of the labeled cultures. The pellets of the labeled virus particles were centrifuged at 234,000 x g for 3 hrs in 15-60 per cent (w/w) linear sucrose gradient. Acid precipitable radioactivity was measured. M_0-RLV purified particles band at 1.16-1.18 g/ml.

Effect of increasing doses of mouse interferon on release of M_0-LV from chronically infected rat cell lines.

Virus release from the different cell lines chronically infected with Moloney Leukemia Virus was assayed by reverse transcriptase activity found in the supernatant fluid of these cells. In rat cell-RE chronically infected with M_0-MuLV, and treated with 20-200 units of mouse interferon, 90-95 per cent inhibition in the release of Moloney Murine Leukemia Virus (M_0-MuLV) was found. This response is shown in mouse cells chronically infected with Moloney Leukemia Virus (M_0-LV).

The RE and RT21 rat cell lines chronically infected with rat Moloney Leukemia Virus (M_0-RLV) responded very poorly to mouse interferon even at concentrations as high as 10^3 or 10^4 units.

Experimental results are demonstrated in Figure 2.

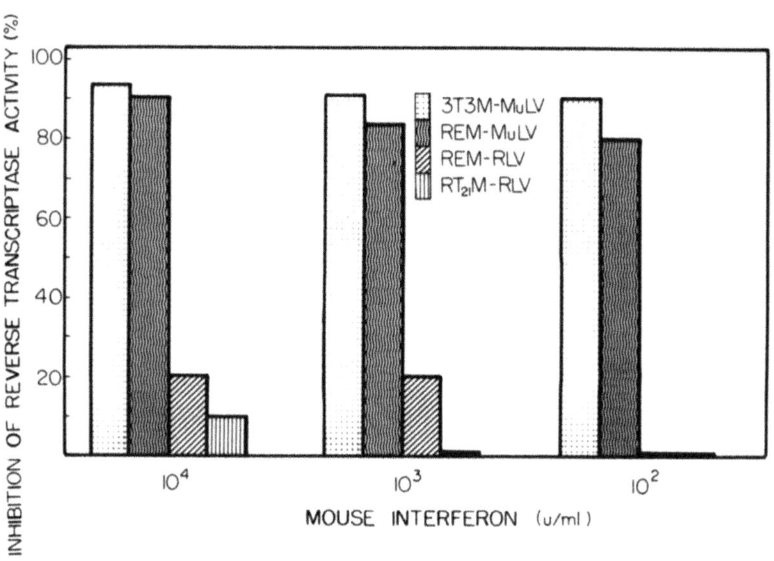

FIGURE 2. Effect of interferon on virus release from rat and mouse cell lines chronically infected with different M_0-LV.
Confluent cultures in 4 cm plastic plates of the rat cell lines and the mouse cell line chronically infected with the different M_0-LV were treated for 24 hrs with 10^4, 10^3 and 10^2 units/ml mouse interferon. Reverse transcriptase activity was assayed in 0.1 ml of supernatant fluids.

DISCUSSION

In this paper we present evidence that three rat cell lines exhibited sensitivity to EMC infection and responded to mouse interferon - as was demonstrated by their antiviral state against VSV and EMC infection - as well as to inhibition of the release of M_O-MuLV.

The acquisition of mouse characteristics was only found in rat cells chronically infected with a Moloney Leukemia Virus from a murine source.

In view of the known facts that retroviruses are RNA viruses which, during their multiplication cycle, produce double-stranded DNA reverse transcripts capable of integrating into the host cells genome, it is most probable that the expression of typical mouse characteristics was due to transfer of a mouse gene or genes carried by the virus from mouse to rat. Chromosome-mediated transfer of murine leukemia virus genetic information to uninfected recipient cell was recently shown (11). This hypothesis is strengthened by the finding that the same three rat cell lines, when infected with Moloney Leukemia Virus from a rat source, do not become sensitive to EMC virus infection and do not respond to mouse interferon.

It may be assumed that the lack of sensitivity of rat cells to mouse interferon is due to the absence in rat cells of proper receptors to mouse interferon. These receptors in mouse are the product of genes located on chromosome sixteen (12). If such a gene is incorporated into M_O-MuLV during its replication in mouse cells, the virus might, upon infection of rat cells, transfer the mouse gene into its genome; its expression might then cause the rat cell to become susceptible to mouse interferon. The results of our experiments support such a possibility.

Further experiments directed towards isolation, characterization and determination of the site of integration of these mouse genes in the rat cell genome, are now in process.

ACKNOWLEDGMENTS

We are grateful to Esther Burstein and Hana Ben-Moshe for their invaluable technical assistance.

REFERENCES

1. Bishop, J.M. Ann.Rev. Biochem. 47:35-88, 1978.
2. Keshet, E. and Temin, H.M. Proc. Natl Acad. Sci. 75:3372-3376, 1978.
3. Duesberg, P.H. Cold Spring Harbor Symposia on Quantitative Biology, Vol. XLIV, 13-29, 1980.
4. Fan, H. and Paskin, M. J. Virol. 14:421-429, 1974.
5. Sherman, L., Eylan, E. and Teitz, Y., Arch. Virol. 66:143-155, 1980.
6. Sherman, L. and Teitz, Y., Expl Cell Biol. 49:185-194, 1981
7. Friedman, R.M., Chang, E.H., Ramseur, M.J., Myers, M.W. J. Virol. 16:509-574, 1975.
8. Naso, R.B., Wu, Y.H.C. and Edbauer, C.A. J. Interferon Res. 2:75-96, 1982.
9. Cremer, N.E., Taylor, D.O.N., Oshiro, L.S. and Teitz, Y. J. Natl Cancer Inst. 45:37-48, 1970.
10. Teitz, Y., Lennette, E.H., Oshiro, L.S. and Cremer, N. J.Natl Cancer Inst. 46:11-23, 1971
11. Levin, J.G., Hughes, B.F., Greater, J.S., Rein, A., Rands, E. and Mukherjee, A.B. J. gen. Virol. 62:227-237, 1982
12. Cox, D.R., Epstein, L.B. and Epstein, C.J. Proc. Natl Acad. Sci. 77:2168-2172, 1980.

11

TRANSFORMING GENES OF RETROVIRUSES AND CANCER CELLS

S. A. AARONSON, National Cancer Institute, Bethesda, Maryland 20205, USA

An approach towards understanding mechanisms involved in processes leading to malignancy has come from studies of acute transforming retroviruses. These viruses have arisen in nature by recombination of replication-competent type C RNA viruses with a limited set of evolutionarily well-conserved cellular genes. When incorporated within the retrovirus genome, such transduced cellular (onc) sequences acquire transforming properties. To investigate the role of onc-related genes in human cancer, we have utilized molecular cloning techniques to isolate the human cellular homologues of a number of retroviral onc genes. These genes are often actively transcribed in human tumor cells, and in some cases in normal cells as well. We have mapped the chromosomal locations of onc-related genes in human cells and shown that in some cases such genes are involved in highly specific translocations associated with certain cancers. These findings raise the possibility that such translocations may affect onc gene expression in a manner that contributes to the neoplastic process. By a different approach involving DNA-mediated gene transfer techniques, dominant transforming genes, or oncogenes, have been detected in human tumors and tumor cell lines. The role of these genes in the malignant process remains to be elucidated. However, many of the oncogenes so far detected are related to the onc genes of bas, has, or kis, a small family of retroviruses comprised of BALB, Harvey, and Kirsten murine sarcoma viruses, respectively. Recent results pertaining to characterization of retroviral transforming genes and their related human oncogenes are presented.

12

CONTROL ELEMENTS FOR THE EXPRESSION OF DNA TUMOR VIRUSES

P. GRUSS
Mikrobiologie, Universitaet Heidelberg, Im Neuenheimer Feld 230,
6900 Heidelberg, Germany

Expression of viral genes is essential for the multiplication of viruses. The early region of many members of the papova virus group encodes a multifunctional protein (T-Ag). One of the functions involves the initiation of DNA replication. DNA replication in turn seems to depend on a certain level of T-Ag. Thus, one important element possibly involved in host-range could be the viral promoter, controlling the expression of T-Ag. In addition to promoter elements, we and others recently described a control region on the viral genome which greatly increases the transcriptional activity of certain genes. This functional element was termed enhancer or activator. It differs from promoter elements in its ability to exert this function relatively independent of distance and orientation with respect to the coding region. The overall activity of the enhancer/activator is host-dependent, suggesting a possible involvement of host specific factors. In an effort to define critical nucleotides we have constructed and analyzed mutants containing multiple nucleotide exchanges within the SV40 72 bp repeat. Our results define nucleotides essential for the enhancer function. We furthermore determined the minimal activity of the A-gene required for a lytic cycle.

13

NON-EXPRESSED TUMOR VIRUS ONCOGENES CAN BE ACTIVATED BY TUMOR PROMOTERS

G. SAUER and E. AMTMANN
Institute for Virus Research, German Cancer Research Center, 6900 Heidelberg, Germany

Papilloma viruses can induce both transformation of cells *in vitro* and tumor formation *in vivo*. The viral genomes reside in the cells as unintegrated episomes. In certain experimental systems such as primary mouse embryo cells, bovine papilloma genomes are not transcribed after infection, neither does replication occur. Eventually, after repeated passages, the viral episomes are lost by segregation from the phenotypically normal cells. However, a brief treatment with a tumor promoting substance induces both replication and transcription of the viral episomes and a transformed phenotype of the cells ensues which is stably maintained. Similarly, when the episome-bearing rodent Mastomys natalensis is exposed to the tumor promoter, skin tumor arise in the treated skin in which the Mastomys papilloma genomes (which are endogenous to this animal species) are amplified. Thus, the synergistic interaction of latent papilloma genomes *in vitro* can also be demonstrated in an animal system.

14

REARRANGEMENT AND ACTIVATION OF A CELLULAR ONCOGENE

E. Canaani, O. Dreazen, M. Horowitz, A. Klar, D. Ram, J. Cohen, G. Rechavi and D. Givol
Department of Chemical Immunology, The Weizmann Institute of Science, Rehovot 76100, Israel

SUMMARY

The c-mos oncogene was shown to be rearranged in the murine plasmacytoma XRPC24. This resulted in transcription of this previously "silent" gene and in its activation as a malignant transforming gene as shown by transfection assays. The rearrangement took place within the 5' region of c-mos coding sequence and it involved the insertion of an endogenous retroviral (intracisternal A-particle) genome. The 5' viral long terminal repeat, placed next to c-mos coding region, probably acts as an enhancer-promoter element.

INTRODUCTION

The nature of the genetic events associated with the malignant transformation of normal cells in natural or induced conditions is little understood. Much of our knowledge of these events has come from studies of the rapidly transforming and slowly transforming retroviruses. It is now well established that the first group of viruses acquired their oncogenic properties by incorporating within their genomes one of a group of cellular onc (c-onc) genes (1,2) which are limited in number (3), evolutionarily conserved (4,5) and probably have an essential physiological role in the normal cell and organism. Once incorporated into the viral genome, the onc gene (now termed v-onc) is controlled by viral signals and expressed in abnormally elevated levels within the infected cell. It is also possible that the viral onc gene product differs in subtle but critical ways from its progenitor, the cellular gene. The slowly tranforming retroviruses, which do not transduce cellular genes, were shown in some cases (6,7) to activate transcription of a particular cellular onc gene by integration in an adjacent position on the cellular DNA.

The situation is more obscure for the non-virally-induced tumours. By using the DNA-mediated gene transfer approach, it was possible to assay tumour DNAs for their ability to transform mouse fibroblasts in culture. This approach led to the identification of transforming genes in chemically- and radiation-induced tumours (8-12) and in spontaneous human and mouse tumours (9,10,13-16). Recently, three transforming genes from human bladder carcinomas (17-19) and one human lung carcinoma (17) were identified as the activated forms of the normal human homologue of the viral onc genes Ha-ras and Ki-ras, respectively. By nucleotide sequencing analysis it was very recently possible to show that the activated genes differed in a single nucleotide from their normal counterparts (20-22). On the other hand, the association of chromosomal aberrations with many murine and human tumours suggested that DNA rearrangements might constitute a general mechanism for tumour induction (23), possibly by their effect on cellular oncogenes (24).

We undertook the present studies to search in non-virally-induced murine myelomas for the possible activation of another cellular oncogene, c-mos. This gene, which is the progenitor of the transforming information in Moloney murine sarcoma virus (M-MuSV) (25,26), was characterized extensively by transfection studies (27-28) and nucleotide sequencing analysis (29,30). These studies established that the cellular information contained within the genome of M-MuSV (denoted v-mos), and presumably transduced out of the mouse genome during generation of this virus, consists of a long open reading frame of 369 codons. This sequence is joined in phase to a segment of five codons of viral helper sequence immediately 5' to the v-mos region (29,30). A 37,000-molecular weight (MW) protein identified within virus-infected cells (31) is presumably encoded by the 374-codons sequence and directly triggers cellular malignant transformation. The cellular homologue of v-mos (c-mos) contains 21 more codons joined in phase 5' to the 369-codon open reading frame described above (30). So far, no proteins encoded by this gene have been found in normal cells. Here we show that the c-mos gene is biologically activated in a particular mouse myeloma. The activation is due to DNA rearrangement at the 5' end of the cellular gene, involving the insertion of a cellular DNA element related to the family of endogenous viral A-particle genomes.

c-mos gene is rearranged and transcribed in a mouse myeloma

We screened the DNA of several BALB/c mouse myelomas by Southern blot analysis using the HindIII-BglI src-specific fragment of v-mos (24) as a probe; one of the tumours - XRPC24 - showed an additional hybridizing band detected by this probe (Fig. 1). In the EcoRI digest, this additional band

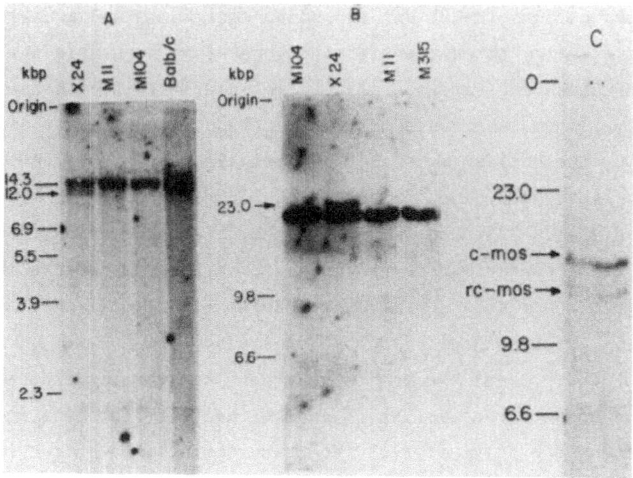

FIGURE 1. Analysis of the c-mos gene in mouse myeloma DNAs. High molecular weight DNA from mouse myelomas MOPC104E, XRPC24, MPC11 and MOPC315 (referred to in the figure as M104, X24, M11 and M315, respectively) was digested with EcoRI (a) or BamHI (b). Aliquots (20 μg) were electrophoresed on 0.7% agarose gels, blotted onto nitrocellulose filters and hybridized to the HindIII-BglI v-mos-specific fragment (24) radiolabelled by nick-translation. After hybridization, the blots were subjected to autoradiography. For molecular cloning, the c-mos and rc-mos EcoRI DNA fragments were enriched from XRPC24 DNA by electrophoresing 500 μg of digested DNA on 0.5% agarose gel, identifying the location of the two fragments on the gel by hybridization (c) and elution of the DNA by adsorption to glass beads. Arrows indicate the rearranged band.

was smaller (12.5 kilobase pairs, kbp) than the 14-kbp c-mos band, whereas in the BamHI digest the additional band was larger (23 kbp) than the 21-kbp c-mos band. These results indicated that the endogenous c-mos gene had undergone some rearrangement in the pristine-induced XRPC24 tumour, therefore we will refer to it as rc-mos. We attempted to determine whether this DNA rearrangement is associated with enhanced transcription of the gene. The poly(A)-containing RNA fraction was extracted from the same plasmacytomas and analysed by the Northern technique for mos-specific transcripts. Figure 2 shows that the probe detected a transcript of 1.2 kilobase (kb) in the RNA of myeloma XRPC24. No hybridization was observed with RNAs from the other tumours tested. Thus, the appearance of rc-mos DNA is correlated with increased transcription of the gene.

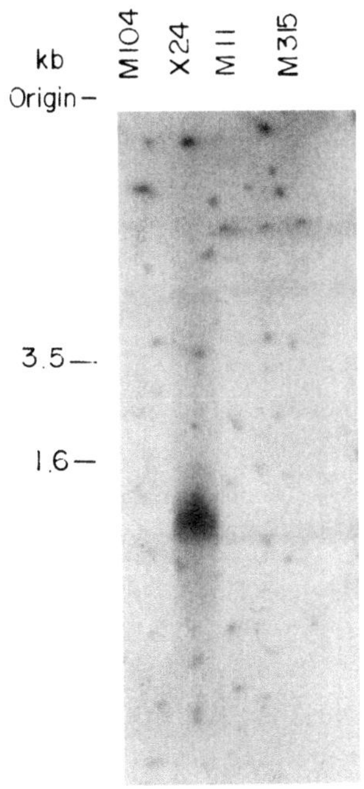

FIGURE 2. Analysis of poly(A)$^{+}$ RNA from mouse myeloma for the presence of c-mos transcripts. Aliquots (50 μg) of poly(A)-selected RNA from myelomas MOPC315, MOPC104E, XRPC24 and MPC11 (M315, M104, X24, M11 in the figure) were heated at 60°C for 10 min in a solution containing 50% formamide, 6% formaldehyde and running buffer (20 mM MOPS pH 7.0, 5 mM NaAc, 1 mM EDTA). The samples were electrophoresed at 100 V for 5 hr in 1% agarose gel containing 6% formaldehyde and 1 x running buffer. The RNA was transferred with 10 x SSC to nitrocellulose filters, fixed by heating at 80°C for 2 hr and hybridized at 43°C for 2 days to 2 x 10^{6} cpm ml^{-1} of v-mos probe in a solution containing 50% formamide, 5 x SSC, 1 x Denhardt's mixture, 20 mM sodium phosphate pH 7.0, and 100 μg ml^{-1} salmon sperm DNA. After washing at 50°C with a solution of 0.1 x SSC, 0.1% SDS, the filters were autoradiographed. Kb, kilobase.

To compare c-mos and rc-mos, we cloned both genes from myeloma XRPC24 DNA digested with EcoRI. The 14-kbp and 12.5-kbp mos fragments were enriched by preparative agarose gel electrophoresis (see Fig. 1c) and cloned into Charon 4A λ phage vector (32). Several clones containing one of the two inserts were purified and analysed in detail.

Biological activity of the rearranged c-mos gene

Previously, it was shown that the molecularly cloned, 14-kbp c-*mos* gene was inactive when assayed for its ability to transform NIH 3T3 cells in transfection experiments (28). To demonstrate biological activity of the gene, it was necessary to ligate a viral long terminal repeat (LTR) upstream from the 5' end of the gene (28). We asked whether the rearranged c-*mos* gene cloned from myeloma XRPC24 is active by itself when assayed by transfection; a typical experiment is shown in Table 1.

The two clones of the rc-*mos* gene that were tested induced a large number of foci having specific infectivities of 3,100 and 4,900 focus-forming units (FFU) per μg DNA, respectively (Table 1). In contrast, clones of the 14-kbp DNA fragment containing the normal c-*mos* gene showed negligible or no activity. As a positive control we used two recombinant phages containing integrated Moloney sarcoma viral DNA (E.C., unpublished) which produced 10,200 and 13,200 FFU per μg DNA (Table 1).

The cells transformed by rc-*mos* DNA were indistinguishable morphologically from those transformed with the viral genome. The transformants consisted of round and spindle-shaped highly refractile cells which grew in multilayers. The foci induced by rc-*mos* DNA could be detected 12 days after transfection in contrast to the 10 days required to detect the viral genome-induced transformants.

Table 1. Biological activity of molecularly cloned c-*mos* and rc-*mos* DNAs

Clone	Insert DNA (ng)	No. of foci	Specific infectivity (FFU per pmol DNA)
λ-MSV 26	50	150	10.200
λ-MSV 3-3	100	220	13,200
λ-c *mos* IA	125	0	67
λ-c-*mos* IB	125	1	<67
λ-rc-*mos* 2B	100	43	3,100
λ-rc-*mos* 2F	100	68	4,900

Intact recombinant phage DNA was transfected into NIH 3T3 fibroblasts by the calcium phosphate method (33). The cultures were kept in medium containing 5% fetal calf serum and were scored for focus formation 2 weeks after transfection. λ-MSV 26 and λ-MSV 3-3 represent M-MuSV molecules integrated within normal rat kidney cell DNA and molecularly cloned in λ Wes B vector. λ-c-*mos* and λ-rc-*mos* are molecular clones of the c-*mos* and rc-*mos* genes, respectively, obtained from XRPC24 myeloma DNA.

c-mos and rc-mos differ at the 5' region

The two cloned genes were analysed by restriction enzymes; the partial physical maps are shown in Fig. 3. The restriction map of c-*mos* is identical to that published by Vande Woude and co-workers (28). rc-*mos* shares with c-*mos* an identical 12-kbp fragment between the *BglI* site at 3.1 kbp on the map and the 3' end of the insert. The two clones differ at the region 5' to the *BglI* site up to the *EcoRI* site. The sequence of this region in the rc-*mos* clone, determined according to the strategy shown in Fig. 3, is shown in Fig. 4 together with the corresponding published sequence of c-*mos* (30).

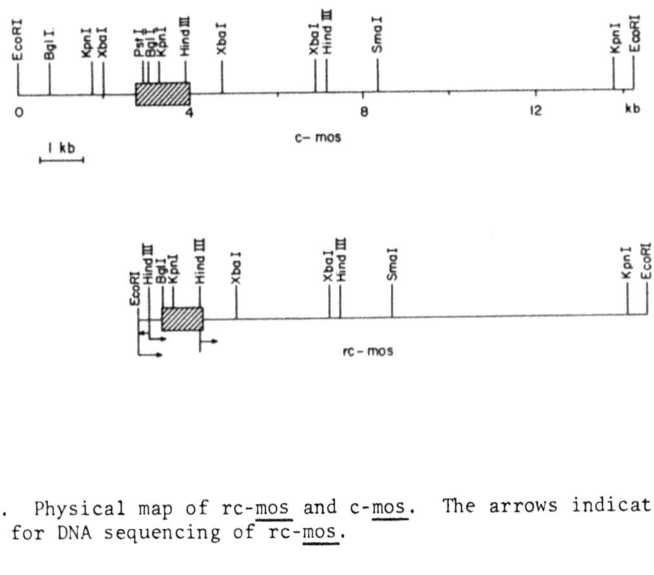

FIGURE 3. Physical map of rc-*mos* and c-*mos*. The arrows indicate the strategy for DNA sequencing of rc-*mos*.

Clearly, rc-*mos* and c-*mos* are identical 3' to nucleotide 513 and completely divergent 5' of this point. The c-*mos* sequence deleted in the rearranged gene includes about one-quarter of the presumed c-*mos* coding sequence (30). The deleted sequence also includes the first 41 codons in the portion of the gene shown to be conserved in both man and mouse (34). The inserted novel DNA fragment contains an open reading frame of 28 codons which joins in phase the rest of the c-*mos* open reading frame at nucleotide 513 in the sequence shown in Fig. 4.

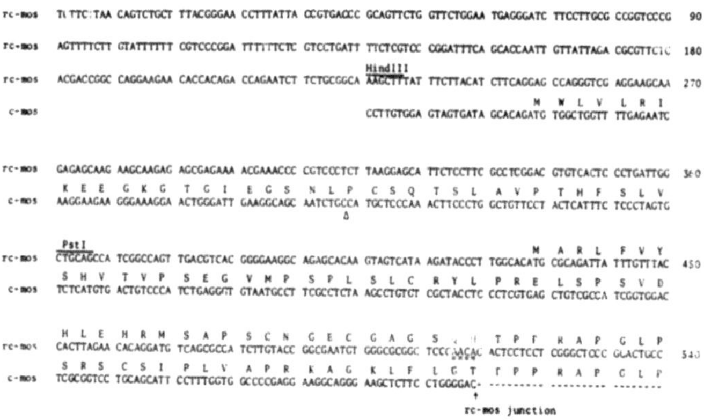

FIGURE 4. Sequence of the novel DNA segment inserted into c-*mos*. The fragment encompassing the *EcoRI-BglI* region at the 5' end of the rc-*mos* gene (Fig. 3) was sequenced according to Maxam and Gilbert (39). The c-*mos* sequence is shown for comparison, beginning with the first ATG of the open reading frame (30). Amino acids translated from the DNA sequence are given in the single letter code. The junction between the novel DNA fragment and the c-*mos* sequence is indicated by a small arrow. The junction (29) between the viral helper DNA and the *mos* information within the M-MuSV genome (v-*mos*) is indicated by an open triangle. The sequence numbers are from the *EcoRI* site shown in Fig. 3.

Analysis of the rc-*mos* sequence immediately 5' to the junction point (Fig. 4) showed that it ended with the tetranucleotide AACA which is identical to the 3' terminus of Drosophila transposable element *copia* (35), yeast Ty 1 mobile element (36,37), and spleen necrosis virus (38).

Promoter-enhancer activity of the inserted DNA element

The appearance of *mos* transcripts in XR24 tumour cells is most likely effected by the insertion of the new cellular DNA element within rc-*mos* coding region. To test a possible transcriptional enhancer-promoter activity of this DNA we subcloned the AvaI-AvaI DNA fragment spanning nucleotides 94-531 (Fig. 4) into the pSVO CAT vector (Fig. 5).

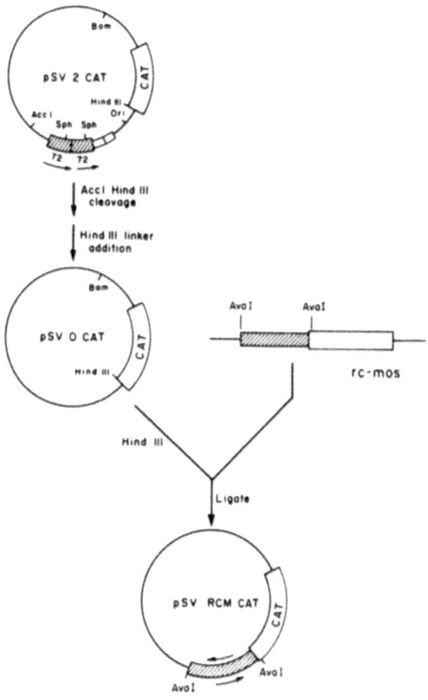

FIGURE 5. Construction of a chloramphenicol acetyltransferase (CAT) expression vector designed to measure transcriptional promoter-enhancer activity of the cellular DNA element (AvaI-AvaI fragment) inserted within rc-mos.

The last vector (40) contains SV40 regulatory sequences, as well as the bacterial gene coding for the enzyme chloramphenicol-acetyltransferase (CAT) which can be accurately and sensitively assayed. A promoter-enhancer sequence placed within this vector will effect synthesis of CAT RNA and consequently the amount of this enzyme synthesized in cells transfected with the vector. Extracts from these cells are assayed in vitro for the conversion of chloramphenicol to acetylchloramphenicol. The extent of the activity in the extracts is proportional to the strength of the enhancer.

The results shown in Fig. 6 indicate that the AvaI-AvaI fragment, when inserted into the CAT vector in orientation similar to the one in which it is present within rc-mos posess a strong enhancer activity (lane C). Only a weak activity is detected when the DNA fragment is inserted in the opposite

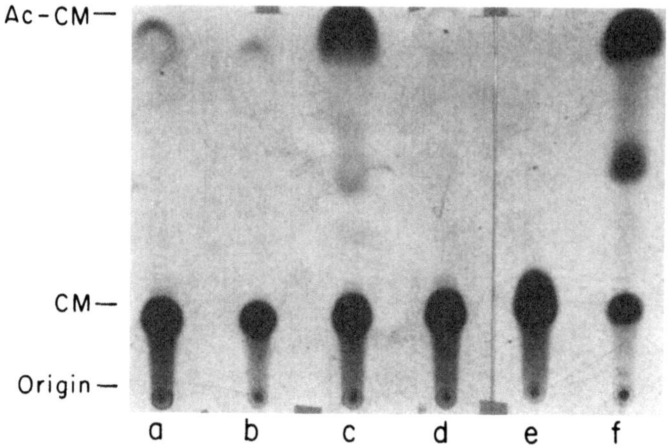

FIGURE 6. Determination of CAT activity in extracts of cells transfected with different plasmids. Plasmids included: pSV2 CAT (SV40 enhancers), (lane a), pSVO CAT (no enhancer) (lance b), pSVO RCM CAT in the native orientation (as in rc-mos) (lane c), pSVO RCM CAT in the reverse orientation (lane d). Activity of nontransfected NIH/3T3 cells is shown in lane e. Markers for chloramphenicol (CM), and acetyl-chloramphenicol (AC-CM) are shown in lane f.

orientation (lane d). For comparison, we also assayed the SV40 enhancer which showed a considerable lower activity (lane a). No activity was detected in extracts from cells infected with pSVO DNA (lane b), or in untreated mouse cells (lane e). These results suggest a strong enhancer activity of the new cellular DNA element within rc-mos. The apparent differential in activity when the AvaI-AvaI fragment is inserted in two orientations could be most easily explained by postulating that the fragment contains both promoter and enhancer signals and their relative orientation with respect to the CAT gene will determine the efficiency of transcription. Finally, inspection of the sequence within the AvaI-AvaI fragment shows two alternating purine-pyrimidine tracks at positions 426-436 and 466-475. These would fit the structures of DNA postulated to be the key elements in transcriptional enhancers (A. Rich, personnel communication).

The inserted DNA element is abundantly expressed in mouse myelomas, and is present within cellular DNA in a high copy number

To begin characterization of the cellular DNA inserted within c-mos we prepared a radiolabelled probe from it, and used it to study expression of this sequence in several mouse myelomas. The results shown in Fig. 7 demonstrate a heavy expression in all tumors examined.

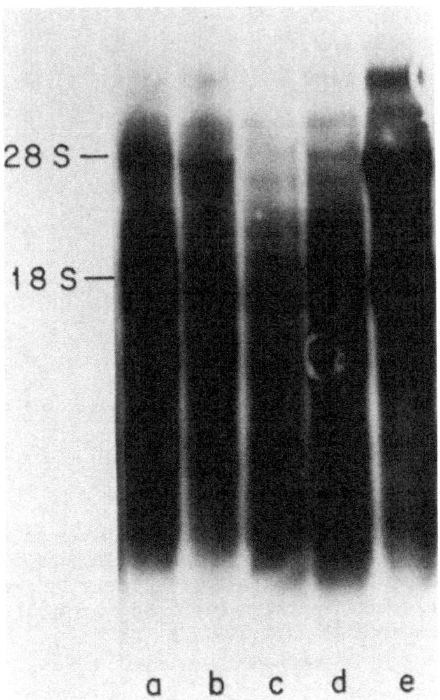

FIGURE 7. Analysis of poly(A)$^+$ RNA from mouse myeloma for the presence of transcripts related to the new DNA element inserted within rc-mos. Radiolabelled fragment was the BglI-EcoRI 0.6 kbp rc-mos DNA (Fig. 3). Hybridization conditions as in Fig. 2. Myelomas included M11, XR24, M104, M315 (lanes a,b,c,d,e, respectively).

Next we analysed genomic DNA for the presence of sequences homologous to the inserted DNA element. DNAs from several mouse strains were digested with EcoRI and examined by Southern blotting. The results as shown in Fig. 8 suggest that the inserted DNA is present within all mouse strains in a large number of copies.

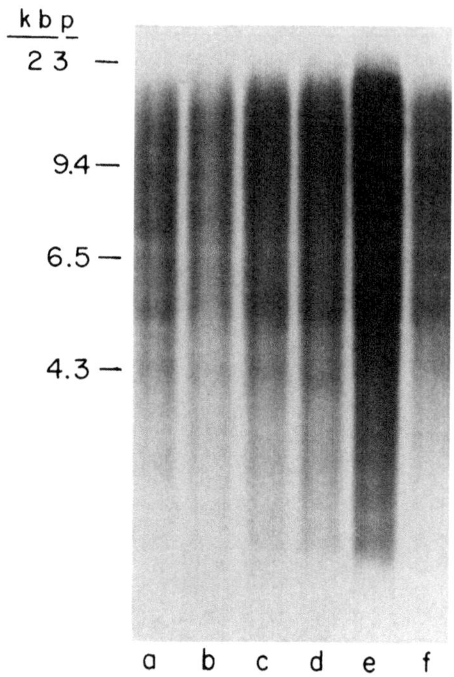

FIGURE 8. Analysis of the rc-mos inserted DNA sequence in total genomic DNA of different mouse strains. Cellular DNAs from livers of BALB/c, AKR, C57BL, SJL, C3H and NZB mice (lanes a-f, respectively) were digested with EcoRI enzyme and analysed by Southern blotting for hybridization to the 0.6 kbp BglI-EcoRI rc-mos DNA fragment (Fig. 3).

Identification and characterization of the cellular sequences 5' to rc-mos

To molecularly clone the cellular sequences 5' to the 12 kbp DNA fragment of rc-mos which were previously cloned, we took advantage of the chance observation that by using total v-mos as a radiolabelled probe one can detect (Fig. 9) on Southern blots of XR24 DNA, an additional new EcoRI fragment of 5.5 kbp (on top of the 12 kbp rc-mos previously detected and cloned). Similarly, two new fragments are detected in a XhoI digest (Fig. 9). These results suggested that the 5' terminal sequence of c-mos previously undetected and considered deleted, is present within XR24 DNA, but is located on a different DNA fragment - EcoRI 5.5 kbp.

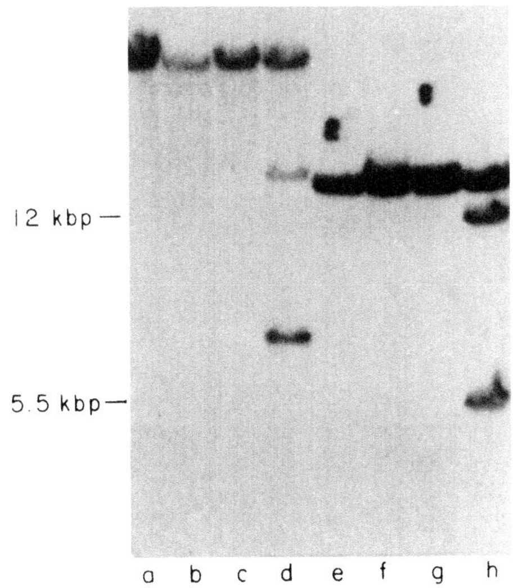

FIGURE 9. Southern blotting analysis of c-mos sequences in mouse myeloma digested with EcoRI (e-h) or XhoI enzyme (a-d) using total v-mos probe. Myelomas tested included M315 (a,f), 104E (b,f), M11 (c,g), XR24 (d,h).

The 5.5 Kbp EcoRI DNA fragment was molecularly cloned into lambda Wes B phage vector. Hybridization analysis demonstrated that it did not anneal to mos sequences, within rc-mos, but reacted strongly with 5' terminal v-mos probe as well as with probe to the inserted DNA element within rc-mos (Fig. 10).

These results indicated that the 5.5 Kbp fragment contained the 5' terminal coding sequences of c-mos, as well as cellular sequences similar to those inserted within rc-mos. Sequencing analysis of a portion of the cloned fragment confirmed these conclusions (Fig. 11).

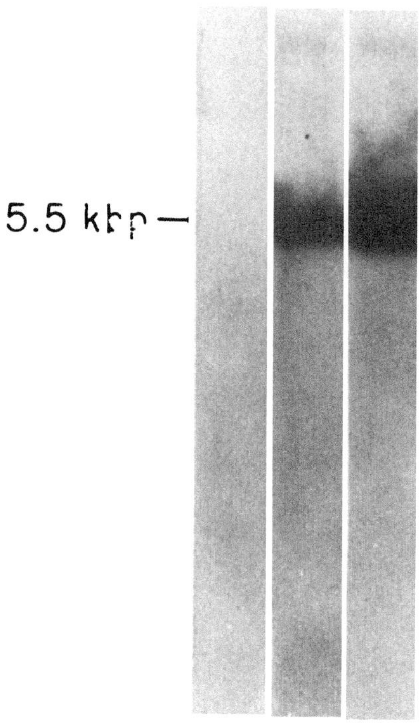

FIGURE 10. Hybridization analysis of molecularly cloned EcoRI 5.5 Kbp mos DNA from XR24 tumour. Probes used (left to right) included: 3' v-mos fragment HindIII-BglI (26), total v-mos fragment HindIII-XbaI (26), AvaI-AvaI DNA insert in rc-mos.

```
                 C  S  I   P  L  V   A  P  R   K  A  G   K  L  F   L  G
.........        CTGCAGCATTC CTTTGGTGG CCCCGAGGA AGGCAGGGA AGCTCTTCC TGGGGACCA
                 PstI

                 c-mos inserted element     20        30        40
5' repeat        CTCCTGTTA TTAGACGC GTTCTCCAC GACCGGCCA GGAAGAACA CCACAGACC
3' repeat            ·····C···  · ········· · ······· ·········· ·AG··A···

                 50        60        70        80        90        100
5' repeat        AGAATCTTC TGCGGCAAA GCTTTATTT CTTACATCT TCAGGAGCC AGGGTCGAG
3' repeat        ········· ········· A·······G ········· ········A ··A··GT·A

                      110       120       130       140       150
5' repeat        GAAGCAAGA GAGCAAGAA GCAAGAGAG AGCGAGAAA ACGAAACCC CGTCCCTCT
3' repeat        ········· ·T·T····· ········· ········· ········· ·······A·

                 160       170       180       190       200
5' repeat        TAAGGAGCA TTCTCCTTC GCGTCGGAC GTGTCACTC CCTGATTGG CTGCAG
3' repeat        ·······A· ··A······ ····A···· ········· ········· .PstI.
```

FIGURE 11. Sequencing analysis of a PstI-PstI fragment (5' repeat) isolated from the molecularly cloned EcoRI 5.5 Kbp DNA - comparison to the sequence of the cellular DNA inserted within rc-mos (3' repeat). mos sequence are present between the left PstI site and nucleotide-1. Six nucleotides underlined with a double solid line, represent a duplication of mos sequence. Four nucleotides underlined with a double interrupted line are an inverted repeat of the terminal tetranucleotide of the DNA element inserted within rc-mos.

Sequence inserted within c-mos is homologous to an endogenous retroviral genome

Our results so far have indicated that in XR24 DNA, c-*mos* was activated by the insertion of a cellular sequence containing a direct terminal repeat of at least 200 nucleotides. This insertion was followed by a duplication of six *mos* nucleotides at the integration site. With the recent appearance of the nucleotide sequence of a portion of an endogenous A-particle viral genome (41), we have found a 90% homology between the cellular sequence inserted in c-*mos* and the 350 nucleotides long terminal repeat of the A-particle viral genome (Kuff *et al*, Nature, in press). These results as well as others emerging from hybridization experiments are schematically shown in Fig. 12.

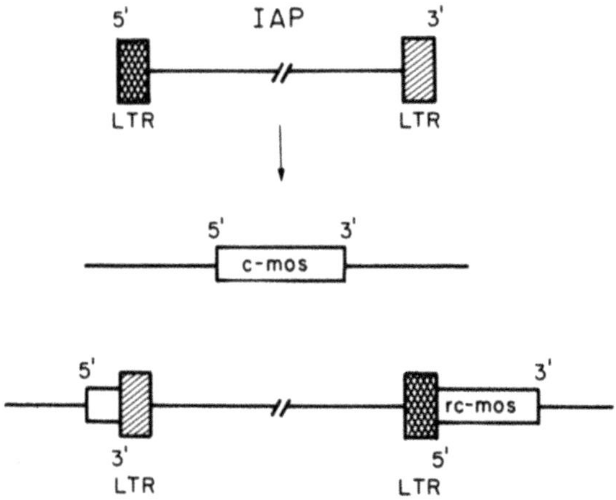

FIGURE 12. Rearrangement of c-*mos* in XR24 tumor by insertion of an endogenous intracisternal A-particle genome (IAP).

CONCLUSIONS

Here we have described a genetic event which led to the activation of a cellular *onc* gene in a non-virally-induced tumour. The normal c-*mos* gene has undergone DNA rearrangements in the murine myeloma XRPC24 and has become highly potent in transforming mouse fibroblast monolayers. The biological activation was associated with the appearance of c-*mos* gene transcripts in the tumour tissue. This is all the more notable as, unlike many cellular *onc* genes which are expressed in various normal and malignant cells (42-45), c-*mos* gene transcripts have not been detected previously in all cells examined (46). The rearrangement resulted from the insertion within c-*mos* of an endogenous retroviral genome (intracisternal A-particle) in a configuration which positioned the 5' viral LTR adjacent to the main coding region of c-*mos*. This 5' LTR, although in the reverse orientation probably acts as a promoter-enhancer element effecting transcription of c-*mos* main coding region downstream to the LTR. This is the first instance where an endogenous retroviral genome has been shown to activate a cellular oncogene. The presence of such endogenous genomes in all higher organisms including humans raise the possibility of finding similar cases in other tumors.

REFERENCES

1. Cold Spring Harb. Symp. Quant. Biol. 44, 1980.
2. Bishop, J.M. Cell 23:5-6, 1981.
3. Coffin, J.M. et al. J. Virol. 40:953-957, 1981.
4. Stehelin, D., Varmus, H.E., Bishop, J.M. and Vogt, P.K. Nature 260:170-173, 1976.
5. Shilo, B.-Z. and Weinberg, R.A. Proc. Natl. Acad. Sci. USA 78:6789-6792, 1981.
6. Hayward, W.S., Neel, G.B. and Astrin, S.M. Nature 290:475-480, 1981.
7. Payne, G.S. et al., Cell 23:311-322, 1981.
8. Shih, C. et al., Proc. Natl. Acad. Sci. USA 76:5714-5718, 1979.
9. Shih, C. et al., Nature 290:261-264, 1981.
10. Lane, M.A., Sainten, A. and Cooper, G.M. Proc. Natl. Acad. Sci. USA 78:5185-5189, 1981.
11. Lane, M.A., Sainten, A. and Cooper, G.M. Cell 28:873-880, 1982.
12. Hopkins, N. et al., Proc. Natl. Acad. Sci. USA 78:7555-7559, 1981.
13. Kronitis, T.G. and Cooper, G.M. Proc. Natl. Acad. Sci. USA 78:1181-1184, 1981.
14. Perucho, M. et al., Cell 27:467-476, 1981.
15. Pulciani, S. et al. Proc. Natl. Acad. Sci. USA 79:2845-2849, 1982.
16. Murray, M.J. et al. Cell 25:355-361, 1981.
17. Der, C.J., Krontiris, T.G. and Cooper, G.M. Proc. Natl. Acad. Sci. USA 79:3637-3640, 1982.
18. Parada, L.F., Tabin, C.J., Shih, C. and Weinberg, R.A. Nature 297:474-478, 1982.
19. Santos, E., Tronick, S.R., Aaronson, S.A., Pulciani, S. and Barbacid, M. Nature 298:343-347, 1982.
20. Tabin, C.J. et al. Nature 300:143-149, 1982.
21. Reddy, E.P. et al. Nature 300:149-152, 1982.
22. Capon, D.J. et al. Nature 302:33-37, 1983
23. Cairns, J. Nature 289:353-357, 1981
24. Klein, G. Nature 294:313-318, 1981.
25. Frankel, A.E. and Fischinger, P.J. Proc. Natl. Acad. Sci. USA 73:3705-3709, 1976.
26. Canaani, E., Robbins, K.C. and Aaronson, S.A. Nature 282:378-383, 1979.
27. Andersson, P., Goldfarb, M.P. and Weinberg, R.A. Cell 16:63-75, 1979.
28. Oskarsson, M. et al. Science 207:1222-1224, 1980.
29. Reddy, E.P. et al. Proc. Natl. Acad. Sci. USA 77:5234-5238, 1980.
30. Van Beveren, J. et al. Cell 27:97-108, 1981.
31. Papkoff, J., Verma, I.M. and Hunter, T. Cell 29:417-426, 1982.
32. Blattner, F.R. et al. Science 196:161-169, 1977.
33. Graham, F.L. and Van Der Eb, A.J. Virology 52:456-467, 1973.
34. Watson, R., Oskarsson, M. and Vande Woude, G.F. Proc. Natl. Acad. Sci. USA 79:4078-4082, 1982.
35. Dunsmuir, P. et al. Cell 21:575-579, 1980.
36. Farabough, P.J. and Fink, G.R. Nature 286:352-356, 1980.
37. Gafner, J. and Philippsen, P. Nature 286:414-418, 1980.
38. Shimotohno, K., Mizutani, S. and Temin, H.M. Nature 285:550-554, 1980.
39. Maxam, A.M. and Gilbert, W. Proc. Natl. Acad. Sci. USA 74:560-564, 1977.
40. Gorman, C., Moffat, L., Howard, B. Mol. Cell. Biol. 2:1044- , 1982.
41. Kuff, E.L. et al. Proc. Natl. Acad. Sci. USA, in press.
42. Spector, D. et al. Cell 13:371-379, 1978.
43. Langebeheim, H., Shih, T.Y. and Scolnick, E.M. Virology 106:292-300, 1980.
44. Witte, O.N., Rosenberg, N.E. and Baltimore, D. Nature 281:396-398, 1979.
45. Eva, A. et al. Nature 295:116-119, 1982.
46. Gattoni, S. et al. Molec. Cell Biol. 2:42-51, 1982.

15

PATHOGENICITY OF VESICULAR STOMATITIS VIRUS IS POTENTIALLY MEDIATED BY WILD-TYPE LEADER RNA WHICH INHIBITS INITIATION OF TRANSCRIPTION

ROBERT R. WAGNER, JOHN J. MCGOWAN, AND BRIAN W. GRINNELL
Department of Microbiology, The University of Virginia, Charlottesville, Virginia 22908 USA

SUMMARY

The highly pathogenic vesicular stomatitis virus shuts off synthesis of RNA, DNA, and protein in infected host cells. Genetic studies with temperature-sensitive mutants and defective-interfering viruses indicate that viral transcription is essential for inhibiting cellular nucleic acid synthesis. Inactivation of viral genome functions by UV-irradiation implicated the 47-nucleotide wild-type leader RNA as the potential inhibitor of cellular RNA and DNA synthesis, but not protein synthesis. Purified preparations of wild-type leader RNA in low concentrations markedly inhibited *in vitro* initiation of transcription of adenovirus-2 late promoter and virus-associated genes as well as SV40 DNA in a Manley-Sharp reconstituted system containing eukaryotic polymerases and co-factors. The leader RNA of non-pathogenic DI virus or viral mRNA at high concentrations had only slight effects on *in vitro* transcription, as did other exogenous synthetic and natural RNAs. The wild-type leader RNA contains TAATA-like and consensus-like sequences, which could potentially inhibit the functions of RNA polymerase II and III, whereas similar sequences are absent in the non-pathogenic DI leader.

INTRODUCTION

Viruses vary greatly in their pathogenicity for host cells, ranging from the most virulent to the completely avirulent or temperate. Rapidly growing viruses tend to be the most virulent but this frequently depends on host range or tissue tropism, which is usually a function of virus recognition of cell-surface receptors. The genomes of virulent viruses may consist of RNA or DNA, single-stranded, or double-stranded; however, the most virulent viruses are generally those that replicate in the cytoplasm and multiply rapidly. Quite clearly, the virulence of a virus does not depend on its replicative strategy, since pathogenic forms exist in all classes of viruses regardless of their nucleic acid content or whether a nuclear or cytoplasmic phase is involved in replication. There is also some evidence that virus-induced cell death is multiplicity-dependent or can, in fact, be caused by large inocula of non-replicating, defective, or inactivated viruses (1). This multiplicity-dependent effect appears to occur by initial interaction of certain viruses with surface membranes resulting in secondary effects on cell functions (2). It is our contention that this interesting phenomenon of multiplicity-dependent, membrane-mediated death of cells is not the principal mechanism by which viruses compromise the physiological functions of host cells.

The cytopathology caused by viruses is generally associated with inhibition of various metabolic functions of the host cell (3), including cellular macromolecular synthesis and membrane permeability (4). The major problem in interpretation of these data is to determine which of the effects is primary and which is secondary. The most reliable and the most direct measurements appear to be the nucleic acid and protein synthesis of cells following infection with a specific virus. Many cytopathogenic viruses simultaneously or sequentially shut off RNA, DNA, and protein synthesis of the target host cell. Here again, the question arises whether inhibition of cellular protein synthesis is caused by prior inhibition of cellular RNA synthesis, or *vice versa*. The temporal relationships of these two events may help to decide whether they are caused by the same viral function or by two independent events. However, *in vivo* systems frequently do not permit one to discriminate between the single- or double-target hypothesis. Investigators have been forced to resort to isolated cell-free systems to determine whether one viral product has one or two effects and whether there are separate cell targets for each inhibitory event.

Biochemical and molecular biological techniques appear to offer the most promising approaches to understanding the viral products and cell targets for inhibition of cellular macromolecular synthesis that lead to cytopathology. The virulent viruses most intensively studied by such techniques are poxviruses, rhabdoviruses, and particularly picornaviruses. Each of these virus groups has a very different replicative strategy, but all inhibit RNA, DNA, and protein synthesis of their host cells (3). Poliovirus has been the most extensively studied, particularly for its inhibitory effect on cellular protein synthesis (5). The evidence seems fairly clear that the genomic plus-strand RNA of poliovirus directly inhibits initiation of translation probably by altering ribosomal factors eIF3 and/or eIF-4B (6). Although less extensively studied, the negative-strand rhabdovirus, vesicular stomatitis virus (VSV), also appears to inhibit protein synthesis by compromising initiation of translation at the level of eIF-3 and/or eIF-4B (7), but, unlike poliovirus, probably by means of a newly synthesized viral double-stranded RNA (8). If these data on poliovirus and VSV are borne out, it would seem that viruses with very different replicative strategies and products may have similar cell targets.

We have used highly virulent VSV as a model virus for studying viral cytopathogenicity as it appears to relate to inhibition of host-cell macromolecular synthesis. This virus has a very wide host range, extending from mammalian to insect cells although with varying cell-dependent degrees of susceptibility (9). VSV also has the great advantage of being fairly simple and well characterized genetically, as well as faithfully transcribing each of the five messengers in vitro by means of its own transcriptase (10). The most dramatic aspect of its cytopathic properties is the capacity of VSV to shut off synthesis of cellular nucleic acids, particularly RNA. Although we are also studying the separate effect of VSV on host-cell protein synthesis (7, 8), we shall confine our remarks in this paper to the biological and molecular events by which VSV shuts off RNA synthesis (as well as DNA synthesis).

BIOLOGICAL PROPERTIES OF VSV RESPONSIBLE FOR INHIBITING CELLULAR RNA SYNTHESIS

Our studies have generally been done with plaque-purified stocks of the Indiana serotype of vesicular stomatitis virus (VSV_{Ind}) free of defective-interfering (DI) particles. VSV_{Ind} rapidly shuts off synthesis of DNA (11, 12), RNA (13), and proteins (14) in a wide variety of cells. Most of our recent studies have been performed with the MPC-11 mouse myeloma cell because it is more susceptible to VSV infection than mouse L cells or BHK-21 hamster cells (13). Inhibition of RNA synthesis occurs very rapidly after infection and proceeds more-or-less linearly, reaching maximum shut off of >90% by 4hr postinfection (13). The New Jersey serotype (VSV_{NJ}) exhibits similar effects on MPC-11 cell but at a somewhat slower rate (B.W. Grinnell, unpublished data). We have also compared the capacity of VSV to shut off RNA, DNA, and protein synthesis in MPC-11 cells as determined by incorporation of radioactive uridine, thymidine, and amino acids (12). These experiments revealed quite similar kinetics for inhibition of RNA and DNA synthesis in MPC-11 cells, suggesting that a single viral function is responsible for shutting off both DNA and RNA synthesis (12). In sharp contrast, protein synthesis in VSV-infected MPC-11 cells was found to be inhibited to a much lesser extent (∿50%); moreover, after an initial decline in cell protein synthesis of ∿40% during the first hour postinfection, there was only limited reduction during the ensuing three hours (12). These divergent effects suggest that different viral functions are responsible for inhibition of cellular protein and nucleic acid synthesis. In fact, we have some evidence that newly synthesized VSV double-stranded RNA specifically inhibits translation in reticulocyte lysates (8), in sharp contrast to the apparent viral product that inhibits cell-free transcription (see below).

Of considerable interest was the observation that fully infectious, cloned VSV B-particles shut off RNA synthesis in MPC-11 cells equally well at multiplicities of 1 or 50 pfu/cell (12). These results suggest that only one infectious virus particle is required to inhibit cellular transcription. It is, of course, still possible that a multiplicity-dependent toxic effect can be produced by even higher multiplicities, but our data strongly suggest that a newly synthesized viral product, not input virions, is the major factor responsible for inhibition of cellular RNA synthesis. An important confirming observation concerns the effect, rather the lack of effect, of highly purified VSV-DI particles, in which two-thirds of the genome is deleted from the 3' end. Infection of MPC-11 cells with DI-T particles at multiplicities up to 1000 particles/cell resulted in minimal inhibition (9-17%) of cellular RNA synthesis (15). This experiment, in which every effort was made to eliminate contaminating standard infectious VSV, provides quite strong evidence that input viral products have little or no effect on cellular RNA synthesis.

Similar results on the effect of VSV on MPC-11 cell DNA synthesis were obtained by McGowan and Wagner (12). In these experiments, it was found that DI-011 (3') VSV had little effect on cellular DNA synthesis, as measured by uptake of [^{3}H]thymidine, whereas DI-LT (5') did retain the capacity to inhibit cellular DNA synthesis. There is an enormous biological difference between DI-011 (3') and DI-LT (5') in that DI-LT retains the capacity to transcribe 4 of the 5 VSV messengers (16); DI particles are transcriptionally inactive except for transcription of a 46-nucleotide leader sequence (17).

Data such as these suggested that the capacity of VSV to inhibit cellular RNA and DNA synthesis depends on intact transcriptional activity of the infecting virion. Earlier studies using temperature-sensitive mutants (14) revealed that VSV transcription is also essential for inhibition of cellular protein synthesis; a complementation group I VSV_{Ind} mutant (*ts*G114) failed to shut off protein synthesis at restrictive temperature (39°C), whereas two other RNA^- mutants not restricted in transcription (18) retained their capacity to inhibit cellular RNA synthesis. Therefore, we also tested the same *ts*G114 mutant for its effect on cellular RNA and DNA synthesis at restrictive and permissive temperature. The data clearly showed that *ts*G114(I) did not inhibit cellular RNA synthesis (15) or DNA synthesis (12) under conditions of restricted VSV transcription but did so at permissive temperature. These experiments provided additional circumstantial evidence that VSV transcription is a sine qua non for shutting off cellular nucleic acid synthesis, as is also the case for VSV inhibtion of protein synthesis (14) and for cell killing (19). It should be kept in mind, however, that these experiments do not identify the inhibitor or cell-killing factor(s) as RNA because VSV protein synthesis is also inhibited under these conditions. A long series of experiments by Phillip Weck (unpublished data), who tested many antibiotics and amino-acid analogues to block viral protein synthesis, failed to determine whether VSV RNA or protein was the inhibitor(s) of cell RNA synthesis; all protein-synthesis inhibitors also inhibited cellular RNA synthesis.

These discouraging technical obstacles led Weck and his colleagues (20) to test UV irradiation as a means to block selectively VSV transcription. Previous investigators (21, 22) had used UV irradiation of virions to map the VSV genome, which is transcribed sequentially from the 3' end without internal initiations by the polymerase moving down the entire length of the genome (23). Therefore, those cistrons at the 5' end of the genome will be more readily inactivated by UV irradiation than cistrons closer to the 3' end. The gene sequence of the VSV genome determined by UV inactivation of transcription is: 3'-leader-N-NS-M-G-L-5' (24). As expected, transcription of the 47-nucleotide leader RNA is extremely resistant to UV irradiation (24). With this information in mind, we tested the UV dose required to inactivate the various biological and transcriptional properties of VSV. Table 1 is a composite of two sets of experiments designed to compare the dose of UV irradiation required to reduce by 63% the infectivity, transcriptase activity, and shut off of RNA and DNA synthesis by VSV infection of MPC-11 cells (12, 20).

Table 1. Comparative effects of UV irradiation on 37% (1/e) survival of VSV functions

VSV functions inactivated	UV dose required (1/e ergs/mm^2)
Infectivity (pfu/ml)[a]	104
Transcriptase (*in vitro*)[a]	1,050
Viral mRNA (*in vivo*)[a]	170
N protein synthesis (*in vivo*)[a]	380
Shutoff of cell RNA synthesis[b]	52,000
Shutoff of cell DNA synthesis[b]	45,000

[a]Data from Weck et al. (20); [b] Data from McGowan and Wagner (12).

These data indicate that relatively small doses of UV irradiation will inactivate the infectivity of VSV as measured by plaque assay and somewhat large doses will produce significant reduction in transcriptase activity. Our findings on gene mapping by UV inactivation of specific transcripts are similar to those or Ball and White (21) and Abraham and Banerjee (22). The most striking finding is the dose of UV light required to inactivate the capacity of VSV to inhibit cellular RNA and DNA synthesis. The only conclusion we could draw from these results was that UV inactivation of the leader RNA was required to reverse the shut off by VSV of cellular RNA and DNA synthesis. These findings are consistent with those of Colonno and Banerjee (24) that the 47-nucleotide leader sequence of the ∿10,000 nucleotide VSV genome would provide such a small target for UV inactivation. Therefore, we postulated that the transcribed VSV leader was the inhibitor of cellular RNA (20) and DNA (12) synthesis. It should also be noted that much smaller doses of UV irradition will inactivate the capacity of VSV to inhibit protein synthesis (25 and Thomas and Carroll, unpublished data), which suggests that the product of the N gene or the NS gene is the inhibitor of protein synthesis.

HOST TARGET SITE AT WHICH VSV COMPROMISES CELLULAR RNA SYNTHESIS

We next turned our attention to the host target to which the VSV-infection product is directed to shut off cellular RNA synthesis. Once again, these studies were done with the highly susceptible MPC-11 mouse myeloma cell (13). After intervals of VSV infection, usually at a multiplicity of 10 pfu/cell, various functions of cells, isolated nuclei, and chromatin were studied and compared with mock-infected cells under otherwise identical conditions. The results of these experiments appear in two publications (13, 26) and can be summarized as follows: 1) Infection of MPC-11 cells with VSV_{Ind} resulted in rapid, marked, and progressive reduction in cellular RNA synthesis reaching maximum inhibition of >90% by 4-5hr pi; this inhibition of RNA synthesis occurred long before cell viability was compromised at 8-12hr pi as measured by exclusion of the supravital dye, trypan blue. 2) This inhibition of cellular RNA synthesis was shown not to be caused by differential membrane permeability to [^{3}H]uridine of uninfected and infected MPC-11 cells. 3) VSV infection did not lead to premature degradation of cellular RNA species, nor was there any significant difference in the polyacrylamide gel electrophoresis patterns of the RNA in infected and uninfected cells. 4) Transport of newly synthesized nuclear RNA to the cytoplasm was not impeded in VSV-infected cells; moreover, infected cells accumulated polyadenylated mRNA at the same rate as did control uninfected cells. 5) Isolated nuclei of VSV-infected MPC-11 cells also exhibited greatly impaired capacity to synthesize RNA despite the absence of cytoplasmic factors; infected-cell cytosol did not inhibit transcription by uninfected cell nuclei, nor did uninfected-cell cytosol reverse viral inhibition of nuclear transcription. 6) Studies with α-amanitin revealed that VSV infection inhibited the activity of RNA polymerases I, II, and III, but only polymerase II was affected progressively throughout infection and to a greater extent than polI and III.

We next attempted to determine whether VSV infection suppressed transcription of nuclear chromatin at the level of chain initiation, elongation, or termination (26). In these experiments we prepared nuclear chromatin from uninfected and VSV-infected MPC-11 cells and tested their capacity to incorporate [^{3}H]UTP or [^{3}H]GTP into acid-precipitable material at intervals after infection. At the time these experiments were done, no reliable techniques were available to study initiation of transcription by reconstitution of polymerases and co-factors with DNA templates. Therefore, all we could examine was chain elongation and number of initiated endogenous DNA templates at intervals postinfection compared to uninfected cell chromatin. In the first studies we looked for the presence of a soluble polymerase inhibitor in infected-cell nuclei and could find none. In crude reconstitution experiments using salt-solubilizing RNA polymerases reacted with endogenous or exogenous (calf thymus DNA) templates, we found that prior VSV infection did not affect the ability of polymerases or chromatin template to function in transcription. Measurement of the number of actively growing RNA chains revealed that infected-cell nuclei contained fewer polymerase units than did uninfected-cell nuclei. However, the rates of RNA chain elongation were the same for nuclei of infected and uninfected cells. There was a steady loss of active polymerase units in nuclei of cells during a 4-hour period of infection to a level 50% that of mock-infected control MPC-11 cells. The α-amanitin-sensitive polymerase II was found to be more severely reduced by viral infection than were polymerases I and III combined. These experiments strongly suggested that VSV infection had little or no effect on RNA chain elongation but appeared to reduce progressively the number of functional RNA polymerases, presumably at the level of initiation of transcription.

COMPARATIVE EFFECTS OF DIFFERENT VSV TRANSCRIPTS ON DNA-DEPENDENT TRANSCRIPTION IN A RECONSTITUTED SYSTEM

The recent availability of reconstituted systems for studying initiation _in vitro_ of eukaryotic transcription led us to design experiments (27) to test our hypothesis that the wild-type leader RNA of VSV is the inhibitor of cellular RNA synthesis. At the same time we hoped to determine whether initiation of transcription was the target of the inhibitor. In these studies we used the system developed by Manley et al. (28) in which disrupted HeLa cells are precipitated with $(NH_4)_2SO_4$ to prepare crude extracts containing all three RNA polymerases and protein co-factors. The DNA templates we chose to use in this reconstituted transcription reaction were a pBR322-Bal-1E clone of the adenovirus-2 late promoter (LP), Pst-1-cleaved SV40 DNA, and a pBR322 clone of the adenovirus-associated RNA gene (VA gene). Under the conditions of these experiments, the LP gene codes for a single mRNA of 560 nucleotides in length, the 3 SV40 DNA segments are templates for RNA species 2051, 1974, and 1530 nucleotides in length (with considerable internal initiations and premature terminations) and the VA gene serves as template for a single RNA 146 nucleotides in length. Polymerase II catalyzes transcription of the LP gene (28) and SV40 (29), whereas polymerase III catalyzes transcription of the VA tRNA-like gene (30).

The strategy of these experiments was to measure RNA synthesis in reconstituted eukaryotic transcription reaction to which are added various amounts of the wild-type VSV leader RNA as well as the DI leader and polyadenylated VSV mRNA free of leader. As additional controls, we added exogenously synthetic poly(A):poly(U), poly(G):poly(C), and yeast RNA. ^{32}P-RNA transcripts synthesized in the presence or absence of these exogenous RNAs were analyzed by electrophoresis on 8M urea-4% polyacrylamide sequencing gels and analyzed by autoradiography and densitometry scanning. Table 2 summarizes the results of these experiments reported by McGowan et al. (27).

Table 2. Inhibition by exogenous RNAs of specific transcription from various DNA templates in a reconstituted system

RNA added	Ad-2 LP Gene		Pst-1 SV40 DNA		Ad-2 VA Gene	
	ng/μl	% Inhibition	ng/μl	% Inhibition	ng/μl	% Inhibition
VSV wt leader	17.4	60	4.2	0	4.2	53
	21.0	86	45.2	91	6.3	72
VSV DI leader	79.5	10	89.0	24	31.8	10
	795.0	53	1361.3	54	843.3	44
VSV mRNA (thio ATP)	362.6	<0	662.7	32	453.3	25
	812.3	45	1450.6	45	812.3	27
VSV poly(A) mRNA	52.7	16	51.2	1	52.7	8
Poly(A):poly(U)	72.3	57	79.5	51	72.3	0
	181.0	91	300.0	93	181.0	90
Poly(G):poly(C)	167.0	56	201.2	37	167.0	8
	301.2	88	687.3	89	602.4	87
Yeast RNA	477.1	62	843.3	0	843.3	5
	954.2	81	2750.0	16	1686.7	77

These data show that relatively low concentrations of the plus-strand RNA made *in vitro* from the 3' end of the wild-type VSV genome inhibits initiation of transcription catalyzed by both RNA polymerases II and III. Polyadenylated VSV messengers and other natural and synthetic RNAs also caused some inhibitory effects on *in vitro* transcription from these DNA templates but only at extremely high concentrations. Compared with the wild-type plus-strand RNA leader, the leader RNA synthesized *in vitro* by defective-interfering VSV showed only limited capacity to inhibit RNA synthesis on adenovirus and SV40 DNA templates and only at concentrations at least 30-times greater than those of the wild-type leader. Of possible mechanistic significance is the fact that the synthetic polynucleotides poly(A):poly(U) and poly(G):poly(C) inhibited the DNA-dependent transcriptions to a greater extent than did the DI leader RNA or the VSV mRNAs. There is also some slight, but not very convincing, evidence that poly(A):poly(U) had a greater effect on polymerase II-catalyzed transcription of LP and SV40 DNA than it did on polymerase III-catalyzed transcription of the VA gene. Conversely, poly(G):poly(C) may have shown a somewhat greater effect on polIII-catalyzed transcription of the VA gene than it did on the polII-transcription reactions. These findings were somewhat unexpected since these extraneous RNAs were meant to serve as negative controls. Thus, far more carefully designed experiments with various types of synthetic polynucleotides must be performed before any significance can be attached to these results. Nevertheless, the existing data may provide interesting leads for investigating the base sequences required for inhibition of transcription.

Figure 1 compares the nucleotide sequences of wild-type leader and DI-T leader of VSV.

Wild-type	5'ppACGAAGACAAACAAACCAUUAUUAUCAUUAAAAGGCUCAGGAGAAAA-OH3'
DI-T	5'ppACGAAGACCACAAAACCAGAUAAAAAAUAAAAACCACAAGAGGGUC-OH3'

FIGURE 1. Comparative nucleotide sequences of wild-type and DI-T VSV leader RNAs. The unique regions of the wild-type leader are highlighted by underlining. Data from wild-type are obtained from results of Colonno and Banerjee (24) and DI-T data are from the results of Schubert et al. (31).

It is tempting to speculate on a possible role played by the one or more nucleotide sequences of wt leader RNA that are distinct from those of the DI leader RNA and whether this sequence (or sequences) could be responsible for inhibition of DNA-dependent RNA synthesis. As shown in Figure 1, wt leader contains a unique sequence (AUUAU) at positions 18 to 30, not present in DI leader, which appears homologous to the TAATA or "Goldberg-Hogness box" (28, 32, 33). There is another unique region of VSV wt leader RNA at positions 32 to 39 (AGGCUCAG) that resembles a deoxynucleotide sequence implicated in RNA polymerase binding (34, 35). The sequence (5' AGGCUCAG 3') is very similar to the consensus sequences (5' AG/GURAG 3') obtained for the 5'-end splice site of intervening sequences (36, 37). All that can be said at this time is that the VSV wt leader RNA contains nucleotide sequences potentially capable of interacting with promoters or host-cell proteins that interact with nucleotide sequences essential for accurate transcription, or perhaps the most intriguing possibility is that VSV wt leader

RNA can serve as a surrogate for other small RNA species found inside the cytoplasm and nucleus of eukaryotic cells (38, 39). Similar sequences do not appear to be present in leader RNAs transcribed from DI particles (Figure 1); DI particles do not possess the capacity to inhibit cellular RNA synthesis (15). It is of some interest that wt VSV transcribes leader RNA in molar amounts higher than that of any of the five mRNAs (24, 40).

The preceding data are far from conclusive for implicating the leader RNA sequence of wild-type VSV as the causative agent for inhibition of cellular RNA synthesis. However, the leader sequence, frequently complexed with viral N protein, is found in VSV-infected cells (41). Of considerable importance is the striking observation that the wild-type leader, but not mRNA, synthesized early in VSV-infected cells rapidly migrates to the nucleus (42). This latter observation provides essential anatomical data to support the hypothesis that the VSV wt leader is at least in the right location to inhibit initiation of chromatin transcription. However, far more data are required before this hypothesis can be entertained seriously. For one, the exact location in the nucleus and the transcription component to which it binds must be determined; the availability of DNA clones of wild-type leader should help greatly in tracing by hybridization techniques the cellular location of the leader RNA. Also of great advantage for design of future experiments is the finding by McGowan (unpublished data) that a pBR322 clone of the cDNA of all but the first six 5' nucleotides of wild-type VSV leader inhibits LP and VA gene transcription as well as does the leader RNA itself. Clones such as these should provide the opportunity to make site-directed deletions in order to identify the critical nucleotide sites necessary for inhibition of transcription catalyzed by polymerase II or III. Another approach for determining the potential inhibitory activities of the wt leader TAATA-like and consensus-like regions is to test the respective transcriptional inhibitory activity of leader segments cleaved by endoribonucleases, such as T1. Perhaps, the most direct way to determine the putative transcriptional-inhibitory activity of specific leader nucleotide sequences is to synthesize and test relevant oligodeoxynucleotides. All these experiments are in progress or on the drawing board. Further down the line are experiments to determine whether leader RNA, leader-like DNA and/or specific oligonucleotide sequences recognize and bind to promoters, the polymerases or protein co-factors required for faithful and efficient transcription.

It is also wise to remember that, even if all or most of these experiments succeed in identifying the specific VSV leader inhibitory oligonucleotides and the sites at which they block transcription, there is still a large hiatus in our knowledge of how or whether viral inhibition of RNA (or DNA and protein) synthesis is really the primary cause for cell death.

REFERENCES

1. Wertz, G.W. and Youngner, J.S. J. Virol. 9:85-89, 1972.
2. Kohn, A. Adv. Virus. Res. 24:223-276, 1979.
3. Bablanian, R. Progr. Med. Virol. 19:40-83, 1975.
4. Carrasco, L. Virology 113:623-629, 1981.
5. Helentjaris, T., Ehrenfeld, E., Brown-Luedi, M.L., and Hershey, J.W. J. Biol. Chem. 254:10973-10978, 1979.
6. Hansen, J. and Ehrenfeld, E. J. Virol. 38:438-445, 1981.
7. Thomas, J.R. and Wagner, R.R. Biochemistry 1983, in press.
8. Thomas, J.R. and Wagner, R.R. J. Virol. 44:189-198, 1982.
9. Robertson, B.H. and Wagner, R.R. In: The Replication of Negative Strand Viruses (Eds. D.H.L. Bishop and R.W. Compans) Elsevier, 1981, pp. 955-963.
10. Wagner, R.R. In: Comprehensive Virology, vol. 4. (Eds. H. Fraenkel-Conrat and R.R. Wagner) Plenum Press, 1975, pp. 1-93.
11. Yaoi, Y. and Amano, M. J. Gen. Virol. 9:69-75, 1970.
12. McGowan, J.J. and Wagner, R.R. J. Virol. 38:356-367, 1981.
13. Weck, P.K. and Wagner, R.R. J. Virol. 25:770-780, 1978.
14. McAllister, P.E. and Wagner, R.R. J. Virol. 18:550-558, 1976.

15. Weck, P.K. and Wagner, R.R. J. Virol. 30:410-413, 1979.
16. Johnson, L.D., Binder, M., and R.A. Lazzarini. Virology 99:203-206, 1979.
17. Emerson, S.U., Dierks, P.M., and Parsons, J.T. J. Virol. 25:708-716, 1977.
18. Hunt, D.M. and Wagner, R.R. J. Virol. 13:28-35, 1974.
19. Marcus, P.I. and Sekellick, M.J. Virology 63:176-190, 1975.
20. Weck, P.K., Carroll, A.R., Shattuck, D.M., and Wagner, R.R. J. Virol. 30:746-753, 1979.
21. Ball, L.A. and White, C.N. Proc. Natl. Acad. Sci. USA 73:442-446, 1976.
22. Abraham, G. and Banerjee, A.K. Proc. Natl. Acad. Sci. USA 73:1504-1508, 1976.
23. Emerson, S.U. Cell 31:635-642, 1982.
24. Colonno, R.J. and Banerjee, A.K. Cell 15:93-101, 1978.
25. Marvaldi, J.M., Sekellick, M.J., Marcus, P.I. and Lucas-Lenard, J. Virology 84:127-133, 1978.
26. Weck, P.K. and Wagner, R.R. J. Biol. Chem. 254:5430-5434, 1979.
27. McGowan, J.J., Emerson, S.U., and Wagner, R.R. Cell 28:325-333, 1982.
28. Manley, J.L., Fire, A., Cano, A., Sharp, P.A., and Gefter, M.L. Proc. Natl. Acad. Sci. USA 77:3855-3859, 1980.
29. Handa, H., Kaufman, R.J., Manley, J., Gefter, M., and Sharp, P.A. J. Biol. Chem. 256:478-482, 1981.
30. Akusjarvi, G., Mathews, M.B., Anderson, P., Vennstrom, B., and Pettersson, U. Proc. Natl. Acad. Sci. USA 77:2424-2428, 1980.
31. Schubert, M., Keene, J.D., Lazzarini, R.A., and Emerson, S.U. Cell 15:103-112, 1978.
32. Baker, C.C. and Ziff, E.B. J. Mol. Biol. 149:189-221, 1981.
33. Gruss, P., Dhar, R., and Khoury, G. Proc. Natl. Acad. Sci. USA 78:943-947, 1981.

34. Breathnach, R., Benoist, C., O'Hare, K., Gannon, F., and Chambon, P. Proc. Natl. Acad. Sci. USA 75:4853-4857, 1978.
35. Jelinek, W.R., Toomey, T.P., Leinwand, L., Duncan, C.H., Biro, P.A., Choudary, P.V., Weissman, S.M., Rubin, C.M., Houck, C.M., Deininger, P.L., and Schmid, C.W. Proc. Natl. Acad. Sci. USA 77:1398-1402, 1980.
36. Lerner, M.R., Boyle, J.A., Mount, S.M., Wolin, S.L., and Steitz, J.A. Nature 283:220-224, 1980.
37. Sharp, P.A. Cell 23:643-646, 1981.
38. Lerner, M.R. and Steitz, J.A. Cell 25:298-300, 1981.
39. Zieve, G.W. Cell 25:296-297.
40. Leppert, M., Rittenhouse, L., Perrault, J., Summers, D.F., and Kolakofsky, D. Cell 18:735-747, 1979.
41. Blumberg, B.M., Leppert, M., and Kolakofsky, D. Cell 23:837-845, 1981.
42. Kurilla, M.G., Piwnica-Worms, H., and Keene, J.D. Proc. Natl. Acad. Sci. USA 79:5240-5244, 1982.

16

STRUCTURAL FEATURE OF PICORNAVIRUS RNA INVOLVED IN PATHOGENESIS: A VERY HIGH AFFINITY BINDING SITE FOR A MESSENGER RNA-RECOGNIZING PROTEIN

RAYMOND KAEMPFER, HAIM ROSEN, GIANFRANCO DI SEGNI and SARAH KNOLLER
Department of Molecular Virology, The Hebrew University-Hadassah Medical School, P.O.B. 1172, 91 010 Jerusalem, Israel.

SUMMARY

Eukaryotic initiation factor 2 (eIF-2) has a dual function in initiation of protein synthesis: it binds Met-tRNA$_f$ to the small ribosomal subunit, and it binds directly to mRNA. The binding of eIF-2 to mRNA is highly specific and occurs in Mengovirus RNA and satellite tobacco necrosis virus RNA at the nucleotide sequences that constitute the ribosome binding sites. A molecule of Mengovirus RNA competes in cell-free translation 35-fold more strongly than (on average) a molecule of globin mRNA. This competition is relieved by excess eIF-2. Mengovirus RNA binds directly to eIF-2 with 30-fold higher affinity than does globin mRNA. Double-stranded RNA (dsRNA) is a powerful inhibitor of translation of cellular mRNA; it acts by causing inactivation of eIF-2. By contrast, initiation of translation on RNA from Mengo and other picorna viruses is totally resistant to such inhibition. Picorna virus RNA protects eIF-2 against inactivation by dsRNA, apparently because it competes more effectively than dsRNA for eIF-2, while cellular mRNA competes less effectively. Our results suggest the following scenario for picorna virus-infected cells: (1) A strong picorna virus RNA template outcompetes cell mRNA for eIF-2, shifting synthesis to viral protein. (2) Early in infection, virus-generated dsRNA acts to decrease the pool size of eIF-2 and thereby to sharpen this competition in favor of viral RNA. (3) Later in infection, when viral RNA is the predominant template for translation, the higher levels of dsRNA fail to inhibit continued initiation of synthesis of viral protein.

INTRODUCTION

The pathogenesis of the picornaviruses is based on their exquisitely high ability to replicate in their host cells, leading to cell lysis and tissue destruction. This property of highly efficient replication is encoded in the viral RNA molecule, which serves both as genome and as messenger RNA. Indeed, the genomic RNA of picornaviruses, in particular of the cardioviruses, such as Mengovirus, ranks among the most efficient mRNA species so far studied (1-4). The molecular basis for this efficiency is not yet known.

Here, we describe molecular properties of the RNA of picornaviruses, in particular of Mengovirus, that explain why this viral RNA is such an effective messenger template in translation, and contribute understanding as to why the synthesis of viral proteins in infected cells can proceed at a high pace, even though the translation of host mRNA is shut off completely. We demonstrate that picornavirus RNA contains a very high affinity binding site for a crucial component of the cell's protein-synthesizing machinery, eukaryotic initiation factor 2 (eIF-2). This high affinity binding site contributes in more than one way to the selective translation of viral messenger RNA.

THE MESSENGER RNA-BINDING FUNCTION OF EUKARYOTIC INITIATION FACTOR 2

Initiation factor eIF-2 is a protein of considerable biological interest, first, because it plays a central role in the recognition of mRNA and its binding to ribosomes -- the rate-limiting step in protein synthesis, as well as the crucial step in the regulation of gene expression at the level of translation -- and second, because the activity of eIF-2 is itself regulated by external signals, including viral ones.

eIF-2 binds with absolute specificity to methionyl-$tRNA_f$ (Met-$tRNA_f$), the initiator species. This binding depends on GTP and leads to the formation of a ternary complex, eIF-2/Met-$tRNA_f$/GTP, that subsequently binds to the small (40 S) ribosomal subunit (5). Only when Met-$tRNA_f$ is bound to the small ribosomal subunit can binding of mRNA take place (6, 7). Thus, the unique property of providing Met-$tRNA_f$ already imparts on eIF-2 a crucial role in the binding of mRNA.

In addition to binding Met-$tRNA_f$, eIF-2 itself can bind to mRNA (8-12). This binding is specific in that all mRNA species tested possess an effective binding site for eIF-2, including mRNA species lacking the 5'-terminal cap or 3'-terminal poly (A) moieties (11), while RNA species not serving as mRNA, such as tRNA (9-13), ribosomal RNA (9,14), or negative-strand viral RNA (11), do not possess such a site.

The mRNA-binding property is a feature of eIF-2 itself. This can be demonstrated in a number of ways, but perhaps most convincingly by the finding that binding of mRNA to eIF-2 can be completely inhibited by competing amounts of Met-$tRNA_f$, provided GTP is present (13). Thus, mRNA and

Met-tRNA$_f$ are mutually exclusive in their binding to eIF-2, suggesting that during protein synthesis, the interaction of a molecule of mRNA with eIF-2 on the 40 S ribosomal subunit displaces the previously bound Met-tRNA$_f$ from this factor.

The importance of the interaction between mRNA and eIF-2 for translational control of gene expression was first suggested by our study of the translational competition between α-globin mRNA and β-globin mRNA, which showed that mRNA competition for eIF-2 is involved in the regulation of α- and β-globin synthesis and suggested strongly that globin mRNA interacts directly with eIF-2 during protein synthesis, with α-globin mRNA binding more weakly than β-globin mRNA (15). This observation showed that individual mRNA species possess different affinities for eIF-2.

The activity of eIF-2 is subject to regulation. Two of the clearest cases of translational control that can be studied in vitro are the block initiation of translation in reticulocyte lysates occurring in the absence of heme or in the presence of (viral) double-stranded RNA (dsRNA). In both cases, protein synthesis is blocked because eIF-2 becomes inactivated (8, 16-18). The translational block induced by interferon also appears to involve dsRNA-dependent inactivation of eIF-2 as a major mechanism of control (19).

THE SITE IN VIRAL MESSENGER RNA RECOGNIZED BY eIF-2

To map the binding site for eIF-2 in mRNA, we first examined the 5' and 3' ends. Globin mRNA molecules from which the 3'-terminal poly (A) tail alone, or together with the 3'-untranslated 90 nucleotides had been removed by controlled, processive phosphorolysis with polynucleotide phosphorylase were compared to native globin mRNA molecules in terms of their binding affinities for eIF-2. All bound with equal affinity, eliminating the sequences downstream from the coding region as essential (12). On the other hand, cap analogs inhibited binding of both mRNA and Met-tRNA$_f$ to eIF-2 (ref. 20). This suggested that the cap might interact with eIF-2. However, the genomic RNA species from Mengovirus or satellite tobacco necrosis virus (STNV) bind extremely well to eIF-2, in fact even better than globin mRNA, yet they do not carry a cap structure (14, 20). This and other observations suggested that binding of eIF-2 to mRNA occurs primarily at an internal sequence and only secondarily through interaction with the cap (20). Since the coding region is unlikely to contain a functional binding site for an initiation factor, this focused our attention on the 5' leader sequence.

STNV RNA is especially suitable for analyzing this sequence, because the RNA, 1,239 nucleotides long (21), has an unmodified 5'-end (22) that makes it particularly amenable for 5'-end labeling with polynucleotide kinase. As can be seen in the gel of Fig. 1c, RNA isolated from STNV virions migrates, after 5'-end labeling, as a heterogeneous collection of fragments, with only a minor amount of label in fully intact viral RNA (arrow). When intact RNA is excised from the gel and digested with pancreatic RNase, a single spot is observed (Fig. 1c) in the position of pApGpU, corresponding to the 5'-terminal nucleotide sequence of the intact viral RNA (23). On the other hand, some 35 different spots are observed in the pancreatic fingerprint of the complete RNA preparation (Fig. 1a),

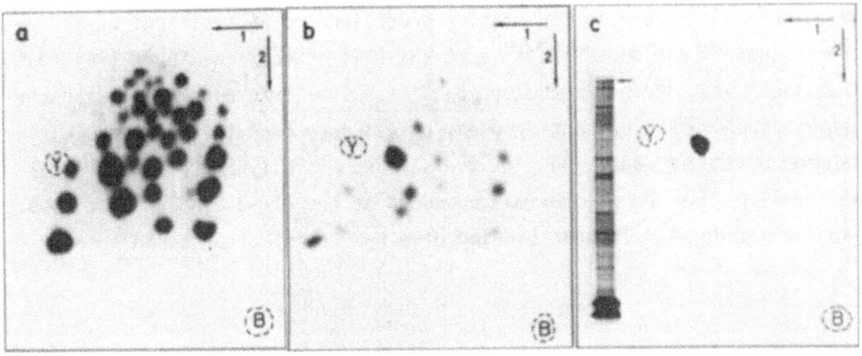

FIGURE 1. Analysis of pancreatic RNase digests of total (*a*), eIF-2-selected (*b*), and purified, intact (*c*) 5'-end-labeled STNV RNA by electrophoresis at pH 3.5 in the first dimension and homochromatography on a DEAE-thin-layer plate in the second dimension (see arrows), followed by autoradiography. (*a*) Total 5'-end-labeled STNV RNA. (*b*) eIF-2-selected RNA. Total 5'-end-labeled STNV RNA was incubated under conditions for RNA binding with increasing amounts of purified eIF-2; bound RNA was isolated by retention on 0.45-μm nitrocellulose filters and eluted with 0.1% $NaDodSO_4$, followed by ethanol precipitation. At saturating amounts of eIF-2, 70% of the labeled RNA could be bound; in the sample analyzed, 7.2% of the input label was bound by a limiting amount (0.3 μg) of eIF-2. (*c*) Intact 5'-end-labeled STNV RNA purified from the total kinase-treated RNA mixture by electrophoresis in a 4% polyacrylamide gel containing 7 M urea [arrow points to intact RNA]. Marker dyes were methyl orange (Y) and xylene cyanol FF (B). Reproduced with permission from ref. 14.

attesting to the presence of many fragments originating from internal regions of the viral RNA. When this RNA was offered to eIF-2 and RNA bound by eIF-2 was isolated and fingerprinted, one major spot was observed, migrating precisely as the 5' end of intact STNV RNA (Fig. 1b). Sequence verification, as well as fingerprinting with RNase T1, confirmed that eIF-2 binds selectively to STNV RNA fragments starting with the 5' end of the intact RNA (14).

To map the eIF-2 binding site more exactly, intact 5'-end-labeled STNV RNA was isolated and digested partially with RNase T1, to generate a nested set of labeled RNA fragments, all containing the 5' end of intact RNA and extending to various points within the molecule. Fragments of discrete size were isolated by gel electrophoresis, and their ability to bind to eIF-2 was studied. Fig. 2 depicts the 5'-terminal sequence of STNV RNA, including the unique AUG translation initiation codon located at positions 30-32. Arrows denote G residues sensitive to RNase T1 attack. We found (14) that eIF-2 does not bind the 32-nucleotide 5'-terminal fragment ending with the initiation codon AUG or shorter ones, but it does bind to the 44-nucleotide fragment or larger ones, and with the same specificity as to intact viral RNA. This places the 3'-proximal boundary of the eIF-2 binding site at or near nucleotide 44. Indeed, binding of eIF-2 to intact STNV RNA greatly

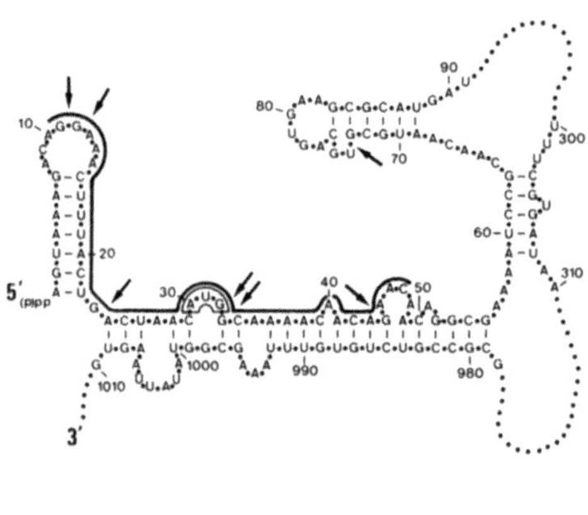

FIGURE 2. Secondary structure model for the 5' end of STNV RNA. The model (21) depicts stable secondary interactions. Line, nucleotides protected by 40 S ribosomal subunits against nucleases (24); arrows, prominent sites of RNase T1 cleavage (14). For eIF-2 binding site, see text. Reproduced with permission from ref. 14.

increases the sensitivity of the RNA to cleavage by RNase T1 at nucleotide 44, attesting to a conformational change induced at this point by the binding of the initiation factor molecule (14). On the 5'-terminal site, the binding of eIF-2 protects positions 11, 12, 23 and 32 against digestion, placing the boundary at or before position 10. Since the G residues at positions 2 and 7 are hydrogen-bonded in the 5'-terminal, 7 base-pair double-stranded stem (23) and thus resistant to RNase T1 attack, it is not certain if the eIF-2 binding site extends to the physical 5' end. The striking aspect of the eIF-2 binding site is, however, that it overlaps virtually completely with the binding site for 40 S ribosomal subunits (24), depicted by the line in Fig. 2. Thus, eIF-2 by itself recognizes the same nucleotide sequence that is bound by 40 S ribosomal subunits carrying eIF-2, Met-tRNA$_f$ and all other components needed for initiation of translation. This finding strongly suggests that the binding of ribosomes to STNV RNA is guided directly by eIF-2.

Fig. 3 depicts the physical map of Mengovirus RNA, constructed by Dr. R. Perez-Bercoff (25). It shows the major oligonucleotides generated by RNase T1 digestion and ordered with respect to their distance from the 3' terminus of the molecule. Mengovirus RNA has a length of about 7,500 nucleotides and contains a poly (C) stretch located within several hundred nucleotides from, but not at, the 5' end (26). The order of oligonucleotides within a given region (demarcated by vertical lines) was not established. Ribosomes were bound to Mengovirus RNA in extracts of L 929 or Krebs ascites cells, to generate 40 S or 80 S initiation complexes that

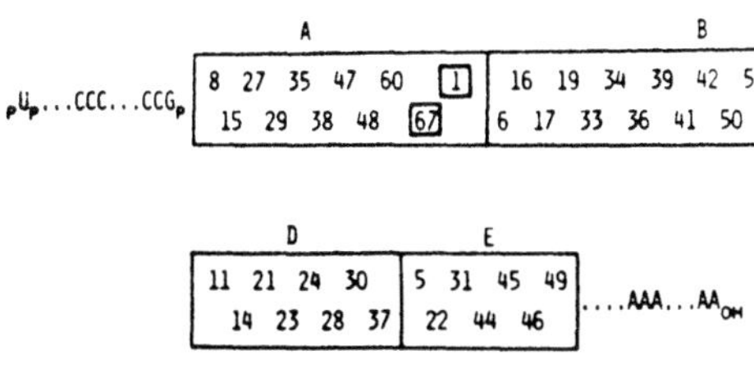

FIGURE 3. Physical map of Mengovirus RNA. The major T1 oligonucleotides (numbers) were ordered relative to the poly (A) stretch at the 3' terminus of the molecule. The order of individual oligonucleotides with a given region (A-E) has not been established. Boxed oligonucleotides Nos. 1, 2, 4 and 67 are protected by ribosomes in initiation complexes, while Nos. 1, 2 and 4 are protected by eIF-2. These four oligonucleotides have been placed at arbitrary positions within their respective regions. Reproduced with permission from ref. 25.

were isolated by sucrose gradient centrifugation. The RNA sequences protected in these complexes were isolated after RNase T1 digestion. In either complex, four unique T1 oligonucleotides were protected: these were oligonucleotides 1, 2, 4 and 67. Although the precise location of these oligonucleotides, each 15 to 28 nucleotides in length, within Mengovirus RNA is not yet known, it is seen from Fig. 3 that they fall into at least two widely separated domains (25). It is not yet clear if these domains form a single site held together by secondary or tertiary interactions in the RNA molecule, or two separate sites for initiation of translation.

Labeled, intact Mengovirus RNA was offered to eIF-2 and the sequences protected by the initiation factor were isolated (25). Three specific T1 oligonucleotides were recovered, and these were oligonucleotides 1, 2 and 4. Most likely, oligonucleotide 67 also interacts with eIF-2, but too weakly to survive the isolation procedure (25). This finding shows that out of a very large sequence, eIF-2 binds, by itself, precisely to those nucleotides in Mengovirus RNA that are protected by ribosomes. The virtual identity of the binding sites in Mengovirus RNA for ribosomes on one hand and for eIF-2 on the other, strongly reinforces the results with STNV RNA and points to the critical role for eIF-2 in recognition of mRNA by ribosomes.

TRANSLATIONAL COMPETITION OF CELLULAR AND VIRAL MESSENGER RNA FOR eIF-2

To study the functional implications of the interaction between Mengovirus RNA and eIF-2 for protein synthesis, we decided to study translational competition between this RNA and cellular mRNA. We chose to work with an mRNA-dependent reticulocyte lysate because it allows the precise quantitation of each mRNA species present during translation and is capable of efficient and repeated initiation. Using this system, we had already revealed that α- and β-globin mRNA compete for eIF-2 in translation (15). In the experiment of Fig. 4, a constant amount of globin mRNA was translated in the presence of increasing amounts of Mengovirus RNA. In these conditions, the number of initiations remained constant (27), while total incorporation of ^{35}S-methionine into protein fell only slightly. Yet, there was a drastic decrease in translation of globin mRNA, accompanied by increasing synthesis of viral protein. At half-maximal inhibition of globin mRNA translation, there were 35 molecules of globin mRNA present for every molecule of Mengovirus RNA. Assuming that equal proportions of

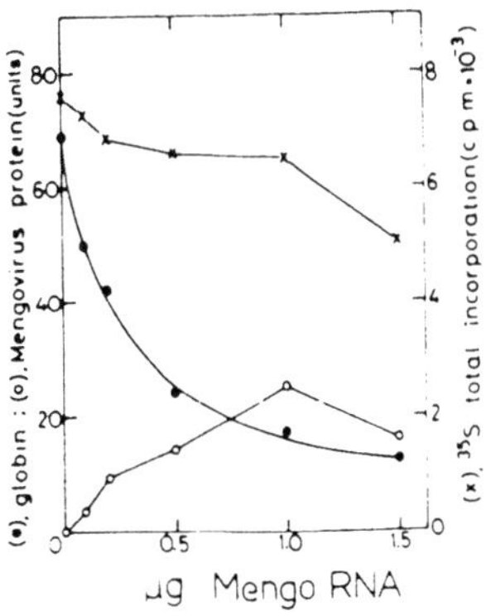

FIGURE 4. Translational competition between globin mRNA and Mengovirus RNA. Reaction mixtures for translation (27) or 25 μl contained micrococcal nuclease-treated rabbit reticulocyte lysate, 1.1 μg of globin mRNA, and the indicated amounts of Mengovirus RNA. After translation, the products were separated by cellulose acetate electrophoresis and the amounts of label in α-globin, β-globin, and Mengovirus proteins were quantitated by densitometry (27). Data are presented in terms of areas under the curves of densitometer scans and are plotted in arbitrary units as total amounts of globin and of Mengovirus RNA-directed products. The total amount of [^{35}S]methionine incorporated into protein is presented as hot CCl_3COOH-precipitable material per 5 μl of reaction mixture. From ref. 27.

these RNA species are translationally active, this means that a molecule of Mengovirus RNA competes 35-fold more strongly in translation than does (on average) a molecule of globin mRNA (27).

Does this competition involve eIF-2? Indeed, we could show that the competition is relieved by addition of highly purified eIF-2. That is, in conditions where globin synthesis is greatly depressed by the presence of Mengovirus RNA, the addition of eIF-2 does not stimulate overall translation, yet restores globin synthesis to the level seen in the absence of Mengovirus RNA (27). Globin synthesis in controls lacking Mengovirus RNA is not stimulated by the addition of eIF-2. We conclude that the addition of eIF-2 allows the more weakly competing, but more numerous, globin mRNA molecules to initiate translation at the expense of the more strongly competing, but less numerous, viral RNA molecules, while the total number of initiations remains constant.

That the translational competition is directly for eIF-2 is supported by the data of Fig. 5, which depicts binding competition between Mengovirus

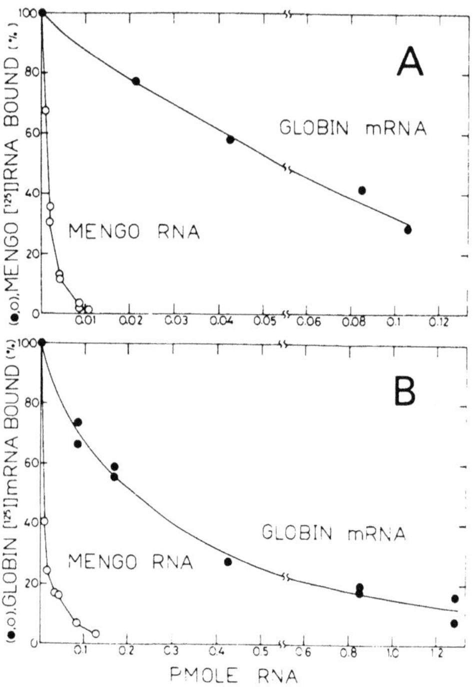

FIGURE 5. Competition between globin mRNA and Mengovirus RNA in direct binding to eIF-2. A, binding of ^{125}I-labeled Mengovirus RNA (2fmol; 4,470 cpm) to a limiting amount of eIF-2 (0.28 ng) was assayed (27) in reaction mixtures containing the indicated amounts of unlabeled Mengovirus RNA and globin mRNA. Binding in the absence of competing RNA (1,900 cpm) was taken as 100%. Background without eIF-2 (70 cpm) was subtracted. B, binding of ^{125}I-labeled globin mRNA (0.15 pmol; 18,000 cpm) to a limiting amount of eIF-2 (20 ng) was assayed in reaction mixtures containing the indicated amounts of unlabeled Mengovirus RNA and globin mRNA. Binding in the absence of competing RNA (6,280 cpm) was taken as 100%. Background without eIF-2 (270 cpm) was subtracted. From ref. 27 with permission.

RNA and globin mRNA for eIF-2. In this experiment, the only macromolecules present were eIF-2 and mRNA. In A, binding of labeled Mengovirus RNA to a limiting amount of eIF-2 is studied in the presence of increasing amounts of unlabeled, competing mRNA. Unlabeled Mengovirus RNA competes as expected, with the same affinity for eIF-2 as the labeled viral RNA. By contrast, 30 times more molecules of globin mRNA must be present before binding of labeled Mengovirus RNA is reduced to 50%. In B, the labels are reversed. In this reciprocal experiment, unlabeled globin mRNA competitively inhibits

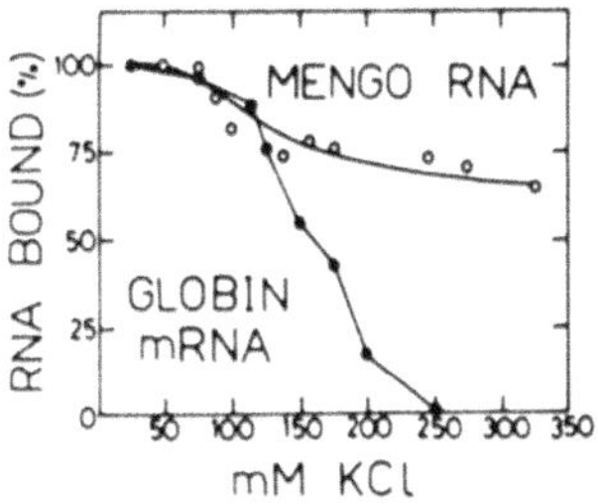

FIGURE 6. Salt sensitivity of complex formation between eIF-2 and globin mRNA or Mengovirus RNA. Binding of ^{125}I-labeled globin mRNA (1.5 x 10^6 cpm/μg; input, 4,270 cpm) or Mengovirus RNA (8 x 10^5 cpm/ g; input 3,100 cpm) to a limiting amount of eIF-2 was assayed (27) in reaction mixtures containing the indicated concentrations KCl. Binding at 25 mM KCl (2,510 cpm for globin mRNA and 1,640 cpm for Mengovirus RNA) was taken as 100%. Background without eIF-2 was subtracted. For globin mRNA, this background was 156 cpm at 25 mM KCl, and 102 cpm at 250 mM KCl. For Mengovirus RNA, the background values were 165 and 160 cpm, respectively. From ref. 27.

the binding of labeled globin mRNA according to expectation, while Mengovirus RNA competes 30 times more effectively. Thus, a molecule of Mengovirus RNA binds to eIF-2 30-fold more strongly than (on average) a molecule of globin mRNA, a result that fits remarkably well with its 35-fold greater ability to compete in translation. As we have seen, the translational competition is relieved by eIF-2.

A further demonstration of the high affinity of Mengovirus RNA is provided in Fig. 6. Binding of globin mRNA to eIF-2 is inhibited progressively at increasing salt concentrations; by contrast, binding of Mengovirus RNA is quite resistant to salt.

These results (27) strongly support the concept that Mengovirus RNA and globin mRNA compete directly for eIF-2 in protein synthesis. More generally, they reveal a direct correlation between the ability of a given mRNA species to compete in translation and its ability to bind to initiation factor eIF-2.

THE SELECTIVE INHIBITORY EFFECT OF DOUBLE-STRANDED RNA ON TRANSLATION

The above results suggest that in the infected cell, mRNA competition for eIF-2 could be a major factor in ensuring the selective translation of picornaviral RNA that is actually observed. Since the replication of the viral RNA generates, as a side-product, double-stranded RNA, and since

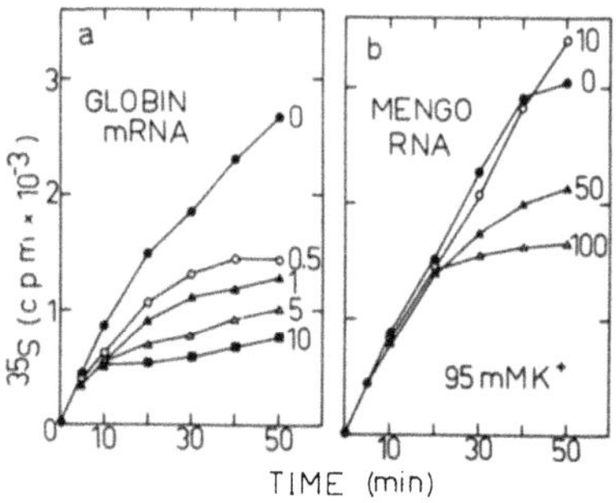

FIGURE 7. Differential sensitivity of Mengovirus RNA and globin mRNA translation to inhibition by dsRNA. A rabbit reticulocyte lysate treated with micrococcal nuclease was incubated with globin mRNA (1.5 μg) (a) or Mengovirus RNA (1.5 μg) (b). DsRNA from *Penicillium chrysogenum* was present at the indicated concentrations in ng/ml. Translation was at 30°C in 50-μl reaction mixtures containing 30 μl of lysate, 6 μg of mouse liver tRNA, 95 mM added KCl, and 0.6 mM added Mg^{2+}. This Mg^{2+} concentration was determined as optimal for both mRNA species. Aliquots of 5 μl were sampled at time intervals, and hot CCl_3COOH-precipitable [^{35}S]methionine was determined. Background without mRNA (300 cpm) was subtracted (28).

dsRNA can cause the inactivation of eIF-2 (8, 7, 18), it was of interest to study the behavior of viral RNA translation in the presence of dsRNA. Indeed, if we assume that dsRNA acts simply by inactivating eIF-2, then the consequent failure of Met-$tRNA_f$ to bind to the 40 S ribosomal subunit -- a necessary prerequisite for the binding of mRNA -- would be enough to explain why initiation of protein synthesis is blocked. This view would predict that dsRNA should equally inhibit the translation of all mRNA species, as each depends on prior binding of Met-$tRNA_f$.

We were surprised to discover that this is not the case (28). As seen in Fig. 7a, the translation of globin mRNA in an mRNA-dependent reticulocyte lysate becomes blocked, after a characteristic lag of 5-10 min, in the presence of as little as 0.5 ng/ml dsRNA, and maximal inhibition is observed when 10 ng/ml dsRNA is present. By contrast, the translation of Mengovirus RNA continues unabated in that condition, and only partial inhibition is seen at 5- or 10-times higher concentrations of dsRNA (Fig. 7b). This holds true also at 125 mM KCl, the optimal salt concentration for Mengovirus RNA translation, where even 1 μg/ml dsRNA gave no greater inhibition than that observed with 0.5 ng/ml in Fig. 7a. Total mRNA from Ehrlich ascites tumor cells behaves as globin mRNA in terms of its sensitivity to dsRNA (28). Thus, in contrast to cellular mRNA, the translation of Mengovirus RNA is resistant to dsRNA.

What is the mechanism of this unexpected resistance? Control experiments (28) with inhibitors show that translation of Mengovirus RNA depends as much on continued initiation as does translation of globin mRNA; thus we are not observing "run-off" synthesis. Second, direct measurement of the number of initiations shows that these remain constant in the presence of dsRNA when the template is viral RNA, but not when it is globin mRNA. Third, the inhibition of globin mRNA translation can be relieved fully by eIF-2, showing that loss of eIF-2 activity is at the basis of the inhibition. In view of the very high competitive ability of Mengovirus RNA and affinity for eIF-2, it was reasonable to suggest that perhaps the translation of Mengovirus RNA can be maintained by traces of eIF-2 left after the bulk of this factor has been inactivated by dsRNA. To test this hypothesis, we translated globin mRNA in the presence of 10 ng/ml of dsRNA and at 15 min, a time when translation has been blocked maximally (cf. Fig. 7a), added Mengovirus RNA. This late addition of Mengovirus RNA completely failed to elicit translation, even though extensive synthesis was observed in these conditions when dsRNA was absent (28). Thus, it appears that once inhibition of translation by dsRNA has been established, the translation of Mengovirus RNA is as sensitive to inhibition as is that of globin or cellular mRNA. This result must mean that Mengovirus RNA acts to prevent the establishment of translational inhibition by dsRNA (Table 1). Indeed, the RNA from picornavirus Coxsackie B5 behaves in the same manner (28).

Mengovirus RNA thus protects the protein-synthetic machinery, that is, eIF-2, against inactivation by dsRNA. The inactivation of eIF-2 is accompanied by phosphorylation of the 38,000 dalton α-subunit (29). Indeed, phosphorylation of this polypeptide was severely decreased in dsRNA-containing lysates when Mengovirus RNA was the template (28). This led us to examine the dsRNA-dependent phosphorylation of eIF-2 more carefully in a crude ribosome system. To our surprise, Mengovirus RNA, when added to this system, effectively inhibited the phosphorylation of eIF-2, while globin mRNA, even in much higher concentrations, did not (Fig. 8). This finding suggested that a competition phenomenon between mRNA and dsRNA might be at the root of the inhibitory mechanism of dsRNA. Since we had already found in 1974 that eIF-2 binds to dsRNA (8), this suggested that mRNA and dsRNA might compete directly for eIF-2.

Table 1. Messenger RNA specificity in the inhibitory effect of dsRNA on translation

Observation	Seen with globin mRNA	Seen with Mengovirus RNA
1. Inhibition of translation by low concentrations of dsRNA	yes	no
2. Inactivation of eIF-2	yes	no
3. Translation depends on continued initiations	yes	yes
4. Sensitivity to dsRNA-activated inhibition	yes	yes
5. Establishment of dsRNA-activated inhibition	yes	no
6. Phosphorylation of eIF-2	yes	no
7. mRNA competitively inhibits dsRNA-dependent phosphorylation of eIF-2	no	yes
8. mRNA affinity for eIF-2, relative to dsRNA	lower	higher

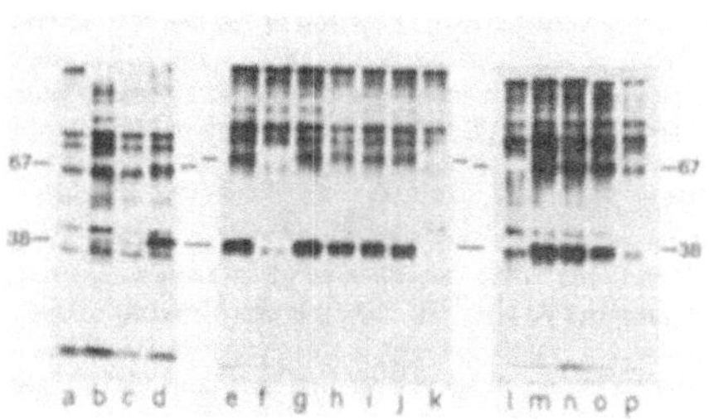

FIGURE 8. Effect of mRNA on dsRNA-dependent phosphorylation of eIF-2. The ribosomal system (28) containing [γ-^{32}P]ATP and crude initiation factors, was incubated without dsRNA (a-c, f, and l), with 50 ng/ml dsRNA (d, e, g-j, and m-p), and with 10 µg/ml dsRNA (k). Globin mRNA was present in (b) (0.5 µg), (g) (0.25 µg), (h) (0.5 µg), (i) (1.0 µg), and (j) (4 µg). Mengovirus RNA was present in (c) (0.5 µg), (n) (1.45 µg), (o) (3.6 µg), and (p) (7.2 µg). The autoradiograms of NaDodSO$_4$-polyacrylamide gels of these reaction mixtures are shown. The positions of 38,000- and 67,000-dalton polypeptides are indicated. Reproduced with permission from ref. 28.

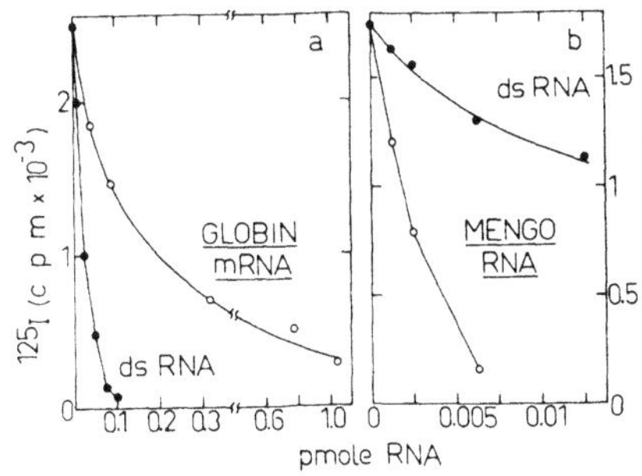

FIGURE 9. Competition of dsRNA, globin mRNA, and Mengovirus RNA for eIF-2. RNA-binding assays in (a) contained ^{125}I-labeled globin mRNA (0.025 pmol; 9500 cpm) and 10.6 ng of eIF-2. In (b), they contained ^{125}I-labeled Mengovirus RNA (0.002 pmol; 4470 cpm) and 0.28 ng of eIF-2. Unlabeled dsRNA, globin mRNA, or Mengovirus RNA was present in the indicated amounts. Control without eIF-2 (about 100 cpm) was subtracted. From ref. 28.

Fig. 9 shows that this is indeed the case. In Fig. 9a, binding of labeled globin mRNA to eIF-2 is competed according to expectation by unlabeled globin mRNA molecules, while dsRNA competes, on a molar basis, about 6-fold more effectively. By contrast, the binding of labeled Mengovirus RNA in Fig. 9b is competed by a molecule of dsRNA some 8-fold more weakly than by Mengovirus RNA. Thus, on one hand dsRNA binds with higher affinity than globin mRNA to eIF-2 and inhibits globin mRNA translation, while on the other hand it binds with lower affinity than Mengovirus RNA to eIF-2 and fails to establish translational inhibition when viral RNA is present.

These results indicate that the rate-determining event in the establishment of translational inhibition by dsRNA involves competition between mRNA and dsRNA and that Mengovirus RNA, because of its high affinity for eIF-2, is able to protect this initiation factor against inactivation by dsRNA.

DISCUSSION

Messenger RNA competition for eIF-2

The concept that mRNA species differ in their efficiency of translation, apparently because of a different affinity for one or more critical components in the initiation step, was first suggested by Lodish (30, 31) who showed that initiation of protein synthesis on a molecule of α-globin mRNA occurs with lower frequency than that on a molecule of β-globin mRNA (32). Lawrence and Thach (2) observed translational competition between host mRNA and encephalomyocarditis (EMC) virus RNA in a cell-free system from mouse ascites cells and concluded that the competition was at the level of initiation. Hackett et al. (3, 4) made similar observations for Mengovirus RNA. Abreu and Lucas-Lenard (33) also found that Mengovirus RNA can suppress the translation of globin mRNA. In searching for a possible target of competition, Golini et al. (34), using a reconstituted cell-free system, suggested that it is an initiation factor, or complex of factors, present in a partially purified preparation of initiation factor eIF-4B. The results reported here, showing that eIF-2 is able to relieve translational competition between Mengovirus RNA and globin mRNA, and that these mRNA species compete directly in binding to eIF-2, cannot be explained by assuming a contamination with eIF-4B. First, eIF-2 prepared by our procedure is at least 98% pure as judged by sodium dodecylsulfate/polyacrylamide gel electrophoresis, and is free of detectable protein at 80,000 daltons, the molecular weight assigned to eIF-4B (13). Second, the purification procedure yields an eIF-2 preparation that contains only a single mRNA-binding component, eIF-2, since binding of mRNA to this preparation is completely sensitive to competitive inhibition by Met-$tRNA_f$, provided GTP is present, but not by uncharged tRNA (13). Third, independent verification that eIF-2 binds 30-fold more tightly to Mengovirus RNA is provided by the observation that Mengovirus RNA is 30 to 40 times more effective than globin mRNA as a competitive inhibitor of ternary complex formation between Met-$tRNA_f$, GTP and eIF-2 (ref. 28). While it is conceivable that eIF-4B may also be a target for mRNA competition in translation, it is nevertheless clear from our studies that the addition of eIF-2 is sufficient to overcome such competition.

It is worth noting that in our translation experiments, the added eIF-2 always was used immediately after purification. Storage of purified eIF-2 leads to a loss of activity. Even though stored preparations can be active in ternary complex formation with Met-$tRNA_f$ and GTP, they contain a considerable proportion of inactive initiation factor molecules that can competitively inhibit the active ones. Moreover, the translation experiments were done in the micrococcal nuclease-treated reticulocyte lysate. This system has two advantages over reconstituted cell-free systems. It is capable of extensive and efficient initiation in conditions more likely to be representative of protein synthesis in intact cells, and except for mRNA, contains all other components for protein synthesis in a proportion much closer to that in the intact cell.

In the lysates used for the translation competition experiments (27), the addition of eIF-2 did not lead to any stimulation of protein synthesis in the presence of globin mRNA alone, nor in the presence of both globin mRNA and Mengovirus RNA. This explains why, when Mengovirus RNA was also present, the increase in globin mRNA translation caused by addition of eIF-2 was concomitant with a decrease in Mengovirus RNA translation. The fact that eIF-2 acts to shift translation in favor of globin synthesis shows clearly that globin mRNA and Mengovirus RNA compete for eIF-2, but does not eliminate the possibility that eIF-2 could act in a nonspecific manner, as by increasing the pool of 40S/Met-$tRNA_f$ complexes. The results of the RNA-binding experiments, however, showing that Mengovirus RNA and globin mRNA compete directly for eIF-2 with an affinity ratio that matches exactly with that observed in translation competition experiments, provide strong evidence for a direct competition of these mRNA species for eIF-2 during translation. While it is conceivable that globin mRNA and Mengovirus RNA compete for free eIF-2 molecules, it is more likely that they compete for eIF-2 molecules located in 40S/Met-$tRNA_f$ complexes (15, 27).

These results point to mRNA affinity for eIF-2 as a new element in translational control. They support the conclusion that translation of different mRNA species is regulated to a large extent by their relative affinities for eIF-2.

<u>The high affinity of Mengovirus RNA for eIF-2</u>

The RNA of picornaviruses is thought to contain only one (35), or at most two ribosome binding sites (36; R. Perez-Bercoff, personal communication). Thus, even though Mengovirus RNA has about a tenfold greater nucleotide length than does globin mRNA, its 35-fold greater ability to compete in translation cannot be explained by a commensurately greater number of ribosome binding sites. The high affinity of Mengovirus RNA for eIF-2, moreover, is not related simply to nucleotide length, for vesicular stomatitis virus (VSV) negative-strand RNA, which is even longer than Mengovirus RNA, binds only very weakly and non-specifically to eIF-2 and lacks the high-affinity binding site found in all mRNA species tested, including the much shorter VSV mRNA (11).

The RNA of Mengovirus, like that of other cardioviruses and of foot-and-mouth disease virus, contains a poly (C) tract. This tract is located close to the 5' terminus, though still several hundred nucleotides removed from it (25, 26). Although the function of the poly (C) tract is not yet known, the affinity of poly (C) for eIF-2 is extremely low, relative to that of other polynucleotides (9). More strikingly, the poly (C) tract is protected neither by ribosomes nor by eIF-2 (25). Instead, as detailed above, from among a sequence of 7,500 nucleotides in Mengovirus RNA, ribosomes in initiation complexes protect only four major T1 oligonucleotides, 15-28 bases in length, against nuclease attack, while eIF-2 recognizes and protects only three, identical to those bound by ribosomes (25). These results show not only that eIF-2 and ribosomes recognize the same sequences in Mengovirus RNA, but also that these sequences constitute only a very small portion of the viral RNA molecule. It is not yet known what molecular features underlie the extremely high affinity of these defined regions in Mengovirus RNA for eIF-2, and the correspondingly high effectiveness of this RNA to compete in translation. Most likely, the binding sites for ribosomes and eIF-2 possess an unusual, highly effective conformation that is not found in cellular mRNA species. In this context, it is worth noting that the related EMC RNA possesses an unusual primary and secondary structure at the start of the coding region (35, 37). It is quite possible that this highly structured site, which is resistant to nuclease treatment, is related directly to the specific site for eIF-2 in Mengovirus RNA.

The extremely high affinity of Mengovirus RNA for eIF-2 may explain why Mengovirus and other picornaviral RNA species lack the 5'-terminal cap structure (38, 39). The existence of a sequence in Mengovirus RNA possessing a very high affinity for eIF-2 would obviate the need for the additional stabilization imparted by binding at the cap.

Implications for the mechanism of selective viral mRNA translation

The present findings bear on the mechanism of shut-off of host protein synthesis by picornaviruses, and possibly also by other virulent viruses. Several mechanisms have been proposed for this shut-off, and they may act in a concerted manner: competition between host and viral mRNA in translation, inhibition by virus-induced dsRNA, increase in intracellular salt concentrations, and loss of initiation factor activity. Indeed, the shut-off of host protein synthesis by poliovirus is accompanied by the

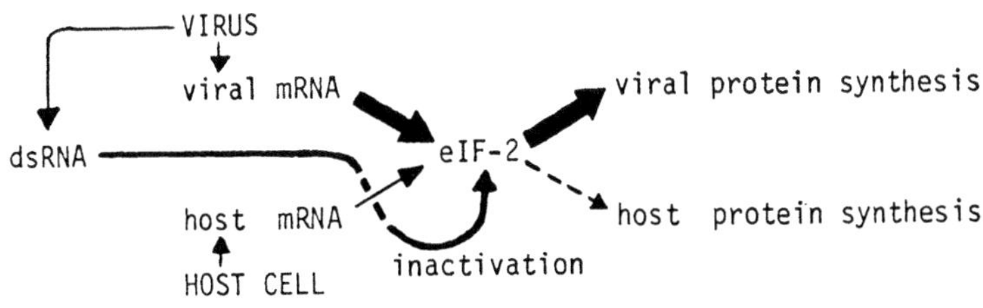

FIGURE 10. Host-virus interactions at the level of translational control that involve eIF-2. See text.

inactivation of cap-binding proteins (40). This mechanism may be particularly important early in infection. As will be seen below, each of the above-listed mechanisms, acting on eIF-2, can contribute to the selective translation of viral RNA at subsequent stages of infection.

Among the host-virus interactions at the level of translational control that involve eIF-2, the ability of a strong viral RNA template to outcompete host mRNA for eIF-2 in translation is most prominent. This ability will lead to the selective translation of viral mRNA that is observed during infection. The high affinity of viral RNA for eIF-2 will result, first, in the displacement of host mRNA from the ribosomes and second, in the highly efficient initiation of translation on viral messenger RNA. Host mRNA, once displaced, eventually may become more prone to limited nuclease attack at sites essential for translation, even though in Mengovirus-infected cells, the physical intactness of host mRNA as a whole does not change perceptibly (41).

A second mechanism, involving dsRNA, may operate during infection: resistance of viral RNA translation to inhibition by dsRNA, coupled with sensitivity of host mRNA translation. Early in infection, dsRNA begins to be formed as a result of viral RNA replication. Initially the level of dsRNA is low, but it increases during infection. A role for dsRNA in shut-off of host protein synthesis was first suggested by Hunt and Ehrenfeld (42), but rejected later when a differential effect of dsRNA on host and viral mRNA translation could not be demonstrated in cell-free systems capable of only limited initiation (43, 44). By using a reticulocyte lysate system capable of initiating translation with an efficiency resembling that of the intact cell, however, we could show that translation of Mengovirus RNA is, in fact, completely resistant to inhibition by concentrations of dsRNA that totally inhibit the translation of globin or ascites tumor cell mRNA. As we have seen, dsRNA causes the inactivation of eIF-2, but fails to do so when Mengovirus RNA is used as template. The most likely explanation of our results is provided by the finding that dsRNA competes with mRNA for eIF-2, binding it more weakly than Mengovirus RNA, but more tightly than globin mRNA. The high affinity of Mengovirus RNA for eIF-2 may thus serve to permit continued viral mRNA translation even in the presence of amounts of virus-generated dsRNA that are sufficient to inhibit host protein synthesis.

The salt optimum for translation of viral RNA appears to be higher than that for host mRNA, for exposure of infected cells to hypertonic salt concentrations leads to a block in initiation of host mRNA translation, while still permitting translation of viral RNA (45). Indeed, picornavirus-infected cells contain higher salt concentrations than do uninfected cells (46, 47), but the change may be too gradual to account for shut-off of host-protein synthesis. Our finding that complexes between Mengovirus RNA and eIF-2 can still form at salt concentrations at which complexes between globin mRNA and eIF-2 are no longer stable (Fig. 9), is consistent with the differential effect of salt on protein synthesis in infected cells.

The high affinity of Mengovirus RNA for eIF-2 may thus contribute in a number of ways to the transition from host to viral protein synthesis (Fig. 10). After infection, when viral RNA begins to accumulate, favorable competition for eIF-2 will lead to displacement of host mRNA from polysomes. Viral dsRNA formed during replication (48) is expected to cause inactivation of some eIF-2 molecules, acting thereby to decrease the pool size of eIF-2 and to sharpen the competition between host and viral mRNA for eIF-2, to the advantage of the stronger template, viral RNA. Likewise, an increase in intracellular salt concentrations will cause a preferential reduction in the binding of host mRNA to eIF-2. These effects will further increase the advantage of the stronger viral mRNA and enhance the shift from cellular to viral protein synthesis. Later in infection, when more dsRNA has accumulated, viral mRNA is the predominant template in translation and protects the residual eIF-2 molecules against inactivation by dsRNA. Viral proteins continue to be synthesized at a high rate.

ACKNOWLEDGMENTS

The work reported here was supported by grants from the National Council of Research and Development (Israel) and the Gesellschaft fuer Strahlenforschung (Muenchen), and from the Israel Academy of Sciences.

REFERENCES

1. Ehrenfeld, E. In: The Molecular Biology of Picornaviruses (Ed. R. Perez-Bercoff), Plenum Publishing Corp., New York, 1979, pp. 223-238.
2. Lawrence, C. and Thach, R.E. J. Virol. 14:598-610, 1974.
3. Hackett, P.B., Egberts, E. and Traub, P. Eur. J. Biochem. 83:353-361, 1978.
4. Hackett, P.B., Egberts, E. and Traub, P. Eur. J. Biochem. 83:341-352, 1978.
5. Levin, D.H., Kyner, D. and Acs, G. Proc. Natl. Acad. Sci. USA 70:41-45, 1973.
6. Schreier, M.H. and Staehelin, T. Nature (London) New Biol. 242: 35-38, 1973.
7. Darnbrough, C.H., Legon, S., Hunt, T. and Jackson, R.J. J. Mol. Biol. 76:379-403, 1973.
8. Kaempfer, R. Biochem. Biophys. Res. Commun. 61:591-597, 1974.
9. Barrieux, A. and Rosenfeld, M.G. J. Biol. Chem. 252:3843-3847, 1977.
10. Barrieux, A. and Rosenfeld, M.G. J. Biol. Chem. 253:6311-6315, 1978.
11. Kaempfer, R., Hollender, R., Abrams, W.R. and Israeli, R. Proc. Natl. Acad. Sci. USA 75:209-213, 1978.
12. Kaempfer, R., Hollender, R., Soreq, H. and Nudel, U. Eur. J. Biochem. 94:591-600, 1979.
13. Rosen, H. and Kaempfer, R. Biochem. Biophys. Res. Commun. 91:449-455, 1979.
14. Kaempfer, R., van Emmelo, J. and Fiers, W. Proc. Natl. Acad. Sci. USA 78:1542-1546, 1981.
15. Di Segni, G., Rosen, H. and Kaempfer, R. Biochemistry 18:2847-2854, 1979.
16. Kaempfer, R. and Kaufman, J. Proc. Natl. Acad. Sci. USA 69:3317-3321, 1972.
17. Kaempfer, R. and Kaufman, J. Proc. Natl. Acad. Sci. USA 70:1222-1226, 1973.
18. Clemens, M.J., Safer, B., Merrick, W.C., Anderson, W.F. and London, I.M. Proc. Natl. Acad. Sci. USA 72:1286-1290, 1975.
19. Kaempfer, R., Israeli, R., Rosen, H., Knoller, S., Zilberstein, A., Schmidt, A. and Revel, M. Virology 99:170-173, 1979.
20. Kaempfer, R., Rosen, H. and Israeli, R. Proc. Natl. Acad. Sci. USA 75:650-654, 1978.
21. Ysebaert, M., van Emmelo, J. and Fiers, W. J. Mol. Biol. 143:273-287, 1980.
22. Wimmer, E., Chang, A.Y., Clark, J.M. Jr. and Reichmann, M.E. J. Mol. Biol. 38:59-73, 1968.
23. Leung, D.W., Browning, K.S., Heckmann, J.E., RajBhandary, U.L. and Clark, J.M. Jr. Biochemistry 18:1361-1366, 1979.
24. Browning, K.S., Leung, D.W. and Clark, J.M. Jr. Biochemistry 19: 2276-2282, 1980.

25. Perez-Bercoff, R. and Kaempfer, R. J. Virol. 41:30-41, 1982.
26. Perez-Bercoff, R. and Gander, M. Virology 80:426-429, 1977.
27. Rosen, H., Di Segni, G. and Kaempfer, R. J. Biol. Chem. 257:946-952, 1982.
28. Rosen, H., Knoller, S. and Kaempfer, R. Biochemistry 20:3011-3020, 1981.
29. Farrell, P.J., Balkow, K., Hunt, T., Jackson, R.J. and Trachsel, H. Cell 11:187-200, 1977.
30. Lodish, H.F. Nature 251:385-388, 1974.
31. Lodish, H.F. Annu. Rev. Biochem. 45:39-72, 1976.
32. Lodish, H.F. J. Biol. Chem. 246:7131-7138, 1971.
33. Abreu, S. and Lucas-Lenard, J. J. Virol. 18:182-194, 1976.
34. Golini, F., Thach, S., Birge, C., Safer, B., Merrick, W. and Thach, R. Proc. Natl. Acad. Sci. USA 73:3040-3044, 1976.
35. Smith, A.E. In: Control Processes in Virus Multiplication (Eds. D.C. Burke and W.C. Russell), Cambridge University Press, Cambridge, 1975, pp. 183-224.
36. Ehrenfeld, E. and Brown, D. J. Biol. Chem. 256:2656-2661, 1981.
37. Porter, A.G., Frisby, D.P., Carey, N.H. and Fellner, P. INSERM Meeting 47:169-176, 1975.
38. Nomoto, A., Lee, Y.F. and Wimmer, E. Proc. Natl. Acad. Sci. USA 73:375-380, 1976.
39. Hewlett, N.J., Rose, J.K. and Baltimore, D. Proc. Natl. Acad. Sci. USA 73:327-330, 1976.
40. Lee, K.A.W. and Sonenberg, N. Proc. Natl. Acad. Sci. USA 79:3447-3451, 1982.
41. Colby, D.S., Finnerty, V. and Lucas-Lenard, J. J. Virol. 13:858-869, 1974.
42. Hunt, T. and Ehrenfeld, E. Nature New Biol. 230:91-94, 1971.
43. Robertson, H.D. and Mathews, M.B. Proc. Natl. Acad. Sci. USA 70:225-229, 1973.
44. Celma, M. and Ehrenfeld, E. Proc. Natl. Acad. Sci. USA 71:2440-2444, 1974.
45. Nuss, D., Opperman, H. and Koch, G. Proc. Natl. Acad. Sci. USA 72:1258-1262, 1975.
46. Carrasco, L. and Smith, A. Nature 264:807-809, 1976.
47. Egberts, E., Hackett, P. and Traub, P. J. Virol. 22:591-597, 1977.
48. Ehrenfeld, E. and Hunt, T. Proc. Natl. Acad. Sci. USA 68:1075-1078, 1971.

17

MOLECULAR APPROACH TO VIRULENCE : ISOLATION AND CHARACTERIZATION OF AVIRULENT MUTANTS OF RABIES VIRUS

P. COULON[1], P. ROLLIN[2], M. AUBERT[3], M. DOLIVO[4], P. KUCERA[4], D. PORTNOI[2], M. KITA[2] and A. FLAMAND [1]

1 Bât. 400, Université Paris-Sud - 91405 Orsay Cedex, France
2 Institut Pasteur, 28 Rue du Dr. Roux, 75724 Paris Cedex 15, France
3 Centre National d'Etudes sur la Rage, BP n°9, 54220 Nancy Malzeville
4 Institut de Physiologie, 7 Rue du Bugnon, CH-1011 Lausanne, Suisse

SUMMARY

Nine anti-G monoclonal antibodies were used to select mutants of the CVS strain of rabies virus resistant to neutralization. Seven mutants were avirulent in adult mice and two others exhibited an attenuated pathogenicity. Both categories were resistant to monoclonal antibodies 194-2 and 248-8. Virulence appears to be associated with a particular configuration of a region of the glycoprotein which is located at the intersection of the epitopes recognized by these two monoclonal antibodies. Our results confirm the role played by the glycoprotein in the neurovirulence of rabies virus.

Penetration in the central nervous system of CVS or its Av01 avirulent derivative has been followed after intraocular inoculation of the virus. The most striking difference is that Av01 has lost its capability to invade the intraocular parasympathetic nerve endings. In addition Av01 but not CVS, is able to invade the lens.

Various parameters of the immune response arising after intramuscular inoculation of CVS or Av01 have also been measured. The major difference is that interferon production and natural killer activity of splenocytes last for 2 or 3 days instead of few hours in the case of Av01 inoculated animals.

INTRODUCTION

Any attempt to understand the molecular basis of rabies virulence has been greatly hampered by the lack of avirulent mutants differing from a virulent strain by a single aminoacid substitution. Since our first discovery that it was possible to select avirulent mutants among survivors to neutralization with appropriate monoclonal antibodies (1-3), it has been demonstrated that such mutants were indeed the result of a single point mutation on the glycoprotein (4).

Systematic search for regions of the glycoprotein linked with virulence has been undertaken, using 9 monoclonal antibodies (kindly provided by Dr. T. Wiktor) which recognized 3 different antigenic sites (5). Results from this study which are presented in part I demonstrated that virulence is associated with a special configuration of the region of the protein which is recognized by monoclonal antibodies n° 194-2 and 248-8. Mutations in sites I, II and IIIa do not affect virulence.

Avirulent mutants have been used for in vivo study of host range spectrum, penetration and propagation of the virus in the organism as well as for investigation of the immune response. Comparison between virulent and avirulent infections are described in part II.

MATERIALS AND METHODS

Viruses and cells

A 5-FU mutagenized stock of the CVS strain of rabies virus was used. Conditions of mutagenesis have been described elsewhere (6). The virus was multiplied in Baby Hamster Kidney (BHK-21) cells and titrated in Chicken Embryo Related (CER) cells (7) as already described (8).

Vesicular Stomatitis Virus (VSV) Indiana serotype was grown in mouse fibroblast (L929) cells.

BHK-21, CER and L929 cells were grown as already described (6). YAC-1 cells were purchased from Flow Laboratories S.A. (France). They were grown in RPMI 1640 + HEPES supplemented with foetal calf serum (10 %), glutamine (30 mg/l) and antibiotics. Mouse ($H.2^a$) neuroblastoma cells (MNB), kindly provided by Dr. T.J. Wiktor (Wistar Institute, Philadelphia), were grown in Minimal Essential Medium (MEM) supplemented with foetal calf serum (10 %), glutamine (30 mg/l) and antibiotics. All the cells were grown at 37°C in a 5 % CO_2 atmosphere.

Production of hybridomas

Hybrid cultures from P_3 x 63 Ag_8 mouse myeloma cells and splenocytes from rabies immune mice were produced and characterized at the Wistar Institute (9).

Selection of avirulent mutants

The selection of mutants was performed according to Koprowski and

Wiktor (10). Briefly, 10 µl of an appropriate dilution of ascitic fluids was mixed with 100 µl of increasing dilutions of the mutagenized CVS stock. Survivors to neutralization were plated onto CER cells. After five days of incubation at 33°C, well isolated plaques were picked up and suspended in saline medium. After multiplication at 33°C in BHK-21 cells, the resistance to the monoclonal antibody was checked. If confirmed, the mutant was stored at -70°C.

The virulent/avirulent phenotype was determined by intracerebral (IC) inoculation of five adult Swiss mice with 1 to 5 x 10^3 pfu of the mutants. Surviving mice were challenged 28 days later by IC inoculation of 40 LD_{50} of CVS virus.

Reactivity towards monoclonal antibodies

The reactivity of the viruses towards antinucleocapsid antibodies was measured as follows : Labtek slides of BHK-21 cells were infected with the virus at a moi of 5 or mock infected. After 18 hrs incubation at 37°C, cells were fixed with acetone for 1 hr at -20°C and washed. Ten µl of anti-N monoclonal antibody was then added. Cells were incubated for 1 hr at 37°C, washed twice with phosphate buffer (0.2 M, pH = 7.2) and incubated with 10 µl of mouse anti IgG fluorescent antibody for 1 hr at 37°C. Preparations were washed several times with phosphate buffer and distilled water, and then observed in a fluorescence microscope.

The reactivity of the viruses towards anti-glycoprotein antibodies was measured as follows : Three to 5 x 10^2 pfu of the virus in 100 µl MEM + 0.1 % bovine serum albumin were incubated with 10 µl of appropriate anti-G monoclonal antibody for 30 min. Survivors were plated onto confluent monolayers of CER cells. The plaques were counted after 5 days incubation at 33°C.

Antibody titration

Blood was taken from anesthesized mice in the axillary hollow, then kept overnight at 4°C and centrifugated. The sera were collected and frozen at -20°C. Before titration of antibody and interferon the sera were decomplemented at 56°C for one half hour. Antibody titration was performed by the RFFIT method in cell culture modified for microtitration (11).

Interferon titration

Interferon was titrated in Balb/c sera and brain extracts by inhibition of the VSV cytopathic effect on L929 cells (12).

Preparation of splenocyte suspension

Mice were sacrificed by cervical dislocation and the spleens were removed.

After dissociation and decantation, the cells were washed twice. The splenocytes were counted, and their concentration was adjusted in RPMI 1640, HEPES, foetal calf serum (10 %), glutamine (30 mg/l), antibiotics.

Infection of MNB cells by CVS

For specific-cell-mediated-cytotoxicity experiments normal and CVS-infected MNB cells were used as target cells. Cells were infected in liquid medium at a moi of 5 to 10 pfu per cell. Twenty four and 48 hrs later, expression of the viral glycoprotein at the surface of the membranes was tested by immunofluorescence (antiglycoprotein conjugate, labelled with fluorescein isothiocyanate).

Labelling by $Na^{51}Cr$

YAC-1 cells. 10^6 cells in 500 µl of RPMI, HEPES + foetal calf serum (10 %) were incubated with 100 µCi of $Na^{51}Cr$ (C.E.A. Saclay, France) for 1 hr. The cells were washed 3 times and counted.

MNB cells. Non-infected or 48 hour-infected MNB cells were labelled as follows : 5×10^6 cells in 500 µl of MEM + foetal calf serum (10 %) were incubated with 200 µCi of $Na^{51}Cr$ for 1 hr. The cells were washed in MEM then twice in RPMI, HEPES + foetal calf serum (10 %).

Cellular-mediated cytotoxicity

"Natural Killer" activity (13, 14) was assayed as follows : 10^4 YAC-1 cells labelled with $Na^{51}Cr$ were laid in wells of a microtitration plaque. 10^6 Balb/c splenocytes were added, and the plaques were incubated overnight at 37°C. After centrifugation for 10 min (200 g), the radiolabelling of 100 µl of the supernatant from each well was counted.

Specific cytotoxic activity was assayed according to Wiktor et al. (15). Briefly, 10^4 non-infected or CVS-infected MNB ($H.2^a$) cells were laid in the wells and the splenocyte suspension of A/J ($H.2^a$) mice were added in effector/target cells ratios of 25 : 1 ; 50 : 1 and 100 : 1. After a 16 hour-incubation at 37°C, the plaques were centrifugated and release of $Na^{51}Cr$ was assayed as for "Natural Killer" activity. Spontaneous and detergent-induced release of $Na^{51}Cr$ was also determined in each experiment. Spontaneous released of $Na^{51}Cr$ was around 20 % for YAC-1 cells, and 30-40 % for MNB cells. For each point, the rate of specific lysis was calculated as follows :

$$\frac{\text{assayed release} - \text{spontaneous release}}{\text{maximum release} - \text{spontaneous release}} \times 100$$

Monoclonal antibody	509-6 (Kelef)°	101-1 (ERA)	162-3 (ERA)	220-8 (CVS)	231-22 (CVS)	719-3 (CVS)	194-2 (ERA)	248-8 (CVS)	507-1 (Kelef)	Total
N° of resistant mutants	15	44	47	19	3	30	39	61	20	278
Frequency in the population	9.8×10^{-4}	5.5×10^{-4}	4.3×10^{-4}	1.4×10^{-4}	10^{-3}	2.8×10^{-3}	3.3×10^{-4}	1.8×10^{-3}	9.8×10^{-4}	-
Mutants inoculated in mice	6	15	15	10	3	15	15	13	20	112
AvO* or AtO* mutants	0	0	0	0	0	0	7	2	0	9

Table 1. Virulence of mutants resistant to neutralization by anti-G monoclonal antibodies.

°Origin of the monoclonal antibodies (17). 719-3 originates from the CVS strain of rabies virus (T.J. Wiktor, personal communication).
*Av = Avirulent ; At = Attenuated ; O = Orsay.

RESULTS

Selection of avirulent and attenuated mutants of the CVS strain of rabies virus

Nine neutralizing monoclonal antibodies were used to select 278 mutants resistant to neutralization. Results presented in table 1 indicate that the frequency of the different mutants in the mutagenized population ranges from 3.3×10^{-4} to 2.8×10^{-3}. Mutants resistant to monoclonal antibody n° 220-8 were selected from a non-mutagenized CVS clone, explaining their lower frequency.

One hundred and twelve mutants were IC injected to adult mice. Most mutants were fully pathogenic and killed the animals between day 7 and day 9, like CVS. Nine mutants had an altered pathogenicity ; all were selected with monoclonal antibodies n° 194-2 or 248-8.

Some of these mutants were totally avirulent in adult mice whatever the inoculation route (Fig. 1) while others were attenuated : they produced transient symptoms, leading only exceptionnally to death.

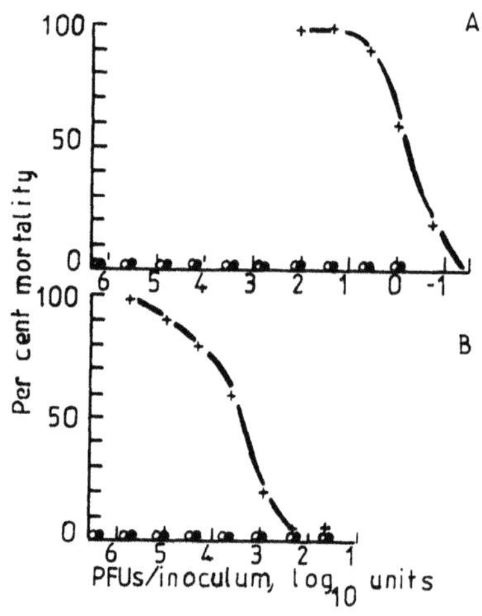

FIGURE 1. Loss of virulence of Av01 (●) and Av02 (o) mutants in adult mice inoculated intracerebrally (A) or intramuscularly (B) compared to the pathogenicity of the parental CVS strain (+).

They were called respectively AvO and AtO mutants (Av = Avirulent, At = Attenuated, O = Orsay). The AvO and AtO mutants were not functionally affected in cell cultures. They grew efficiently in BHK-21, CER cells and

	Hybridoma		Virus				
Specificity	Origin	N°	CVS	Av01	Av02	HEP*	Virulent derivative of HEP*
Nucleocapsid	ERA	c12	+	+	+	-	-
	CVS	237-7	+	+	+	-	-
	CVS	222-2	+	+	+	-	-
	ERA	102-27	-	-	-	+	+
	ERA	103-17	+	+	+	+	+
	ERA	104-7	+	+	+	-	-
	ERA	120-2	+	+	+	+	+
	CVS	239-10	+	+	+	-	-
	Kelef	377-10	-	-	-	+	+
	Kelef	515-3	+	+	+	+	+
	Kelef	590-2	+	+	+	+	+
	Mokola	422-7	-	-	-	-	-
Glycoprotein	CVS	231-22	+	+	+	+	+
	CVS	248-8	+	-	-	-	+
	Kelef	507-1	+	+	+	+	+
	Kelef	509-6	+	+	+	+	+
	ERA	162-3	+	+	+	+	+
	ERA	194-2	+	-	-	-	+
	CVS	220-8	+	+	+	-	-
	CVS	240-3	+	+	+	-	-
	CVS	719-3	+	+	+	ND°	ND°

Table 2. Reactivity of antinucleocapsid and antiglycoprotein antibodies against virulent and avirulent strains of rabies virus.

*The reactivity of HEP and its virulent derivative towards antinucleocapsid and antiglycoprotein antibodies was measured by radioimmunoassay, as described in Flamand et al. (17), using 200 ng of purified virus per assay.

°ND, Not Done.

Virus		CVS	Avirulent mutants	Attenuated mutants	Other mutants		
	509-6 (Kelef)	+°	+	+	+	+	+
	101-1 (ERA)	+	+	+	+	+	+
	162-3 (ERA)	+	+	+	+	+	+
Anti-G monoclonal antibodies	220-8 (CVS)	+	+	+	+	+	+
	231-22 (CVS)	+	+	+	+	+	+
	719-3 (CVS)	+	+	+	+	+	+
	194-2 (ERA)	+	-°	-	+	-	-
	248-8 (CVS)	+	-	-	-	+	-
	507-1 (Kelef)	+	+	+	+	+	+
N° of mutants		-	7	2	10	8	1

Table 3. Sensitivity to neutralizing monoclonal antibodies of mutants selected with antibodies n° 194-2 and 248-8.

°+, Neutralization ; - no neutralization.

neuroblastoma (NS 20) cells ; They were not thermosensitive. They probably multiply in the nervous system since they killed suckling mice after IC inoculation. An immunofluorescence test on thin sections of the brain of the moribund animals with antinucleocapsid fluorescent antibody clearly demonstrate the multiplication of the virus in the central nervous system of young mice.

SDS-polyacrylamide gel electrophoresis shows no difference in the migration of the structural proteins of Av01, Av02 mutants and CVS. The reactivities of the nucleocapsid proteins of Av01 and Av02 towards anti-N monoclonal antibodies were compared to CVS by indirect immunofluorescent staining of infected, acetone-treated BHK-21 cells. The results show that the pattern of reactivity towards anti-N antibodies was similar in the three viruses (Table 2) indicating that the two mutants possess antigenic determinants similar to the CVS N protein. Both observations indicate that Av01 and Av02 retain the general characteristics of the CVS strain.

The study of the reactivity of Av01 and Av02 towards anti-G neutralizing monoclonal antibodies shows that they were resistant to antibody n° 194-2 in addition to antibody n° 248-8 used for their selection (Table 2).

Avirulent and attenuated mutants selected with monoclonal antibody n° 194-2 (Table 1) were also resistant to monoclonal antibody n° 248-8 (Table 3). On the contrary, 18 mutants which were only neutralized by one of these two antibodies retained their pathogenicity for adult mice (Table 3). An analog situation was observed with the HEP strain of rabies virus which is avirulent in adult mice (16). It is possible to select a virulent derivative of HEP after one passage in suckling mice brain. Comparison between the reactivities of anti-G monoclonal antibodies against avirulent and virulent HEP indicates that the virulent derivative acquired the sensitivity to antibodies n° 194-2 and n° 248-8. No other difference could be detected (Table 2).

Monoclonal antibodies n° 194-2 and n° 248-8 were used to select double resistant mutants from independant CVS clones. Injected to adult mice, these independant mutants exhibited an altered pathogenicity, demonstrating the efficiency of antibodies n° 194-2 and n° 248-8 to select avirulent mutants.

Virulence seems to be associated with a special configuration of the antigenic site(s) recognized by these two monoclonal antibodies. Correlation between avirulence and resistance to specific anti-G monoclonal antibodies is not absolute ; One mutant selected with antibody n° 194-2 was resistant to antibody n° 248-8 but its pathogenicity was unmodified.

The nine anti-G neutralizing monoclonal antibodies used in these experiments recognize three antigenic sites on the glycoprotein of the CVS strain (5). Figure 2 presents the antigenic map of the glycoprotein, where ovals delineate epitopes recognized by the different monoclonal antibodies and crosses within ovals represent mutants resistant to the corresponding monoclonal antibodies. Our results indicate that the site of virulence is located within site n° III at the intersection of the epitopes recognized by monoclonal antibodies n° 194-2 and n° 248-8.

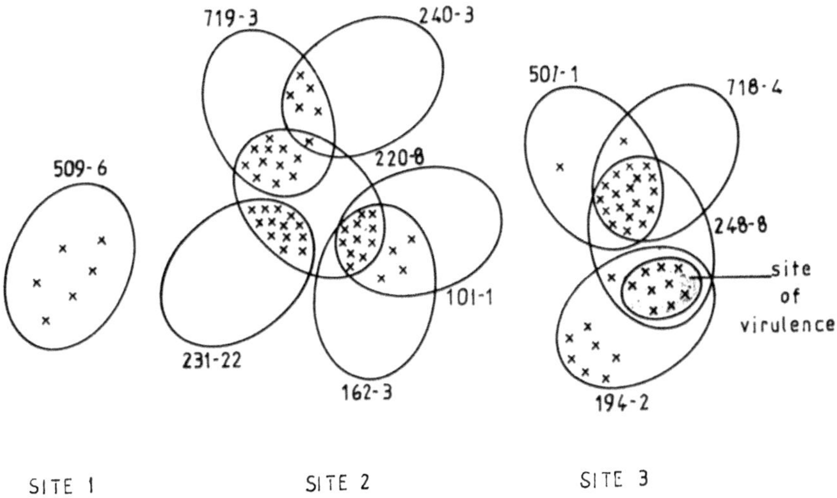

FIGURE 2. Localization of the site of the virulence . The antigenic map of the CVS glycoprotein is done according to Lafon et al. (5).

Immune response of animals injected with AvO1 or CVS

Various parameters of the immune response were investigated after intramuscular (IM) inoculation of 10^5 pfu of AvO1 or CVS to series of Balb/c adult mice.

Production of circulating antibodies. Until day 7, where CVS-inoculated animals died, no difference could be detected in the titre of circulating antibodies (Fig. 3). This result is different from what was previously found in the case of IC inoculation : the titre of circulating antibodies was much higher in the case of AvO1-inoculated animals.

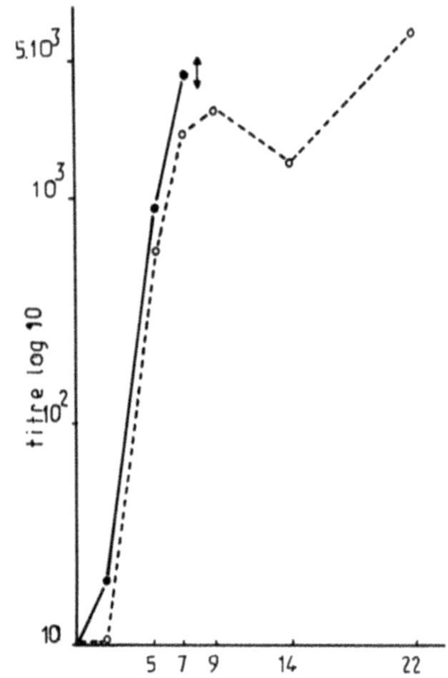

FIGURE 3. Neutralizing antibodies titre of sera.
Adult swiss mice were injected intramuscularly with 10^5 pfu of CVS (●—●) or Av01 (o---o) (100 µl of viral suspension in one of the hind feet). Results are expressed as the dilution of sera reducing to 50 % the number of rabies fluorescent focus in cell culture (using an antinucleocapsid - fluorescein conjugate).

Production of interferon. The production of interferon was assayed in the sera and in the brains. The injection of CVS and Av01 was followed by an early production of interferon in the sera within the first 8 hrs. This initial production lasted longer with Av01 than with CVS (Fig. 4). With CVS, a second period of production was detected between day 4 and day 7 while there was none with Av01. At that time there was also a peak of interferon in the brain of animals inoculated with CVS, and to a lesser extent with Av01.

Cytotoxic activity of splenocytes.

- "Natural Killer" cell-mediated cytotoxicity of splenocytes. Assays of "Natural Killer" (NK) activity of splenocytes gave results similar to those of interferon production. In the first 24 hrs after IM inoculation, an important NK activity was observed with Av01 : Again, this activity persisted longer than with CVS (Table 4). Correlatively with the second

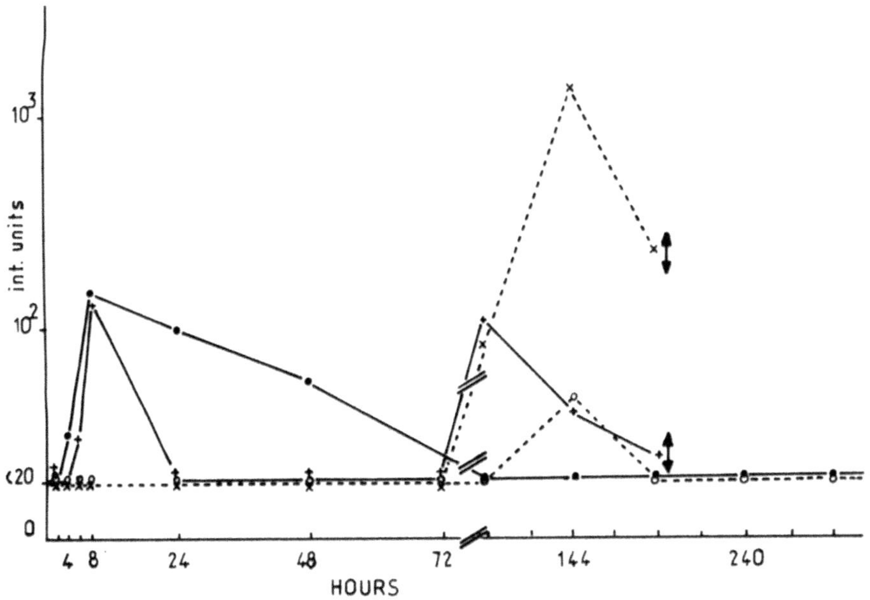

FIGURE 4. Interferon production.
Interferon activity was assayed in CVS-injected mice sera (+—+) or on brain extracts (x.....x) and in Av01-injected mice sera (●—●) or brain extracts (o---o). Interferon titres are expressed in terms of international units.

Time after infection (days)	% Cytotoxicity CVS mean* + SD (range)	% Cytotoxicity Av01 mean* + SD (range)
0	30.8 ± 2.1 (28.5 - 32.6)	30.8 ± 2.1 (28.5 - 32.6)
1	49.4 ± 1.2 (48.3 - 50.6)	60.7 ± 3.3 (57.0 - 63.5)
2	38.8 ± 3.1 (36.6 - 42.3)	49.8 ± 4.3 (45.3 - 53.8)
3	35.3 ± 3.4 (31.9 - 38.7)	42.3 ± 2.2 (40.2 - 44.6)
4	52.4 ± 8.9 (45.3 - 62.3)	39.3 ± 1.2 (38.5 - 40.7)
6	30.5 ± 3.5 (27.1 - 34.0)	29.7 ± 2.8 (26.9 - 32.5)
8	22.4 ± 3.2 (19.2 - 25.6)	27.3 ± 3.8 (23.1 - 30.3)
10		34.1 ± 1.9 (32.7 - 36.3)
12		33.7 ± 1.1 (32.5 - 34.7)
14		32.6 ± 2.7 (30.7 - 34.5)

Table 4. Kinetics of the NK activity in the course of CVS and Av01 infection.

* Average of 3 values.

production of interferon, another period of increased NK activity was detected on days 4-5 in the CVS-injected mice.

- T cytotoxic activity of splenocytes. In preliminary experiments, susceptibility of normal and CVS-infected MNB cells to NK activity was determined. MNB cells were susceptible to NK activity, and the virus infection enhanced this susceptibility, leading to a situation where NK activity could be confused with T cytotoxic activity of splenocytes. In A/J mice, syngenic to MNB cells, a lytic activity of the splenocytes of Av01-injected mice was found. It was maximum at days 6 and 7 where no NK activity could be detected (Fig. 5). Results obtained with splenocytes from CVS-inoculated animals were extremely variable probably because of the poor condition of the animals.

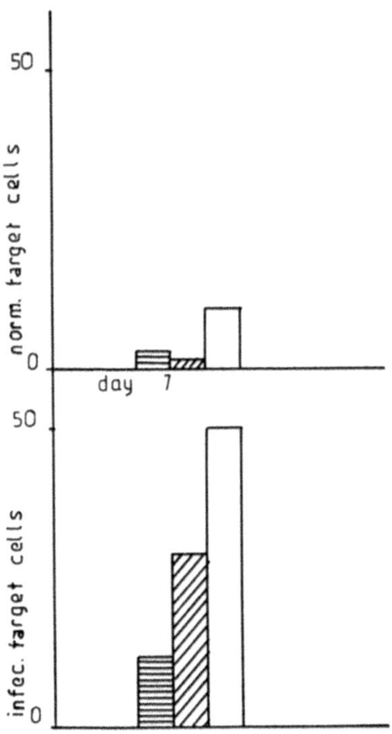

FIGURE 5. Specific cell-mediated cytotoxicity of splenocytes from Av01-injected mice.

Effector/target ratio 25 : 1 ▤ 50 : 1 ▨ 100 : 1 □

Penetration of AvO1 and CVS in the Central Nervous System (CNS)

Both strains were inoculated into the anterior chamber of the eye (4.10^5 pfu). As the nerve connections between the eye and the CNS are relatively well-known the entry and subsequent propagation of the viruses can be studied in thin sections of the eye and brain stained with anti-nucleocapsid fluorescent antibodies. Since the fluorescence can be hardly detected in the infected cells before 16-18 hrs a bright fluorescence found at 24 hrs indicates the primary target cells for the virus.

With CVS, a characteristic fluorescence has been found in the ciliary (Fig. 6a) and Gasser's ganglia at 24 and 48 hrs after inoculation indicating the intraocular parasympathetic and trigeminal endings as the primary sites of penetration.

On the contrary, the mutant AvO1 has shown a strong affinity for the lens epithelium (Fig. 6b) but has lost the capacity to invade the parasympathetic intraocular endings. The penetration of the trigeminal system was not impaired (Fig. 6c) but the slowly following intracerebral infection was rather attenuated and cleared up after 8 days.

Surprisingly, neither the CVS nor the AvO1 infect primarily the intraocular adrenergic terminals (no fluorescence found in the superior cervical ganglion) or the retina and the optic nerve (although a part of inoculum was constantly found in the posterior chamber of the eye). In the case of CVS, the retinal ganglion cells do become fluorescent but only very lately when the infection has disseminated throughout the brain.

Despite the specific differences between both strains these results support the idea that the rabies viruses are conveyed by the retrograde axoplasmic transport.

DISCUSSION

We have demonstrated that appropriate anti-G neutralizing monoclonal antibodies allow an efficient selection of avirulent mutants of rabies virus. From 112 mutants which were studied, 10 were resistant to both antibodies : 194-2 and 248-8, 9 of which were attenuated or totally avirulent. Using the same technique, 2 avirulent mutants of CVS and ERA strains of rabies virus have been isolated at the Wistar Institute (4). The nucleotide sequence of the non-pathogenic mutants indicates that both were mutated at position 333 of the glycoprotein, an arginine being replaced by a glutamine or an isoleucine. We are now currently investigating

a
b
c

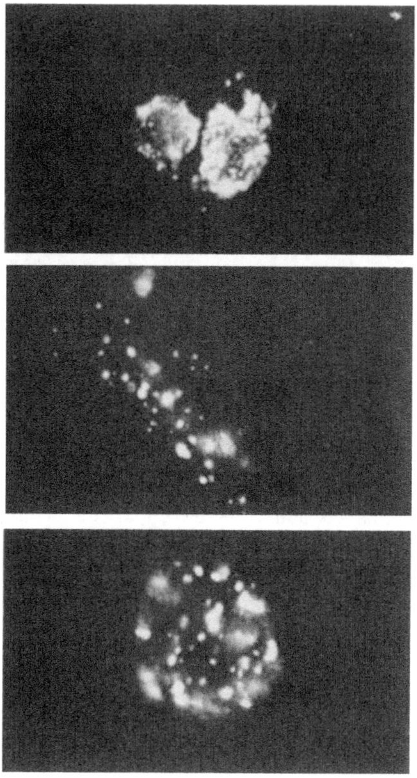

FIGURE 6. Primary targets of AvO1 and CVS after intraocular inoculation (Immunofluorescence).

a) CVS : infected neurons in the ciliary ganglion, 24 hrs after inoculation.
b) AvO1 : anterior epithelium of the lens, 24 hrs after inoculation.
c) AvO1 : infected neuron in the ganglion Gasseri, 48 hrs after inoculation.

Bar : 20 μm.

which AA substitution corresponds to our attenuated and avirulent mutants, in order to relate a mutation to a particular phenotype.

In addition we would like to investigate if other regions of the glycoprotein are implicated in virulence. Continuation of this exploration necessitates the isolation of monoclonal antibodies which recognize other antigenic determinants of the protein. Several anti-G monoclonal antibodies have been already isolated in our laboratory and will be studied in the next future.

Why are the mutants avirulent ? Direct study of penetration and propagation of the virus in the brain of infected rats demonstrate that the mutation completely modifies the host-range spectrum of the virus. Anyhow it is likely that modification of penetration is not sufficient to explain the loss of virulence : nude and suckling mice died after inoculation of Av01, although with different symptoms and after a longer incubation period.

Intervention of the immune response is probably necessary for the survival of animals inoculated with avirulent or attenuated viruses. The most striking difference concerns the production of interferon and Natural Killer splenocytes which is strongly stimulated in Av01-injected animals. Injection of mice with anti-interferon immunoglobulins should demonstrate whether this stimulation is sufficient to explain the lack of pathogenicity. Why is production of interferon lasting more than 2 days in Av01-inoculated animals, and less than 1 day in CVS-injected mice is also under investigation. Preliminary studies concerning in vitro stimulation of macrophages with Av01 or CVS showed no difference in the production of interferon by the infected cells. Continuous multiplication of the virus at the site of inoculation in Av01 (but not CVS)-infected animals could also explain this prolonged stimulation.

Our hope is that careful comparison of the various aspects of an infection with a virulent strain and its avirulent derivatives could lead to a better understanding of the mechanism of viral pathogenicity.

REFERENCES

1. Coulon, P. and Flamand, A. In: Abstracts of the fifth international congress of virology, Strasbourg, 1981, p. 159.
2. Coulon, P., Rollin, P., Aubert, M. and Flamand, A. J. Gen. Virol. 61 : 97-100, 1982.

3. Coulon, P., Rollin, P.E. and Flamand, A. J. Gen. Virol. 64 : 693-696, 1983.
4. Dietzschold, B., Wunner, W.H., Wiktor, T.J., Lopes, A.D., Lafon, M., Smith, C.L. and Koprowski, H. Proc. Natl. Acad. Sci. U.S.A. 80:70-74, 1983.
5. Lafon, M., Wiktor, T.J. and MacFarlan, R.I. J. Gen. Virol. 64:843-851, 1983.
6. Bussereau, F. and Flamand, A. In: Negative Strand Viruses and the host cell (Eds. B.W.J. Mahy and R.D. Barry). Academic Press, 1978, pp. 701-708.
7. Smith, A.L., Tignor, G.H., Mifune, K. and Motohashi, T. Intervirology, 8 : 92-99, 1977.
8. Bussereau, F., Flamand, A. and Pese-Part, D. J. Virol. Meth. 4:277-282, 1982.
9. Wiktor, T.J. and Koprowski, H. Proc. Natl. Acad. Sci. U.S.A. 75 : 3938-3942, 1978.
10. Koprowski, H. and Wiktor, T.J. In: Monoclonal antibodies (Eds. R.H. Kenneth, T.J. MacKearn and K.B. Bechtol). London and New-York. Plenum Press, 1980, pp. 335-351.
11. Sureau, P., Rollin, P.E. and Zeller, H. Comp. Immun. Microbiol. Infect. Dis. 5 : 143-150, 1982.
12. Imanishi, J., Hoshino, S., Hoshino, A., Oku, T., Kita, M. and Kishida, T. Biken Journal. 24 : 103-108, 1981.
13. Kiessling, R., Klein, E. and Wigzell, H. Eur. J. Immunol. 5 : 112-117, 1975.
14. Kiessling, R., Klein, E., Pross, H. and Wigzell, H. Eur. J. Immunol. 5 : 117-121, 1975.
15. Wiktor, T.J., Doherty, P.C. and Koprowski, H. Proc. Natl. Acad. Sci. U.S.A. 74 : 334-338, 1977.
16. Koprowski, H. Bull. W.H.O. 10:709-724, 1954.
17. Flamand, A., Wiktor, T.J. and Koprowski, H. J. Gen. Virol. 48:105-109, 1980.

ACKNOWLEDGMENT

We thank Dr P. Vigier for critical discussion of this work . The excellent technical assistance of Mrs Benejean , Thiers , Gagnat and Alexandre is gratefully acknowledged . This work was supported by the Centre National de la Recherche Scientifique through the L.A. 040086 and by la Fondation pour la Recherche Médicale .

18

STUDIES ON THE STRUCTURE AND ACTIVITIES OF INFLUENZA VIRUS HAEMAGGLUTININ

J.J. SKEHEL[+], R.S. DANIELS[+], A.R. DOUGLAS[+], M. KNOSSOW[++], J.C. PAULSON[+++], G.N. ROGERS[+++], M.D. WATERFIELD[++++], I.A. WILSON[++] and D.C. WILEY[++].

\+ National Institute for Medical Research, Mill Hill,London, NW7, UK; ++ Department of Biochemistry and Molecular Biology, Harvard University, Cambridge, Massachusetts, USA; +++ Department of Biological Chemistry, UCLA School of Medicine, Loss Angeles, California, USA; ++++ Imperial Cancer Research Fund, Lincolns Inn Fields, London, WC1, UK.

The initial events in infection by influenza virus involve binding of virus to sialic acid residues of glycosylated membrane receptors, endocytosis, and fusion of the membranes of virus and endosome. Virus adsorption has been known for some time to be a property of the haemagglutinin glycoprotein (HA) of the virus membrane and the recent direct implication of HA in membrane fusion (1) confirms previous conclusions from a variety of observations that HA is also involved in this activity (2,3). As a consequence of these functions and of its location in the virus membrane, antibodies which affect HA neutralize virus infectivity and the frequent epidemics of influenza are caused by viruses with antigenically novel HAs.

This article is a summary of recent results on the molecular basis of receptor binding, membrane fusion and antibody binding by the haemagglutinin of X-31 (H3N2) influenza virus.

RECEPTOR BINDING

The interaction of HA with sialic acid-containing receptors is known to vary in detailed specificity for different influenza viruses. A comparison of viruses of the H3 subtype has revealed at least three distinct specificities based on preferential binding to either one or both of the sequences NeuAcα2-6Gal α1-4GlcNAc- or NeuAcα2-3Galα1-3GalNAc- commonly found to terminate glycoprotein oligosaccharides linked to asparagine and to threonine or serine, respectively (4). Preferential

Table 1. Correlation of receptor specificities in haemagglutination tests and amino acid substitutions at residue 226 in mutants of X-31 and A/Memphis/102/72 viruses.

Virus	Receptor specificity as judged by binding to		Amino acid at 226
	NeuAcα2-3Gal	NeuAcα2-6Gal	
X-31	-	+	Leucine
X-31 (Horse serum selected)	+	-	Glutamine
X-31 (Guinea pig serum selected)	+	-	Glutamine
M1/HS10 (Plaque selected in MDCK cells)	-	+	Leucine
M1/HS7 "	+	+	Methionine
M1/HS8 "	+	-	Glutamine

binding to the NeuAcα2-6Gal linkage was also found to correlate with high sensitivity to neutralization of infectivity by glycoprotein (γ) inhibitors present in horse serum (5) and this sensitivity has allowed selection of receptor specific variants from X-31 virus. Thus while X-31 preferentially binds the NeuAcα2-6Gal- linkage and is very sensitive to inhibition by horse serum, the selected mutants bind to NeuAcα2-3Gal- linkages and haemagglutination by them is insensitive to horse serum components. In addition to these viruses receptor specific variants of A/Memphis/102/72(H3N2) were obtained after growth of virus in MDCK cells in the presence or absence of horse serum. The variants obtained have similar binding properties to the X-31 variants except for one virus which bound well to both NeuAcα2-6Gal- and NeuAcα2-3Gal- linkages.

The amino acid sequences of the HAs of these viruses were deduced from the nucleotide sequences of their RNA genes. These were determined using dideoxynucleotide chain terminators (6) in a primer extension system containing total virus RNA mixtures, reverse transcriptase, and 5'-^{32}P-labelled synthetic oligodeoxynucleotide primers (7). The results indicated single amino acid substitutions at residue 226 as a result of base changes in the triplet 754-756, CTG-leucine to CAG-glutamine or ATG-methionine, Table 1. Variants which bound the

NeuAcα2-6Gal- linkage contain leucine at position 226, those which bound NeuAcα2-3Gal- contain glutamine at 226 and the variant which bound both linkages contains methionine at 226.

These observations on the modifications of sialic acid binding specificity as a consequence of amino acid substitutions at residue 226 directly support the tentative identification (8) of the sialic acid binding site as a surface pocket at the distal end of the molecule illustrated in the α-carbon tracing of the X-31 HA monomer shown in Fig. 1. This proposal was originally based on the presence of conserved tyrosine 98, histidine 183, glutamic acid 190, tryptophan 153, and leucine 194 amino acid residues in this pocket. Amino acid 226 is in this site which is consistent with a direct role for this residue in receptor binding.

MEMBRANE FUSION

Evidence that the HA is involved in influenza virus-mediated fusion includes the observations that post-translational cleavage of a precursor HA, HA_0 to HA_1 and HA_2 is required for both virus infectivity (9,10) and in vitro virus-mediated fusion (11,12) and that the hydrophobic amino-terminal sequence of HA_2 is analogous to that of the amino terminus of the F_1 component of Sendai virus fusion glycoprotein (13,14,15). Furthermore, the findings that the amino-terminal sequence of HA_2 consists of 10 uncharged hydrophobic amino acids (13) and is the most highly conserved sequence in the haemagglutinin suggest that the terminal region may be directly involved in the membrane fusion reaction. Analysis of the three-dimensional structure (8) of bromelain-released HA which lacks the carboxyl-terminal hydrophobic region through which the complete HA is associated with the lipid membrane of the virus particle suggested that a conformational change is required before this can occur. In vitro membrane fusion and haemolysis mediated by influenza viruses are maximal at about pH 5.0 (16,17,18) a pH close to that in endosomes and we have observed that at this pH the conformation of HA is in fact different from that at neutrality. In studies of the structure of bromelain released HA the observed consequences of this low pH structure transition are that the molecule acquires the ability to bind to lipid vesicles, to bind detergent, to aggregate in lipid and detergent-free solutions and it becomes susceptible to proteolysis (19). In studies to define the regions of the HA involved in membrane fusion activity we have observed that following incubation at pH 5.0 aggregated X-31 HA is susceptible to tryptic digestion and the products obtained can be separated into aggregate and soluble fractions. The aggregate contains the HA_2 glycopolypeptide disulphide bonded to the 27 amino-terminal residues of HA_1, and the self-association site. Aggregate prepared in this way was rendered soluble by digestion with thermolysin. Analyses of the products obtained indicate that incubation of the aggregate with 2% thermolysin for 90 minutes resulted in a decrease in the apparent molecular weight of HA_2 of about 2000 and that about 90% of the HA_2 component was released from the aggregate as judged by sedimentation in sucrose density gradients. The amino-terminal sequence of HA_2 solubilized in this way was found to be heterogeneous in agreement with polyacrylamide gel electrophoresis results which indicate more than one polypeptide product of thermolytic digestion. However, the identities of the predominant residues at each cycle of the sequence analysis indicate that the major amino acid sequences correspond to those beginning at residues 23 and 27 of HA2, Fig. 1. These results suggest a direct involvement of the amino terminal region of HA_2 in membrane fusion since the self association of BHA which is induced specifically at pH 5.0, the optimum pH for fusion, is reversed by modification of the amino-terminus by thermolytic digestion. Furthermore similar experiments in which HA_2 is released from its pH 5.0 induced association with lipid vesicles by thermolysin have indicated that soluble products of the same size as those obtained from aggregated BHA are produced. Further analyses of these products and the residues involved in lipid association are in progress.

ANTIBODY BINDING

The results of studies of the amino acid sequences of HAs of natural and laboratory selected antigenic variants of H_3 subtype influenza viruses led to the proposal that four or five regions of the HA interact with antibodies and undergo structural modification during antigenic drift. Evidence for three of these sites A, C and D included the location of specific amino acid alterations detected in the HAs of monoclonal antibody selected variants (20,21) and we have extended these analyses to sites B and E. This is based on the amino acid substitutions listed and illustrated in Fig. 1. In studies on the structure of HA in the low pH conformation it has been observed that the antigenicity of the molecule is specifically modified in the antigenic sites designated B and D. Furthermore it was observed that antibodies which bind to site B can be divided into two groups on the basis of their reaction with HA in the pH 5.0 conformation. Antibodies which recognize amino acids 157 and 193 which extend towards site A react with HA at both pH 7.0 and pH 5.0; antibodies which recognize residues 156, 158, 198 and 199 react only with HA at pH 7.0. This delineation of sequentially adjacent amino acid residues into different antibody binding sites and the locations of the sites in the molecule are similar to those described for viruses of the H_1 subtype (22).

The three monoclonal antibodies used to define site E all recognize amino acid 63 and variants selected by growing X-31 in their presence contained asparagine instead of aspartic acid at 63. This substitution resulted in the formation of the glycosylation site -Asn63-Cys64-Thr65- and we have shown by analyzing the HA_1 polypeptides of the variants that carbohydrate side chains are attached at this site. We have also observed by binding antibody to variant HA produced in the presence of tunicamycin that glycosylation at residue 63 is required to prevent antibody binding to the variants at this site, thus demonstrating a direct effect of glycosylation on antigenicity.

Initial considerations of the amino acid sequences of HAs of different antigenic subtypes indicated that oligosaccharide attachment sites are found in areas equivalent to the antibody binding sites of the H_3 subtype HA. The suggestion was therefore made that oligosaccharide chains may alter the antigenic nature of these sites by sterically blocking access to the protein surface. We have analyzed the HAs of the H_3 subtype with particular reference to the role of oligosaccharide side chains in antigenic variation and have made the two following observations. 1) In A/Victoria/3/75 the attachment site at 81 is lost and new sites are gained at residues 63 and 126 (23). Assuming that oligosaccharide at 63 could occupy nearly the same location as that at 81 this change, tentatively labelled 'E?' was considered antigenically neutral (24). A refined crystallographic model, however, indicates that the carbohydrate side chain at 81 extends parallel to the polypeptide chain at residues 119-124 in contact through N-acetyl glucosamine residues with Glu 119, Phe 120 and Ile 121 and through mannose residues with Thr 122, Glu 123 and Gly 124. This orientation is not toward residue 63 and, therefore, the amino acid variations in region E may not be antigenically neutral. This conclusion is of course supported by the analyses of monoclonal antibody selected variants at amino acid 63 described above. 2) Nucleotide sequence analyses of H_3 subtype viruses isolated in 1980, 1981 and 1982 indicate that changes at residue 124 in A/Belgium/2/81 and at 248 in A/Shanghai/31/80, A/Hong Kong/1/82 and A/Netherlands/246/82 provide extra sites for glycosylation of these HAs. These residues are located in antigenic sites A and B respectively and in these recent H_3 viruses it may be that three of the antibody binding sites, A, B and E are modified by glycosylation when compared with earlier H_3 isolates. The epidemiological significance of these modifications and their effects on subsequent antigenic variation remains to be determined.

ACKNOWLEDGEMENTS

We thank D. Stevens, R. Gonsalves, G. White, G. Scrace and N. Totty for assistance and acknowledge support from NIH AI 13654 (DCW), NSF PC-771398 (DCW - computing hardware) and NIH AI 16165 (JCP). GNR received a fellowship DRG-494F from the Damon Runyon-Walter Winchell foundation and JCP an American Cancer Society Faculty Research Award.

REFERENCES

1. White, J., Helenius, A. and Gething, M.J. Nature 300: 658-659, 1982.
2. Rott, R. and Klenk, H.-D. In: Cell Surface Reviews (Eds. G. Poste and G.L. Nicolson), North Holland, Amsterdam, 1977, pp. 47-81.
3. Compans, R.W. and Klenk, H.-D. In: Comprehensive Virology (Eds. H. Frainkel-Conrat and R.R. Wagner) Plenum Press, New York, 1979, pp. 293-407.
4. Rogers, G.N. and Paulson, J.C. Fed. Proc. 41: 5880, 1982.
5. Rogers, G., Pritchett, T. and Paulson, J.C. Fed. Proc., in press, 1983.
6. Sanger, F., Nicklen, S. and Coulson, A.R. Proc. Natl. Acad. Sci. USA 74: 5463-5467, 1977.
7. Patel, T.P., Millican, T.A., Bose, C.C., Titmas, R.C., Mock, G.A. and Eaton, M.A.W. Nuc. Acids Res. 10: 5605-5620, 1982.
8. Wilson, I.A., Skehel, J.J. and Wiley, D.C. Nature 289: 368-373, 1981.
9. Klenk, H.-D., Rott, R., Orlich, M. and Blodorn, J. Virology 68: 426-439, 1975.
10. Lazarowitz, S.G. and Choppin, P.W. Virology 68: 440-454, 1975.
11. Huang, R.T.C., Rott, R. and Klenk, H.-D. Virology 110: 243-247, 1981.
12. White, J.M., Matlin, K. and Helenius, A. J. Cell Biol. 89: 674-679, 1981.
13. Skehel, J.J. and Waterfield, M.D. Proc. Natl. Acad. Sci. USA 72: 93-97, 1975.
14. Gething, M.J., White, J.M. and Waterfield, M.D. Proc. Natl. Acad. Sci. USA 75: 2737-2740, 1978.
15. Scheid, A. and Choppin, P.W. Virology 80: 54-66, 1977.
16. Maeda, T. and Ohnishi, S. Febs Lett. 122: 283-287, 1980.
17. Huang, R.T.C., Rott, R. and Klenk, H.-D. Virology 110: 243-247, 1981.
18. White, J.M. and Helenius, A. Proc. Natl. Acad. Sci. USA 77: 3273-3277, 1980.
19. Skehel, J.J., Bayley, P.M., Brown, E.B., Martin, S.R., Waterfield, M.D., White, J.M., Wilson, I.A. and Wiley, D.C. Proc. Natl. Acad. Sci. USA 79: 968-972, 1982.
20. Laver, W.G., Air, G.M., Webster, R.G., Gerhard, W., Ward, C.W. and Dopheide, T.A.A. Virology 98: 226-237,1979.

21. Laver, W.G., Air, G.M., Dopheide, T.A., Ward, C.W. Nature 283: 454-457, 1980.
22. Caton, A.J., Brownlee, G.G., Yewdell, J.W. and Gerhard, W. Cell 31: 417-427, 1982.
23. Both, G.W. and Sleigh, M.J. J. Virol. 39: 663-672, 1981.
24. Wiley, D.C., Wilson, I.A. and Skehel, J.J. Nature 289: 373-378, 1981.

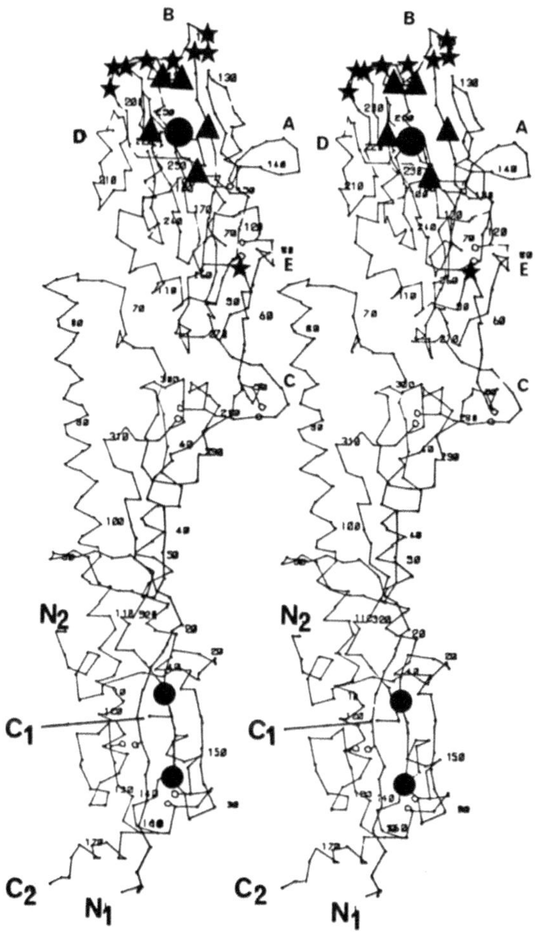

FIGURE 1. Stereo drawing of the α-carbon tracing of a haemagglutinin monomer to show the antigenic sites A, B, C, D and E; the conserved residues of the sialic acid binding pocket, ▲; residue 226 the residue found to vary with changes in receptor binding specificity, ●; the amino- and carboxy-terminal residues of the HA_1 and HA_2 glycopolypeptide components, N_1, C_1, N_2 and C_2; and the sites of thermolytic cleavage of HA aggregate at pH 5.0 which render the protein soluble, ●. The amino acids in sites B and E recognized by specific monoclonal antibodies are shown by ★.

19

ON THE ROLE OF VIRAL ENVELOPE PROTEINS IN PATHOGENESIS

PURNELL W. CHOPPIN
The Rockefeller University, New York, New York 10021

SUMMARY

The paramyxovirus envelope contains two surface glycoproteins (HN and F) and an internal non-glycosylated protein (M). The F protein is involved in viral pathogenesis in three different ways, each related to its ability to cause membrane fusion, which mediates virus penetration, cell fusion, and hemolysis. 1. Membrane fusion is activated by a specific proteolytic cleavage yielding two polypeptides (F_1 and F_2), and because this is accomplished by a host enzyme, the tropism and virulence of the virus is significantly determined by the F protein and its susceptibility to cleavage by a protease available in the host tissue. 2. In some virus-cell systems cell death has been shown to be due to the membrane-damaging effects of the F protein. 3. Lack of antibodies to F has been implicated in an immunopathological response in patients receiving formalin-inactivated measles virus vaccine. The membrane fusion reaction has been characterized using intact cells and liposomes as target membranes, and both intact virus and membranes reconstituted with purified F protein as the fusogenic agents. Requirements for fusion have been determined, including a conformational change in the F protein upon cleavage, the lipid composition of the target membrane, and the effect of pH. Extensive evidence has been obtained indicating that the N-terminus on F_1 generated by the activating cleavage is involved in membrane fusion reaction, probably through a hydrophobic interaction with the target membrane. Synthetic oligopeptides with sequences that mimic this N-terminal region specifically inhibit membrane fusion, and thereby virus infectivity at the level of penetration. The oligopeptides have been found to act at the cell membrane, presumably competing with the F_1 polypeptide for sites on the cell membrane.

Lack of the expression of the M protein of measles virus is involved in the persistent infection leading to the chronic neurological disease subacute sclerosing panencephalitis (SSPE), and evidence has been obtained which suggests that there is a host restriction in the expression of this protein in brain cells. In vitro studies have provided structural evidence supporting the role of the M protein in virus assembly at the cell membrane.

A new protein has been identified in cells infected with influenza B virus that has not been found with influenza A virus, the first instance of a protein unique for influenza B. The protein (NB) is glycosylated and is encoded in the same viral genome RNA segment as the viral neuraminidase (NA). A single mRNA codes for both proteins, with NB translated from the first initiation codon and NA translated from the second.

INTRODUCTION

Paramyxoviruses comprise a large group that includes parainfluenza virus types 1 to 5, mumps, measles, canine distemper, and Newcastle disease viruses. Our laboratory has for many years studied the structure, replication, and mechanisms of pathogenesis of these viruses, with emphasis on the structure and biological activities of the viral membrane proteins. These results and experimental methods are described in detail in a number of primary publications cited below, which represent the work of many individuals in our laboratory. The earlier studies have been reviewed previously (1,2), and this work will be summarized briefly here, emphasizing the more recent results.

Three proteins are associated with the membrane of paramyxoviruses, two glycoproteins (HN and F) that form spike-like projections from the viral membrane, and a non-glycosylated protein (M) associated with the inner surface of the lipid bilayer of the viral membrane (reviewed in 1,3). These proteins have been isolated and purified in biologically active form, and their functions identified (4-9). The HN protein has both receptor binding (hemagglutinating) and receptor-destroying (neuraminidase) activities. The F protein is involved in virus penetration and virus-induced cell fusion and hemolysis; it will be a major focus of this paper. The M protein has long been thought to play an important role in the assembly of the virus particle by budding from the plasma membrane (3,10,11).

THE F PROTEIN

Structure

The F protein is synthesized as a biologically-inactive precursor protein (F_o) with a mol wt of $\sim$ 65,000, which is cleaved by a host enzyme to yield two disulfide-bonded polypeptides (F_1 and F_2) with mol wts of $\sim$ 50,000 and ∿ 15,000, respectively (4-6,12-14). The F_1 polypeptide has the C-terminus of F_0, which is anchored in the viral membrane, and the F_2 polypeptide contains the original N-terminus. The cleavage of F_0 generates a new N-terminus on F_1 (14). This N-terminus is highly hydrophobic (9, 15,16); in Sendai virus there is a stretch of 25 uncharged amino acids. The sequence of this region is highly conserved among different paramyxoviruses. As shown below, there are only 5 positions at which differences occur in the first 18 amino acids of the F_1 polypeptide of Sendai, SV5, and Newcastle disease viruses (9,16), and in each case one hydrophobic amino acid is exchanged for another.

Sendai	Phe-Phe-Gly-Ala-Val-Ile-Gly-Thr-Ile-Ala-Leu-Gly-Val-Ala-Thr-Ala-Ala-Gl
SV5	Phe-Ala-Gly-Val-Val-Ile-Gly-Leu-Ala-Ala-Leu-Gly-Val-Ala-Thr-Ala-Ala-Gl
NDV	Phe-Ile-Gly-Ala-Ile-Ile-Gly-Gly-Val-Ala-Leu-Gly-Val-Ala-Thr-Ala-Ala-Gl

If a paramyxovirus is produced by cells that lack a suitable protease to cleave F_0, virions are produced that are non-infectious and cannot induce cell fusion or hemolysis (6,12,13). These functions, each of which involves fusion of the viral membrane with the cell membrane, can be activated by treatment of the virus _in vitro_ with the appropriate protease, e.g., trypsin in the case of wild type Sendai virus (6,12-14). The activities of the F protein involve fusion of membranes, and therefore F has no activity as an isolated protein, however if reconstituted into a membrane with lipid (17), the F protein can induce membrane fusion, provided a mechanism is present to deliver it to the target membrane. In the virus, this is provided by the HN protein, however it can be done with reconstituted membranes containing F protein with a lectin such as wheat germ agglutinin (17).

The role of the F protein in pathogenesis

Tropism, spread in the host, and virulence. It was found that some cells (e.g., MDBK cells or L cells in the case of Sendai virus) lack the appropriate protease to cleave the F protein (6,12,13) and activate virus penetration by fusion of its membrane with the cell membrane. Virus produced by such cells cannot undergo multiple cycle infection, spread in the host, and cause disease. These results indicated that the host range, tissue tropism, and virulence of the virus was dependent on the susceptibility of the F protein to cleavage by an available host protease. This concept was substantiated by the isolation of mutants of Sendai virus that required different proteases to cleave and activate F, and which consequently exhibited a different host range with respect to cultured cells and the chick embryo (13,18). Based on these results in cell culture and the chick embryo, we postulated that the pathogenesis of paramyxoviruses in these natural animal hosts would depend on the susceptibility of the F protein to cleavage by host enzymes, citing as possible examples virulent and avirulent strains of NDV (19). The importance of F protein cleavage in determining viral virulence was soon clearly demonstrated with NDV in chickens (20,21).

The finding of proteolytic activation of the F protein, and that protease activation mutants occur, has other important implications in addition to elucidating the mechanism of viral penetration through viral and cell membrane fusion and demonstrating the role of the F protein in pathogenesis. These have been discussed previously (1,7) and will be only briefly mentioned. When a paramyxovirus is isolated from an infected individual in a laboratory host (cell culture, chick embryo, or animal) the F protein of the virus isolated must be cleaved by an enzyme present in that host. This may result in the selection of a virus that is not representative of the major virus population in the infected individual, and which may not be efficiently cleaved by the proteases available in the infected human or other natural host. Such variant selection may provide an explanation for the often observed rapid attenuation for humans of viruses by a single or few passages in the laboratory. We have also correlated the loss on serial passage on cells such as monkey kidney cells to support paramyxovirus replication with the loss of a specific activation protease (18). Finally, the generation in the laboratory of mutants activated by different proteases with different host ranges (13) raises the possibility that such mutants may arise in nature, and may explain the occasional infection of an organ or species not usually infected by wild type virus.

Cell injury due to F protein action. Since the middle of the 19th century pathologists have noted that paramyxoviruses such as measles caused the formation of giant cells in infected tissues, and the finding of such cells has diagnostic significance in certain cases (22,23). Early studies with cultured cells indicated that cell fusion occurred in infected cells *in vitro* (24,25), and it was shown that concentrated, ultraviolet light-inactivated virus could induce cell fusion (25,26), and this activity was related to the ability of paramyxoviruses to cause hemolysis (25-28). As discussed above, these activities were subsequently shown to be associated with the F protein (6,12-14). One of the characteristics of many paramyxoviruses is their ability, in the appropriate host cell system, to cause persistent infection without cell death, and this property made it possible to evaluate the correlation between cell damage and the membrane fusing action of the virus. With the parainfluenza virus SV5, we found earlier that the killing of cells (e.g., the BHK21 line of hamster kidney cells) or the survival of cells (e.g., primary rhesus monkey kidney cells) in apparently normal condition without inhibition of cell metabolism, was dependent on the susceptibility or resistance of the plasma membrane to the fusing activity of the virus (30-32). The role of the membrane fusing activity of the F protein in early cell death of paramyxovirus-infected cells was subsequently demonstrated clearly in studies with a specific oligopeptide inhibitor of the action of the F protein (32). It was shown that measles virus infected cells which were prevented from fusing by the inhibitor survived for many days without cytopathology while producing virus, whereas infected cells not treated with the inhibitor fused and died within a day (32). Thus the direct action of the F protein on cell membranes is an important factor in cell death induced by paramyxoviruses. It should be recognized that other factors also play a role in cell injury by paramyxoviruses, and that the action of the F protein on cell membranes can lead to lysis and death before or in the absence of recognizable cell fusion, as exemplified by the hemolysis reaction.

The role of the F protein in an immunopathological reaction. The ability to isolate the HN and F protein in biologically active form (4-6,17) made it possible to prepare monospecific antibodies and to evaluate the importance of immunity to each in the spread of infection in the host (33,34). As expected from the known activities of the protein, anti-HN antibodies inhibited hemagglutination and neuraminidase activities, and also neutralized infectivity in a conventional neutralization test in which virus and antibody were mixed before inoculation of cells. Anti-F antibody had no effect on hemagglutination or neuraminidase activity and neutralized the virus due to inhibition of penetration. However, an important finding was that if a cell in a population was infected before the virus came in contact with antibody, anti-HN antibody could not inhibit the spread of infection. Even though such antibody could neutralize released virus in the medium, it could not prevent the spread of infection to adjacent cells due to the membrane fusing action of the F protein (33). On the other hand, anti-F antibody could neutralize both released virus, at the level of penetration, and prevent spread to adjacent cells through fusion of membranes. These results thus clearly demonstrated that for a paramyxovirus vaccine to be effective it must stimulate immunity to the F protein. In addition, when coupled with the work of Norrby and coworkers (35-37) indicating that formalin inactivates antigenicity of the F protein of some paramyxoviruses, our results provided an explanation (33) for not only the failure of previous formalin inactivated vaccines, but also for the atypical and severe infections that occurred in some patients who received formalin-inactivated measles vaccine, a condition that was thought at the time to involve immunopathological reaction (38-40). Our hypothesis (33) to explain this atypical measles syndrome is as follows. The inactivated vaccine induces antibodies to the hemagglutinin protein and other internal viral proteins, but not to the F protein. When such individuals were subsequently exposed to the virus, some cells in the respiratory tract could be infected and the infection could spread to adjacent cells by membrane fusion due to the lack of immunity to F. As the infection spreads, viral antigens are produced and released, stimulating a secondary immune response to H and the other viral proteins except F. Thus, the situation would exist in which there is continuing production of viral antigens in the presence of a hyperimmune response to these antigens, a situation which sets the stage for a pathological immune reaction. It has been shown that patients with atypical measles have very high levels of antibodies to measles virus proteins (41). A similar explanation could apply to the severe respiratory infections that occurred in children receiving a formalin-inactivated respiratory syncytial (RSV) virus vaccine (42).

The failure of these inactivated paramyxovirus vaccines in the 1960's effectively halted attempts to develop inactivated vaccines for these viruses, and no effective vaccine exists for parainfluenza viruses 1, 2, and 3 and RSV. Now that there is an explanation for the previous failure of formalin inactivated vaccines, and it is known to what protein immunity must be developed, it appears appropriate to reopen the question of purified protein vaccines, possibly incorporated into lipid vesicles, for these viruses (2,17,33). The potential use of recombinant DNA techniques to produce large quantities of viral proteins inexpensively provides another stimulus for reconsideration of purified paramyxovirus protein vaccines.

The mechanism of action of the F protein

Structure-activity relationships. The fact that the membrane fusion mediated by the F protein that is reflected in the biological properties of viral penetration, virus-induced cell fusion, and hemolysis is activated by the proteolytic cleavage that generates a new N-terminus on the F_1 polypeptide (14) suggested that this N-terminus might be directly involved in the fusion reaction. This was supported by the findings that the primary structure in this region is highly conserved among different paramyxovirus type and mutants of Sendai virus that are cleaved by different proteases (9,16). This suggested a requirement for a specific amino acid sequence at the N-terminus. Furthermore, this N-terminal region of F_1 is extremely hydrophobic, suggesting that a hydrophobic interaction could occur between this region of the protein and the target membrane (9,15,16,43,44). This could result in the F_1 polypeptide facilitating fusion by bringing the bilayers of the cell and viral membrane together, since its N-terminus would be inserted in the cell membrane and its C-terminus anchored in the viral membrane. Further support for this hypothesis has been obtained by the demonstration of a conformational change in the F protein upon cleavage (44). This was shown by a change in the circular dichroism spectrum with an increase in α-helical content, and by an approximate doubling of the amount of the detergent Triton X-100 bound to the cleaved protein as compared to the uncleaved protein (i.e., 67 vs 27 molecules of detergent per protein molecule), indicating an increased hydrophobicity compatible with exposure of the N-terminus of F_1. Following our demonstration of a conformational change in the F protein of Sendai virus a similar finding was obtained on the hemagglutinin protein of influenza virus in which cleavage and acid pH expose a hydrophobic region (45).

The effect of pH on membrane fusion by paramyxoviruses. It should be emphasized that it has long been known that paramyxoviruses can cause cell fusion and hemolysis at membrane pH as well as over a wide pH range (pH 5-9). Thus acid pH is not required for fusion of paramyxoviruses with membranes as it is for Semliki forest virus, vesicular stomatitis virus, and influenza virus (46-48). With these viruses the evidence suggests that virus penetration is accomplished by fusion of the viral membrane with the cell membrane in an endocytic vesicle at acid pH, rather than by fusion at the plasma membrane as occurs with paramyxoviruses. It has been found in our laboratory that paramyxovirus fusion activity not only does not require acid pH, it is enhanced at alkaline pH (49). An irreversible enhancement of activity of the F protein was found upon exposure of Sendai virus to alkaline pH (optimum pH 9.0) followed by returning the virus to neutral pH before assaying cell fusing or hemolysing activities. The increased activity was correlated with an irreversible conformational change in the protein as shown by its circular dichroism spectrum (49). The results with different enveloped viruses suggest that a similar hydrophobic interaction could mediate the fusion of viral membranes with cell membranes with the variation among viruses in the site of membrane fusion being dependent on the pH at which the viral protein involved in membrane fusion acquires the conformation for fusion activity.

Liposomal membranes with defined composition as targets for F protein action. The compositional requirements of the target membrane for fusion by the F protein have been investigated (50) using unilamellar liposomes of defined composition for fusion by Sendai virus and two different methods to quantitate fusion, one involving proteolytic cleavage of viral proteins by protease trapped within the liposomes, and the other based on centrifugal separation of fused and unfused liposomes. A Sendai virus mutant (pa-cl) with an uncleaved F, and therefore fusion inactive, provided an ideal control in these studies. Fusion of the virus with liposomes occurred in the absence of any membrane protein when ganglioside was incorporated in the membrane to serve as a virus receptor, and, to a lesser extent, in the absence of ganglioside. A definite requirement for fusion was the presence of cholesterol in the membrane, with an optimal mole fraction of cholesterol to phospholipid of 0.3-0.4. Cleavage of the F protein was found to be required for fusion of the virus with liposomes, just as it is required for fusion with cell membranes, indicating the appropriateness for membrane fusion studies of this model system with defined membrane components. The results obtained thus far are compatible with a hydrophobic interaction between the cleaved F protein and the target membrane.

Specific inhibition of F protein action

The findings summarized above suggested strongly that the N-terminal region of the F_1 polypeptide was involved in membrane fusion. We therefore reasoned that it might be possible to inhibit specifically the action of the F protein with oligopeptides synthesized to mimic this region of the polypeptide. We were encouraged in this approach because we noted the similarity between the N-terminus of F_1 and a peptide (carbobenzoxy-D-Phe-L-Phe-L-(NO_2) ARG) that had been found in an antiviral screening program to inhibit measles virus replication (51) and was later shown to interfere with cell fusion and hemolysis by measles (52) and SV5 (53). We therefore synthesized a large number of oligopeptides using the N-terminus of the Sendai virus F_1 polypeptide as a model and varying a number of parameters, including the sequence and steric configuration of the amino acids and substitutions at the termini of the peptides. These oligopeptides were then tested for their ability to inhibit the fusion activity of the F protein, i.e., infectivity at the level of penetration, and cell fusion and hemolysis induced by measles, Sendai, SV5 and canine distemper viruses. Oligopeptides with the proper structure inhibited these activities by each of the viruses, however because measles virus was the most sensitive it was used in extensive structure-activity studies and dose response curves were determined using a plaque assay to quantitate inhibition of infectivity. These studies are described in detail elsewhere (16,54,55) and will be briefly discussed below. It should be emphasized that in these assays, the activity of the cleaved F protein is being inhibited, not the cleavage that activates the protein. The peptides are not protease inhibitors, but rather inhibitors of the membrane fusing activity of the protein.

Our earlier studies with these oligopeptides (16,43,54) can be summarized as follows. Oligopeptides with the correct amino acid sequence are highly active, specific inhibitors. The longer the peptide, the greater its activity; the most active peptide examined to date is a heptapeptide with the sequence of the first 7 amino acids of the Sendai virus F_1 N-terminus (50% effective concentration, 0.02 μM). A carbobenzoxy (Z) group on the N-terminal amino acid increased activity significantly, however other hydrophobic additions (e.g., a dansyl group) at the N-terminus also increased inhibitory activity as compared to oligopeptides with no modification of the N-terminus. The steric configuration of the N-terminal phenylalanine also significantly affected activity, e.g., Z-D-Phe-L-Gly was more active than Z-L-Phe-L-Phe-Gly. Esterification of the C-terminal amino acid decreased activity. Table 1 illustrates the 50% effective doses obtained with some of the many oligopeptides tested.

Table 1. Inhibition by oligopeptides of plaque formation by measles virus

Peptide	50% effective concentration (μM)
Z-D-Phe-L-Phe-Gly-D-Ala-D-Val-D-Ile-Gly	0.02
Z-D-Phe-L-Phe-Gly	0.20
Z-D-Phe-L-Phe	28
Z-D-Phe-L-Phe-Gly(chloromethylketone)	0.20
Z-D-Phe-L-Phe-L-(Azido-Phe)	0.28
Z-D-Phe-L-Phe-Gly(methyl ester)	20
DNS-D-Phe-L-Phe-Gly	0.34
Z-L-Phe-L-Ser	141
Z-Gly-L-Phe-L-Phe	530
D-Phe-L-Phe-Gly	180

Z indicates a carbobenzoxy group and DNS a dansyl group

These results with a large number of different oligopeptides have clearly shown the importance of the correct amino acid sequence for maximum inhibitory activity. The effects of the substitutions at the termini of the oligopeptides may be related to the positioning of the inhibitor at its site of action. The carbobenzoxy or dansyl groups add hydrophobicity to the N-terminus, and esterification of the C-terminus decreases the polarity of the peptide; such changes could affect its orientation. The explanation for the greater activity of peptides with an N-terminal D-phenylalanine rather than the naturally occurring L-amino acid is also not clear. This could involve protection of the peptide from proteolytic activity, however this is probably not the only factor, because similar results were obtained in hemolysis assays, in which proteolytic activity is less likely to be a factor than in plaque assays. Thus a steric effect on activity seems to be involved.

The importance of the correct amino acid sequence for inhibitory activity was further demonstrated by the selection of a mutant of measles virus that is resistant to the action of a tripeptide (Z-D-Phe-L-Phe-L-(NO_2)Arg) by repeated passage of the virus in the presence of this peptide (16), but which remains sensitive to a peptide (Z-D-Phe-L-Phe-Gly) which differs only in the third amino acid. It is of interest that the sensitivity of this mutant to these oligopeptides resembles that of canine distemper virus, which, like measles virus, is a member of the morbillivirus subgroup of paramyxoviruses and is immunologically related to measles virus.

These oligopeptides with specific antiviral activity do not appear to have significant toxic effects on the cells used in these studies. The cells survive for days with no detectable cytopathic changes and can multiply normally in the presence of the peptides in concentrations many fold higher than those causing inhibition of virus replication.

The site of action of the oligopeptide inhibitors. To investigate the site of action of these oligopeptides several experimental approaches have been used (55). Oligopeptides have been synthesized with ^{3}H or ^{125}I labels to determine their site of binding. Radioactively-labeled oligopeptides were reacted with purified virus, with mock-infected cells, or with infected cells at 4°C and, after washing, specific binding was calculated. Such studies suggested that the oligopeptides bind to the cell and not to the virus (55). The inhibitory action of the oligopeptides is reversible at 37°C; removal of the oligopeptides from the medium and washing the cells greatly reduces activity. To further investigate their site of action, an oligopeptide was synthesized with a chloromethylketone (CK) group attached to Z-D-Phe-L-Phe-Gly (abbreviated ZPPGCK). This oligopeptide binds irreversibly as do the protease inhibitors, N-p-tosyl-L-phenylalanine-chloromethylketone (TPCK) and N-p-tosyl-L-lysine-chloromethylketone (TLCK), which can alkylate histidine, serine, methionine, or cysteine residues near the active sites of proteases. The oligopeptide with the chloromethylketone derivative (ZPPGCK) retained virus inhibitory activity. Whereas TPCK and TLCK, which were included as controls, had no activity. This irreversible inhibitor was used to investigate its site of action (55). Z-PPGCK was preincubated with monkey erythrocytes (RBC) or purified measles virus and, after extensive washing, hemolysis assays were performed. ZPPGCK irreversibly inhibited hemolysis when it was preincubated with RBC, or RBC plus virus, but not when it was preincubated with virus alone. These results provide further evidence that the oligopeptides act on the cell and not on the virus.

A third experiment to investigate the site of action of the peptides involved the use of an oligopeptide inhibitor containing an azido group as a photoaffinity probe, i.e., Z-D-Phe-L-Phe-(Azido)Phe (Z-APPP!) (55). RBC or virus were preincubated with this peptide, exposed to UV-light for various times, washed extensively, and hemolysis assays were then carried out. Pre-treatment of virus with Z-APPP! had no effect, but pre-treatment of RBC inhibited hemolysis. These studies with irreversible inhibitors indicate that the site of action of the oligopeptides is on the target cell membrane, not on the virus. The results suggest that the inhibitors compete with the N-terminus of the F_1 polypeptide for a site on the cell membrane, however the exact location and nature of this presumed "receptor" for the F_1 protein is under investigation. If such a receptor exists, there may be several related, but different, classes of receptors, because different oligopeptides inhibit different viruses to different extents and, as discussed above, resistant mutants of measles virus can be isolated which differ markedly in sensitivity to specific oligopeptides as compared to wild type virus.

Significance of the oligopeptide inhibitors. Whether these inhibitors will have any use in the chemotherapy or chemoprophylaxis or virus infections *in vivo* is not yet clear. However, they provide an example of an important principle, i.e., that specific inhibitors of virus replication have been designed utilizing knowledge gained from basic research on the structure and biological function of individual virus proteins and evidence suggesting that a particular region of a viral protein was involved in activity. In this case, the step in virus replication that is inhibited is virus penetration, which occurs by fusion of the membrane of the virus with that of the cell. In addition to demonstrating that specific inhibition of viral functions other than viral enzymes can be designed on the basis of knowledge of the proteins, these studies have shed light on the mechanisms of virus penetration and virus-induced cell fusion and hemolysis, and membrane fusion in general. The availability of these inhibitors with specific probes attached is also facilitating further studies on the precise site and mechanism of action of viral proteins.

THE ROLE OF THE NON-GLYCOSYLATED MEMBRANE PROTEIN (M) OF MEASLES VIRUS IN SUBACUTE SCLEROSING PANENCEPHALITIS (SSPE)

SSPE is a chronic fatal neurological disease caused by a persistent, abortive infection with measles virus (56-59). Isolation of the virus from the brains of patients with the disease has been possible only by co-cultivation of brain cells with cell lines, not by direct isolation of released infectious virus (58,59). Studies in many laboratories have attempted to define biochemical markers that would distinguish between SSPE isolates and wild type measles virus. Although some minor differences have been observed when comparing two given strains, no consistent biochemical difference has been found that allows an unequivocal distinction between SSPE and other measles virus strains. Thus the existence of specific mutants that cause SSPE has not been demonstrated. In addition, the epidemiological evidence suggests that there are not "SSPE strains" circulating in the population; there are not clusters of cases indicating that a certain population was infected with such a strain.

In recent years much evidence has been accumulated in this laboratory and others that indicates that a lack of the expression of the M protein of the virus is involved in SSPE. In 1979 we found very high titers of antibodies to the other measles virus proteins in the serum and cerebrospinal fluid of patients with SSPE, but in striking contrast, little or no antibody to the M protein (41). This was rapidly confirmed in other laboratories (60,61). Several possibilities for the lack of antibodies to M were considered initially (41), including lack of synthesis of M in brain cells, rapid degradation of M before it became available to the immune system, and a specific immunological defect in which M, but not the other viral proteins, was not recognized by the immune system. No evidence has been found to support the last possibility, and it was shown that the lack of antibody to M was not due simply to antigenic differences among SSPE strains; an SSPE patient's serum failed to precipitate the M protein of the virus isolated from that patient's brain tissue. It has also shown that the same pattern of high titers of antibody to the other proteins but no antibody to M persisted over a period of 7 years in a single patient, indicating that this was not a feature confined to a certain stage of the disease (41, and subsequent follow-up of that patient).

Additional evidence that the M protein was not expressed normally in SSPE brain cells was obtained in an explant culture of SSPE brain cells in which the synthesis of the other virus proteins but not M was detected (62). Subsequently, a technique was developed to detect virus proteins directly in the brain, and no M protein could be detected in the brain of SSPE patients, although the other viral proteins were found (63). In addition, cell lines have been described that were derived by co-cultivation of brain cells from SSPE patients with cell lines in which the M protein was not detected but other measles virus proteins were found (61,64).

The findings described above lead us to hypothesize that there was a host restriction of the synthesis or processing of the M protein in brain cells (2,63,65,66). It was suggested that this host control of M expression is not limited to the brains of SSPE patients, but is a property of certain cells on the brain. Several additional points support this hypothesis. It should be recalled that measles virus can be isolated only after co-cultivation of brain cells with other cells, indicating that the entire viral genome is present in the brain cells, but cannot be expressed until introduced into another cell; this in itself indicates host control. In addition, in cases of acute measles encephalitis, many unsuccessful attempts have been made to isolate virus from the brain, but the successful attempt involved co-cultivation (67), suggesting that in acute measles encephalitis also there is host restriction of virus replication. We have also examined the serum and CSF of a patient 24 years after acute measles encephalitis at age 35. This patient had severe neurological deficits and very high antibody titers to all the other measles virus proteins but no antibody to M (66). This observation strongly suggests that measles virus genetic information was still being expressed in the patient's central nervous system many years after acute encephalitis, but that M was not being synthesized. In addition to supporting the concept of host-cell restriction in the brain, these results suggest that persistent infection of the nervous sytem with measles virus is not limited to the SSPE syndrome. Finally, in England acute severe measles virus infections are an important problem in immunosuppressed children, such as those with leukemia, and several cases have been examined in which acute encephalitis occurred and measles virus antigen and nucleocapsid were observed in the brain, but the virus could not be isolated (68), suggesting again a restricted infection. In our view the most straightforward explanation for these observations is a host restriction of measles virus replication by brain cells, and it is not necessary to evoke a mutant virus or a defect in the patient with SSPE. Whether a patient develops SSPE may be determined by the extent of the seeding of the brain with virus at the time of the initial infection and the state of the brain at that time. In the latter regard, it should be noted that SSPE patients frequently have had their acute measles infection before the age of 2, thus the stage of the development of the central nervous system could play a role. The explanation for the lack of expression of the M protein in brain cells is not yet clear, i.e., whether it is at the level of transcription of the mRNA for M, its translation, or rapid degradation of the protein after synthesis. These questions are now being approached in many laboratories.

Whatever the explanation for the lack of M, its absence explains many of the features of SSPE (2,41,63,65). Because the M protein plays a primary role in the assembly of the virus at the plasma membrane (3,10, 11), lack of M would be expected to result in an abortive infection in which no infectious virus is assembled, but viral nucleocapsids accumulate in the cytoplasm and the viral glycoproteins appear at the cell membrane to stimulate high levels of antibodies. These are the findings that characterize SSPE.

Some recent evidence has been obtained with regard to the role of the M protein in the virus assembly process (69). It was suggested in 1971 (3,10,11), that the M protein served as the recognition site at the plasma membrane between the viral nucleocapsid in the cytoplasm and the glycoproteins in the membrane. Much evidence has been accumulated subsequently to support this hypothesis, including the association of M with nucleocapsid when virus is fractionated or the proteins are mixed in vitro (70,71), the finding of M on the inner surface of the plasma membrane (72), cross-linking of M and NP in cells (73), and failure of assembly of virus in cells infected with a mutant of Sendai virus with a lesion in the M protein (74). Recently purified M protein has been assembled in vitro into ordered structures (69), one of which consists of a sheet formed of strands of M protein subunits with a spacing of 7.2 nm. Tubular structures were also seen which appeared to be formed by rolling of the sheet into tubes in which the strand of subunits were inclined with a left-handed orientation. The arrangements and dimensions of these structures formed in vitro by M protein are strikingly similar to the ordered arrays seen on the inner surface of the plasma membrane of infected cells that were thought to be M, although no direct evidence for this was available (75,76). These arrays of M protein formed in vitro have arrangements that are compatible with their interaction with the helical nucleocapsid of the virus (69). These results support the concept that ordered arrays of the M protein are involved in the assembly of the virus by budding from the cell membrane, providing specific recognition sites for the viral nucleocapsid at the cytoplasmic surface of the plasma membrane.

A NEW INFLUENZA B VIRUS PROTEIN

The nucleotide sequences of RNA segment 6 of influenza B virus.

The sequences of RNA segments 4,6,7, and 8 of influenza B virus resemble those of influenza A virus at both the nucleotide and amino acid levels (77-81). Segment 6 in both viruses codes for the neuraminidase (NA) protein (82-84). The nucleotide sequence of the influenza B/Lee/40 NA gene differs from that of the other influenza virus genes in that its translation is initiated at the second AUG codon from the 5' end of the mRNA (81). The second usual feature of the influenza B RNA segment 6 is that the first AUG codon, which is separated from the second by only 4 nucleotides, is followed by an open reading frame coding for 100 amino acids that overlaps the NA reading frame by 292 nucleotides (81). This reading frame (designated NB) has not been found in influenza A NA genes.

The NB protein.

After finding this second open reading frame, we searched for a polypeptide product translated from it in infected cells. Such a protein would contain 100 amino acids with a mol wt of 11,242. The nucleotide sequence was used to select appropriate labeled amino acids, i.e., cysteine and isoleucine, to detect such a protein. The NB protein has been detected in infected cells (85). The deduced amino acid sequence contains four glycosylation sites, and the protein has been found to be glycosylated in infected cells. The amino acid sequence contains stretches of hydrophobic amino acids at residues 1-16, 18-41, and 83-96, which might act as membrane insertion sites, however the two uncharged regions near the ends of the molecule contain potential glycosylation sites that would not be expected to be glycosylated if that region is inserted into a membrane. The NB protein has not been detected in virions.

Sucrose gradient sedimentation analyses and analyses of the mRNA structure by S_1 mapping and primer extension sequencing indicated that the NB and NA proteins are translated from a single, bicistronic mRNA (85).

A protein analogous to NB has not been found with influenza A virus, and this represents a major difference betwen the two virus types. The function of NB is at present unknown, however there are several biological differences between influenza A and B viruses. Genetic reassortment does not occur between influenza A and B viruses, and influenza B virus does not undergo the periodic major antigenic shifts as does influenza A virus, possibly due to a lack of an animal resevoir for influenza B virus to provide strains for recombination with human viruses. Influenza B virus may be a more "toxic" virus, e.g., the incidence of Reye's syndrome is greater after influenza B infection than influenza A. There is no evidence to implicate the NB protein in any of these aspects of influenza B virus replication and pathogenesis, but it is of great interest as the only protein unique to influenza B virus.

ACKNOWLEDGMENTS

Research described was supported by research grants AI-05600 from the National Institute of Allergy and Infectious Diseases, PCM80-13464 from the National Science Foundation, RG-1216 from the National Multiple Sclerosis Society and CA-18213 from the National Cancer Institute.

REFERENCES

1. Choppin, P. W. and Scheid, A. Rev. Infect. Dis. 2:40-61, 1980.
2. Choppin, P. W., Richardson, C. D., Merz, D. C., Hall, W. W. and Scheid, A. J. Infect. Dis. 143:352-363, 1981.
3. Choppin, P. W. and Compans, R. W. In: Comprehensive Virology (Eds. H. Fraenkel-Conrat and R. R. Wagner), Plenum Press, New York, vol. 4, 1975, pp. 95-178.
4. Scheid, A., Caliguiri, L. A., Compans, R. W. and Choppin, P. W. Virology 50:640-652, 1972.
5. Scheid, A. and Choppin, P. W. J. Virol. 11:263-271, 1973.
6. Scheid, A. and Choppin, P. W. Virology 57:475-490, 1974.
7. McSharry, J. J., Compans, R. W., Lackland, H. and Choppin, P. W. Virology 67:365-374, 1975.
8. Tozawa, H., Watanabe, M. and Ishida, N. Virology 55:242-253, 1973.
9. Scheid, A., Graves, M. C., Silver, S. M. and Choppin, P. W. In: Negative Strand Viruses and the Host Cell (Eds. B.W.J. Mahy and R.D. Barry), Academic Press, London, 1978, pp. 181-193.
10. McSharry, J. J., Compans, R. W. and Choppin, P. W. J. Virol. 8: 722-729, 1971.
11. Choppin, P. W., Compans, R. W., Scheid, A., McSharry, J. J. and Lazarowitz, S. G. In: Membrane Research (Ed. C. F. Fox), Academic Press, New York, 1972, pp. 163-179.
12. Homma, M. and Ohuchi, M. J. Virol. 12:1457-1465, 1973.
13. Scheid, A. and Choppin, P. W. Virology 69:265-277, 1976.

14. Scheid, A. and Choppin, P. W. Virology 80:54-66, 1977.
15. Gething, M. J., White, J. M. and Waterfield, M. D. Proc. Natl. Acad. Sci. USA 75:2737-2740, 1978.
16. Richardson, C. D., Scheid, A. and Choppin, P. W. Virology 105: 205-222, 1980.
17. Hsu, M.-C., Scheid, A. and Choppin, P. W. Virology 95:476-491, 1979.
18. Silver, S. M., Scheid, A. and Choppin, P. W. Infection and Immunity 20:235-241, 1978.
19. Scheid, A. and Choppin, P. W. In: Proteases and Biological Control (Eds. E. Reich, D. B. Rifkin, and E. Shaw), Cold Spring Harbor Laboratory, 1975, pp. 645-659.
20. Nagai, Y., Klenk, H.-D. and Rott, R. Virology 72:494-508, 1976.
21. Nagai, Y. and Klenk, H.-D. Virology 77:125-134, 1977.
22. Virchow, R. Virchow's Arch. 14:1, 1858.
23. Roizman, B. Cold Spr. Harbor Symp. Quant. Biol. 27:327-340, 1962.
24. Enders, J. F. and Peebles, T. C. Proc. Soc. Exp. Biol. Med. 86:277-286, 1954.
25. Henle, G., Deinhardt, F. and Girardi, A. Proc. Soc. Exp. Biol. Med. 87:386-393, 1954.
26. Okada, Y. Biken's J. 1:103-110, 1958.
27. Morgan, H. R., Enders, J. F. and Wagley, P. F. J. Exp. Med. 88:503-514, 1948.
28. Kohn, A. Virology 26:228-245, 1965.
29. Holmes, K. V. and Choppin, P. W. J. Exptl. Med. 124:501-520, 1966.
30. Compans, R. W., Holmes, K. V., Dales, S. and Choppin, P.W. Virology 30:411-426, 1966.
31. Choppin, P. W., Klenk, H.-D., Compans, R. W. and Caliguiri, L. A. In: From Molecules to Man (Ed. M. Pollard), Perspectives in Virology VII, Academic Press, New York, 1971, pp. 127-158.
32. Graves, M. C., Silver, S. M. and Choppin, P. W. Virology 86: 254-263, 1978.
33. Merz, D. C., Scheid, A. and Choppin, P. W. J. Exptl. Med. 151: 275-288, 1980.
34. Merz, D. C., Scheid, A. and Choppin, P. W. Virology 109:94-105, 1981.
35. Norrby, E., Enders-Ruckle, G. and ter Meulen, V. J. Infect. Dis. 132:262-269, 1975.
36. Norrby, E. and Gollmar, Y. Infect. Immun. 11:231-239, 1975.
37. Norrby, E. and Pettinen, K. J. Infect. Dis. 138:672-676, 1978.
38. Fulginiti, V. A., Eller, J. J., Downie, A. W. and Kemp, C. H. J.A.M.A. 202:101-106, 1967.
39. Buser, F. N. Engl. J. Med. 277:250-251, 1967.
40. Scott, T. F. M. and Bonnano, D. E. N. Engl. J. Med. 277:248-250, 1967.
41. Hall, W. W., Lamb, R. A. and Choppin, P. W. Proc. Natl. Acad. Sci. USA 76:2047-2051, 1979.
42. Chanock, R. M., Parrott, R. H., Kapikian, A. Z., Kim, H. W. and Brandt, C. D. In: Perspectives in Virology VI (Ed. M. Pollard), Academic Press, New York, 1968, pp. 125-139.
43. Choppin, P. W., Richardson, C. D., Merz, D. C. and Scheid, A. In: Adhesion and Microorganism Pathogenicity (Ed. M. Oconnor), Ciba Foundation Sumposium, Pitman Medical, London, 1981, pp. 252-263.
44. Hsu, M.-C., Scheid, A. and Choppin, P. W. J. Biol. Chem. 256: 3357-3363, 1981.
45. Skehel, J. J., Bayley, P. M., Brown, E. B., Martin, S. R., Waterfield, M. D., White, J. M., Wilson, I. A. and Wiley, D. C. Proc. Natl. Acad. Sci. USA 79:968-972, 1982.

46. Helenius, A., Kartenbeck, J., Simono, K. and Fries, E. J. Cell Biol. 84:404-420, 1980.
47. Lenard, J. and Miller, D. K. J. Cell Biol. 84:430-437, 1980.
48. White, J., Matlin, K. and Helenius, A. J. Cell Biol. 89:674-679, 1981.
49. Hsu, M.-C., Scheid, A. and Choppin, P. W. Proc. Natl. Acad. Sci. USA 79:5862-5866, 1982.
50. Hsu, M.-C., Scheid, A. and Choppin, P. W. Virology 146:361-369, 1983.
51. Miller, F. A., Dixon, G. J., Arnett, G., Dice, J. R., Rightsel, W. A., Schabel, F. M., McLean, J. W. Appl. Microbiol. 16:1489-1496, 1968.
52. Norrby, E. Virology 44:599-608, 1971.
53. Choppin, P. W. Unpublished experiment, 1971.
54. Richardson, C. D., Scheid, A. and Choppin, P. W. In: The Replication of Negative Strand Viruses (Eds. D. H. L. Bishop and R. W. Compans), Elsevier-North Holland, New York, 1981, pp. 509-515.
55. Richardson, C. D., Scheid, A. and Choppin, P. W. Virology, to be submitted, 1983.
56. Bouteille, M., Fontaine, C., Vedrenne, C. and Delarue, J. Rev. Neurol. (Paris) 113:454-458, 1968.
57. Connolly, J. H., Allen, I. V., Hurwitz, L. I. and Millar, J. H. Lancet I:542-544, 1967.
58. Payne, F. E., Baublis, J. V. and Itabashi, H. H. N. Engl. J. Med. 281:585-589, 1969.
59. Horta-Barbosa, L., Fuccillo, D. A., Sever, J. L. and Zeman, W. Nature 221:974, 1969.
60. Wechsler, S. L., Weiner, H. L. and Fields, B. N. J. Immunol. 123:884-889, 1979.
61. Machamer, C. E., Hayes, E. C. and Zweerink, H. J. Virology 108:515-520, 1981.
62. Hall, W. W. and Choppin, P. W. Virology 99:443-447, 1979.
63. Hall, W. W. and Choppin, P. W. N. Eng. J. Med. 304:1152-1155, 1981.
64. Lin, F. H. and Thormar, H. Nature 285:490-492, 1980.
65. Choppin, P. W. (Editorial). Annals Neurology 9:17-20, 1981.
66. Hall, W. W., Sahgal, V., Harter, D. H. and Choppin, P. W. Lancet II:967-968, 1979.
67. ter Meulen, V., Muller, D., Kackell, Y. and Katz, M. Lancet II:1172-1175, 1972.
68. Marshall, W. C. In: Problems of Antiviral Chemotherapy (Ed. C. Stuart-Harris), Academic Press, London and New York, 1983, in press.
69. Heggeness, M. H., Smith, P. R. and Choppin, P. W. Proc. Natl. Acad. Sci. USA 79:6232-6236, 1982.
70. Shimizu, K. and Ishida, N. Virology 67:427-437, 1975.
71. Yoshida, T., Nagai, Y., Yoshii, S., Maeno, K., Matsumoto, K. and Hoshino, M. Virology 71:143-161, 1976.
72. Bowen, H. A. and Lyles, D. S. J. Virol. 37:1079-1082, 1981.
73. Markwell, M. A. K. and Fox, C. F. J. Virol. 33:152-166, 1980.
74. Yoshida, T., Nagai, Y., Maeno, K., Iinuma, M., Hamaguchi, M., Matsumoto, T., Nagayoshi, S. and Hoshino, M. Virology 92:139-154, 1979.
75. Büchi, T. Virology 106:41-49, 1980.
76. Büechi, M. and Büchi, T. Virology 120:349-359, 1982.
77. Briedis, D. J., Lamb, R. A. and Choppin, P. W. Virology 112:417-425, 1981.
78. Briedis, D. J. and Lamb, R. A. J. Virol. 42:186-193, 1982.
79. Briedis, D. J., Lamb, R. A. and Choppin, Virology 116:581-588, 1982.

80. Krystal, M., Elliot, R. M., Benz, E. W., Jr., Young, J. F. and Palese, P. Proc. Natl. Acad. Sci. USA 79:4800-4804, 1982.
81. Shaw, M. W., Lamb, R. A., Erickson, B. W., Briedis, D. J. and Choppin, P. W. Proc. Natl. Acad. Sci. USA 79:6817-6821, 1982.
82. Palese, P. and Schulman, J. L. Proc. Natl. Acad. Sci. USA 73:2142-2146, 1976.
83. Ueda, M., Tobita, K., Sugiura, A., and Enomoto, C. J. Virol. 25:685-686, 1978.
84. Raccaniello, V. R. and Palese, P. J. Virol. 29:361-373, 1979.
85. Shaw, M. W., Choppin, P. W. and Lamb, R. A. Proc. Natl. Acad. Sci. USA 80:, in press.

20

ON THE PATHOGENICITY OF AVIAN INFLUENZA VIRUSES

RUDOLF ROTT, HANS-DIETER KLENK, and CHRISTOPH SCHOLTISSEK
Institut für Virologie, Justus-Liebig-Universität Giessen, Frankfurter Strasse 107, D-6300 Giessen, FRG

INTRODUCTION

Influenza A viruses cause different diseases in mammals and birds. While the clinical manifestations of influenza virus infection in mammals is virtually always a result of localized infection of the respiratory tract epithelium, the clinical disease in birds infected with highly pathogenic avian influenza viruses is commonly associated with systemic infection and death. The majority of avian influenza viruses, however, induce asymptomatic infection which is restricted to local sites in mucous membranes frequently in the gut and the respiratory tract.

Our group in Giessen has concentrated in the last few years mainly on obtaining information about the prerequisites for avian influenza virus pathogenicity, whereby pathogenicity is defined as the capacity of a virus strain to cause overt disease.

Avian influenza viruses occur with at least 13 different hemagglutinin (HA) subtypes and in many HA-neuraminidase combinations (1,2). In addition to variability of the genes coding for HA and neuraminidase in one subtype, there are also considerable differences in base sequence homologies or oligonucleotide fingerprints of the other genes (references in 3). This suggests that avian influenza viruses arise in nature in a continuous process of segmented gene reassortment. Furthermore, these viruses show differences in their pathogenicity for chicken. Highly pathogenic virus strains exist among the H7 (Hav1) and H5 (Hav5) subtypes. So far as is known, virus strains of all the other subtypes are nonpathogenic (4). Therefore, for studies on pathogenicity of influenza viruses the avian system offers the advantages that a variety of naturally occurring different virus strains can be used in their natural host. Furthermore, this virus-host system enables us also to analyze the contribution of the individual virus genes to pathogenicity.

PROTEOLYTIC CLEAVAGE OF THE HEMAGGLUTININ AS A DETERMINANT FOR PATHOGENICITY

The HA which is located as spike-like protrusions on the surface of influenza virus particles is of great significance for the uptake of virions by host cells and therefore for infectivity. It is responsible for the attachment of the virus to sialic acid containing receptors on the host cell and it is involved by acting in concert with the neuraminidase in virus penetration through fusion of the viral envelope with cellular membranes (references as in 5). The capacity for penetration depends on posttranslational proteolytic cleavage of the precursor HA into the subunits HA_1 and HA_2. Cellular proteases are involved in this activation reaction and, depending on the presence of an appropriate enzyme in a given cell, virus particles with cleaved or with uncleaved HA may be formed (references in 6). It has been claimed that neuraminidase alters the cleavability of HA (7). However, naturally occurring strains or our laboratory reassortants never show this effect, although both types of virus isolates contained different HA-neuraminidase combinations (8-10).

It has become increasingly evident that the vast majority of influenza viruses have a very restricted host range for cells within an organism and for specific cell cultures. The reason for this is that, in the majority of cell types, the HA of most influenza viruses is not cleaved. In the case of natural isolates of avian influenza viruses, there is a strict correlation between susceptibility for proteolytic cleavage and pathogenicity. While the HA of pathogenic strains is cleaved in all cell types tested, the HA of nonpathogenic strainsis cleaved only in certain host cells (9).

The differences in HA cleavability can be explained by the structural variation at the cleavage site of the HA molecule (Fig. 1). On the basis of amino acid sequence analyses it could be shown that with influenza viruses nonpathogenic for chickens a single arginine residue is eliminated at the site of the cleavage reaction (references in 11). In contrast, elucidation of the primary structure of HA of the highly pathogenic fowl plague virus revealed that HA_1 and HA_2 are linked by several arginine or lysine residues (12,13). This intervening sequence is also removed in the cleavage reaction.

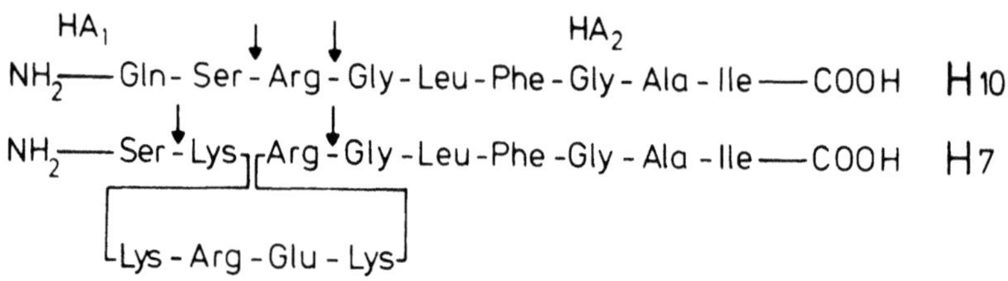

FIGURE 1. Part of the amino acid sequences around the cleavage site of H10 (Hav2) and H7 (Hav1) hemagglutinins.

The H10 hemagglutinin is of the nonpathogenic A/chick/Germany/49 strain (15) and H7 hemagglutinin is of the pathogenic A/FPV/Rostock strain (13). Arrows indicate the established C-termini of HA_1 and the NH_2-termini of HA_2.

That this presence or absence of arginine or lysine residues may be a common property of pathogenic influenza viruses, is suggested by recent observations. A series of pathogenic and nonpathogenic viruses,all of which carry HA of the same serotype, were found to differ in HA cleavability. The loss of basic residues is paralleled by a shift in charge which can be demonstrated by analysis of HA using isoelectric focusing. It was therefore of interest to find that the pathogenic strains show a larger charge shift than the nonpathogenic strains (14). Thus, it appears that the size of the basic intervening sequence is an essential structural determinant for the susceptibility of HA to cleavage.

Two enzymes are involved in this reaction. First a trypsin-like endoprotease cleaves the arginine-glycine linkage; subsequently, the arginine or the entire intervening basic peptide is removed by a carboxypeptidase B (15). It has long been assumed that the trypsin-like enzyme is of host origin, since the host cell determines whether or not cleavage occurs. There is evidence that differences exist in the enzymatic specificity involved in the cleavage of pathogenic and nonpathogenic viruses. It appears that HA of pathogenic strains is activated by a trypsin-like protease present in all host cells that requires a pair of basic amino acids at the cleavage site. On the other hand, HA of the nonpathogenic strains, which is activated in a few host cells, is cleaved by an enzyme which can act at a single arginine (13). The caboxypeptidase B is found in purified virus particles, but it appears also to be host cell-derived. This is indicated by a host-dependent variation of the isoelectric point of this enzyme (Garten, personal communication).

It was interesting to find that endoproteases with specificities different from those of trypsin, such as thermolysine or chymotrypsin, cleave but do not activate the hemagglutinin. After treatment of HA with these enzymes the cleavage site is shifted, but only by one or three amino acids towards the carboxy terminal direction. From these findings one can conclude that a specific structure at either the carboxy terminus of HA_1 or the amino terminus of HA_2 is necessary for activity (15). The finding that elimination of the arginine can be prevented by inhibitors of carboxypeptidase B, and that this treatment does not influence activity of HA, indicate that a specific structure at the amino terminus of HA_2 is required for the fusing capacity of HA (Garten, personal communication).

The findings discussed so far demonstrate that activation of HA requires cleavage of a specific bond and that the amino acid sequence of the connecting peptide between HA_1 and HA_2 determines the range of cells in which cleavage can occur. Since pathogenic strains contain HA which is activated in a broad spectrum of different host cells, these viruses are able to undergo multiple cycle replication and spread in the host organism. In contrast, spread of nonpathogenic virus is inhibited as soon as the virus infects cells which do not possess appropriate HA activating enzymes.

This hypothesis could be confirmed convincingly using the chicken chorioallantoic membrane as a model organ system in studies with influenza viruses pathogenic or nonpathogenic for the chicken. It was found that the hemagglutinin of the pathogenic strains was activated in all three germinal layers (the endoderm, the mesoderm, and the ectoderm), whereas the hemagglutinin of the nonpathogenic viruses was cleaved only in the endodermal cells. Accordingly, there are differences between the pathogenic and the nonpathogenic viruses in their spread through the membrane as indicated by immune histology. Multiplication of nonpathogenic strains was restricted to the cell layer which was inoculated. Spread of newly synthesized virus was inhibited as soon as the virus reached the barrier of nonpermissive mesodermal cells. On the other hand, the pathogenic virus spread through the whole membrane and gained entrance into the blood vessels regardless of the route of infection (16).

A similar mechanism could be shown in the course of infection of the mature avian organism. Pathogenic as well as nonpathogenic avian influenza viruses are produced in infectious form, that is with a cleaved hemagglutinin, in epithelial cells which line the respiratory- and intestinal tract of the bird. Spread of nonpathogenic viruses is inhibited as soon as the virus reaches the lamina propria which is nonpermissive for these viruses. Only the very few pathogenic avian influenza viruses, which possess a cleavable hemagglutinin, may pass this barrier. The resulting generalized infection leads to the lethal disease (9,17).

Our results support the concept that differences in pathogenicity are the result of structural variations in the hemagglutinin. The available evidence indicates that these structural variations are confined to the cleavage site, i.e. to a small, but functionally important part of the molecule.

GENE CONSTELLATION AND PATHOGENICITY

In contrast to natural influenza virus isolates, reassortants obtained by a mixed infection _in vitro_ have revealed that in addition to HA an appropriate gene constellation is important for pathogenicity. Experiments on replacement of RNA segments by reassortment have shown that the source from which the RNA segment is derived, as well as the RNA segment which is exchanged, are important for pathogenicity. Investigations with a large number of different reassortants lead to the following conclusion: 1) There is no general rule which RNA segment(s) has to be replaced in order to increase or lower pathogenicity. 2) The more RNA segments of pathogenic strains that are replaced by nonpathogenic strains, the higher is the probability of obtaining nonpathogenic reassortants. 3) Especially important for pathogenic properties is the gene constellation of the polymerase complex, i.e. the genes coding for the three P proteins and the NP protein (references in 18).

Comparing the results obtained with influenza virus reassortants isolated _in vitro_, and those obtained from naturally occurring avian influenza viruses, there is ample evidence that in nature such avian influenza viruses are selected whose genomes possess an optimal composition. If in addition to the optimally functioning genome, the viruses possess a hemagglutinin which is cleaved in many different host cells, thereby becoming activated, the viruses are truly pathogenic (references in 19). Evidently, a major factor for this selection is the body temperature of the bird at 41° C, since only pathogenic reassortants grow efficiently at this elevated temperature. As a consequence of double infection _in vitro_ with two different influenza viruses at the elevated temperature, reassortants are selected which are exclusively pathogenic for the chicken. If on the other hand, mixed infection is performed at 37° C, most progeny are nonpathogenic (10). Preliminary studies on the temperature sensitive block of the nonpathogenic reassortants indicate that growth inhibition at 41° C can occur at different levels within the multiplication cycle (unpublished results).

In view of these findings one may suspect that in vitro produced nonpathogenic reassortants are temperature-sensitive because of missmatching of the individual viral components, especially of the polymerase complex. If this assumption is correct, it should be possible to convert the appropriate genes by mutation and selection to allow function at the body temperature of the bird. To achieve this goal we (20) passaged nonpathogenic reassortants serially at high multiplicities (von Magnus passages) at 41° C. It has been shown by others (references in 21) that, under these experimental conditions, deletions occur preferentially on the genes coding for the polymerase complex.

As can be seen from Fig. 2, the infectivity declined rapidly as expected under these conditions. However, with most reassortants tested, the infectivity thereafter increased with further passages, the number of which depended on the reassortants used and varied from experiment to experiment. Most interestingly, all isolates obtained after von Magnus passages at 41° C were pathogenic for chicken although the degree of pathogenicity varied. After intramuscular inoculation, some isolates killed the birds two to three days after infection like the highly pathogenic fowl plague virus. With some isolates, chickens died only after a prolonged incubation or became sick, but did not die in a 14-day period of observation (Table 1).

When the von Magnus passages were carried out at 37° C only one of ten isolates could be obtained in which pathogenicity was reactivated. The reason for this inefficiency presumably is that there is no selective pressure at 37° C

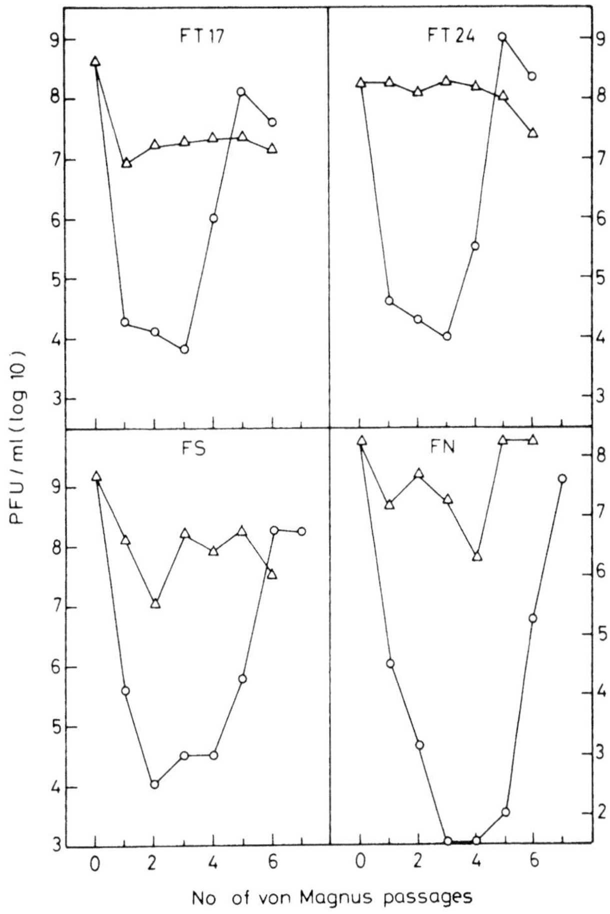

FIGURE 2. Behavior of infectivity of different influenza virus reassortants during von Magnus passages at 37° C (▲) or 41° C (O).

The "0" von Magnus passage represents the infectivity titer of the input virus which was prepared at 37° C. For detail see (20).

and a mutant with new properties has no chance of overgrowing the original parental type.

Table 1. Pathogenicity for chicken of influenza virus reassortants before and after von Magnus passages.

Reassortants[a]	Gene constellation[b]	Pathogenicity[c] before	after von Magnus passages at 37° C	after von Magnus passages at 41° C
FT1	TTFFTTTF	-	-	+(4)
FT3	FTTFFTFT	-	-	±
FT15	FTFFTTTT	-	-	±
FT17	TTFFTTTF	-	-	+(6)
FT22	FTTTTFTT	-	-	+(2)
FT23	FFTFFFFF	-	+(7)	+(2)
FT24	FFTFFFFF	-	-	+(3)
FS	FSFFFSFF	-	-	+(5)
FE	FFFFEEFF	-	-	+(7)
FN	FFNFFNNN	-	-	±

[a] The reassortants were obtained after mixed infections with the parent viruses F = A/fowl plague/Rostock/34 (H7N1), T = A/turkey/England/63 (H7N3), N = A/chick/Germany/N/49 (H10N7), S = A/swine/1976/31 (H1N1), and E=A/equine/1/63 (H3N8).

[b] The letters indicate the parental source of the genes and are arranged according the electrophoretic migration rates of the RNA of A/fowl plague/Rostock. The order designates the genes in the following sequence: PB1, PB2, PA, HA, NP, NA, M, and NS (3).

[c] Chickens die (+) after infection (No. of days), or become sick (±), or showed no signs of illness (-) during a 14-day observation period.

For detail see (20).

CONCLUSIONS

In the course of these investigations, it became obvious that pathogenicity of avian influenza viruses depends on an otpimal gene constellation which permits virus replication in the bird. The individual combination of genes which achieves this pathogenic constellation is not unique and is not restricted to any particular RNA segment in the virus genome. The reassortment of any possible combination which guarantees a rapid replication of the virus in the host organism will exhibit the best chance of being pathogenic. These events must regularly occur in natural infection. The experiments using high multiplicity passages under selection at 41° C show the dynamic response at the biological level of the genetic system composed of the eight influenza virus genes.

There is one specific restriction on pathogenicity, however, which can be defined with certainty: in order to become pathogenic (i.e. to induce a systemic infection), a virus strain must have a hemagglutinin which is cleavable in a broad range of different host cells. Such a situation permits rapid production of a great many virus particles which can flood the host organism before defense mechanisms, in particular a potent immune response, can block the spread of the virus. If a certain HA gene product is only cleaved in a restricted number of cell types, the infection will be confined to localized areas in the bird. This type of infection is likely to be clinically inapparent.

ACKNOWLEDGMENT

This work was supported by the Deutsche Forschungsgemeinschaft (Sonderforschungsbereich 47).

REFERENCES

1. Hinshaw, V.S., Webster, R.G. and Rodriguez, R.J. Arch. Virol. 67:191-201, 1981.

2. Hinshaw, V.S., Air, G.M., Gibbs, A.J., Graves, L., Prescott, R. and Karunakaran, D. J. Virol. 42:865-872, 1982.
3. Scholtissek, C. In: Genetics of Influenza Viruses, Virology Monographs (Ed. P. Palese) Springer-Verlag, Wien, 1983, in press.
4. Easterday, B.C. In: The Influenza Viruses and Influenza (Ed. E.D. Kilbourne) Academic Press, New York, 1975, pp. 449-481.
5. Rott, R., Huang, R.T.C., Wahn, K. and Klenk, H.-D. In: Replication of Negative Strand Viruses (Eds. D.H.L. Bishop and R.W. Compans), Elsevier/North Holland, New York, 1981, pp. 269-276.
6. Klenk, H.-D. and Rott, R. Curr. Top. Microbiol. Immunol. 90:19-48, 1980.
7. Schulman, J.L. and Palese, P. J. Virol. 24:170-176, 1977.
8. Klenk, H.-D., Rott, R., Orlich, M. and Blödorn, J. Virology 68:426-439, 1975.
9. Bosch, F.X., Orlich, M., Klenk, H.-D. and Rott, R. Virology 95:197-207, 1979.
10. Rott, R., Orlich, M. and Scholtissek, C. Virology 120: 215-224, 1982.
11. Ward, C.W. Curr. Top. Microbiol. Immunol. 94:1-74, 1981.
12. Porter, A.G., Barber, C., Carey, N.H., Hallewell, P.A., Threlfall, G. and Emtage, J.S. Nature (London) 282:471-477, 1979.
13. Garten, W., Linder, D., Rott, R. and Klenk, H.-D. Virology 122:186-190, 1982.
14. Bosch, F.X., Garten, W., Klenk, H.-D. and Rott, R. Virology 113:725-735, 1981.
15. Garten, W., Bosch, F.X., Linder, D., Rott, R. and Klenk, H.-D. Virology 115:361-374, 1981.
16. Rott, R., Reinacher, M., Orlich, M. and Klenk, H.-D. Arch. Virol. 65:123-133, 1980.
17. Webster, R.G., Campbell, C.H. and Granoff, A. Virology 44:317-328, 1971.
18. Rott, R. and Scholtissek, C. In: Basic and Applied Influenza Research (Ed. A.S. Beare), CRC Press, Inc., Boca Raton, 1982, pp. 189-210.
19. Rott, R. Arch. Virol. 59:285-298, 1979.
20. Rott, R., Orlich, M. and Scholtissek, C. Virology 126: 1983, in press.
21. Sivasubramanian, N. and Nayak, D.P. Virology 124:232-237, 1983.

21

GLYCOPROTEINS OF VENEZUELAN EQUINE ENCEPHALITIS (VEE) VIRUS: MOLECULAR STRUCTURE AND FUNCTION IN VIRUS PATHOGENICITY AND HOST IMMUNITY

DENNIS W. TRENT,[1] JOHN T. ROEHRIG,[1] JOHN R. BELL,[2] JAMES H. MATHEWS,[1], RICHARD M. KINNEY[1], and JAMES H. STRAUSS[2]
[1]Division of Vector-Borne Viral Diseases, Centers for Disease Control, Fort Collins, Colorado, USA and [2]Division of Biology, California Institute of Technology, Pasadena, California, USA

SUMMARY

The VEE virus particle contains a ssRNA genome enclosed in an icosahedral capsid and an envelope containing two glycoproteins, E2 (56k daltons, gp56) and E1 (50k daltons, gp50). The amino terminal sequences of the envelope proteins have been determined and compared with data for other alphaviruses. The alphavirus glycoproteins share regions of homology, particularly in the conservation of proline and cysteine residues, implying that the three dimensional conformation is conserved. Deletions and amino acid replacements can occur at most locations with the alphavirus glycoproteins indicating that variation within the proteins can be accommodated without destroying biological function. Tryptic map analysis of the two virus glycoproteins shows differing degrees of diversity within the VEE complex, the larger protein being the more variable. Viral neutralization (N) and hemagglutination inhibition (HAI) determinants have been placed on the gp56 VEE/TC-83 protein using monoclonal antibody to purified proteins. Antigenic analysis indicates the presence of at least three antigenic epitopes on gp56 and four on gp50. Biological functions of N and HAI are primarily associated with gp56^{c}, the critical N site. <u>In vivo</u> protection from virus infection is provided by passive immunization with antisera to gp56 and gp50 as well as highly avid monoclonal antibodies to gp56^{c}, gp50^{b}, gp50^{c}, and gp50^{d}. Small amounts of gp56^{c} antibodies were needed to provide protection, whereas less avid antibodies to other determinants on gp56 and gp50 were less protective. Protection from VEE virus infection appears to be determined by the specificity of antibody to the virion glycoprotein epitope, avidity of the antibody and the ability of the antibody to react with antigenic epitopes proximal to the critical N site.

1. INTRODUCTION

Venezuelan equine encephalitis (VEE) virus was first isolated in 1938 by Kubes and Rios (1) from the brain of a horse which died during an epizootic of a previously unrecognized disease in Venezuela. VEE-related viruses were subsequently isolated during the period of 1943-1963 in Venezuela, Colombia, Peru, Trinidad, Brazil, Surinam, Argentina, Panama, Mexico, and the United States (2). Shope et al. (3) first defined the viruses in the VEE complex by showing serological relationships between classical VEE, Mucambo, and Pixuna viruses. Young and Johnson (2) serologically characterized a variety of VEE isolates and proposed that the complex be divided into four subtypes (I, II, III, and IV). Viruses in subtype I were divided into five variants designated IA through IE. During 1969-1971 a VEE epizootic-epidemic occurred in South America, Central America, and the United States involving a subtype IAB virus which caused high mortality among equines and human disease (4).

Venezuelan equine encephalitis viruses are alpha-togaviruses which contain a positive strand ribonucleic acid genome enclosed in an icosahedral nucleocapsid. The virion has an envelope which contains two glycoproteins: E2 of 56,000 daltons (gp56) and E1 of 50,000 daltons (gp50) (5,6). Viral neutralization (N) and hemagglutination (HA) sites have been placed on E2 by the use of monospecific rabbit antisera and monoclonal antibodies specific for purified viral structural proteins (7-10). Only anti-E2 antisera neutralized virus infectivity or blocked virus hemagglutination. The E2 envelope glycoprotein, therefore, possibly plays a central role in eliciting a protective immune response following either immunization or natural infection with VEE virus (11).

In this report, the biochemical and immunological characterization of the VEE E1 and E2 envelope glycoproteins will be presented, and the function of these proteins in providing protective immunity will be discussed.

2. MOLECULAR WEIGHT COMPARISONS

2.1. Antigenic analysis of the VEE virus structural proteins in the Venezuelan encephalitis complex of alphaviruses using polyclonal antisera

2.1.1. Structural proteins. Young and Johnson (2) serologically classified Venezuelan equine encephalitis (VEE) viruses by kinetic HAI into four subtypes (I to IV), with subtype I exhibiting five antigenic variants (IA to IE; Table 1). The serological reactivities demonstrated in HAI and N indicate that the antigenic determinants responsible for type-specificity and induction of host protective immunity are located on the two glycoproteins present on the surface of the virus particle (7-14).

TABLE 1. Venezuelan equine encephalitis complex viruses. Antigenic classification, geographic origin, and year of isolation.

Prototype Virus	Antigenic Classification[a]	Origin	Year Isolated
Trinidad donkey	I-A	Trinidad	1943
TC-83[b]	I-A		
PTF-39	I-B	Guatemala	1969
P676	I-C	Venezuela	1963
3880	I-D	Panama	1961
Mena II	I-E	Panama	1962
Everglades Fe3-7c	II	U.S.A. (Florida)	1963
Mucambo BeAn 8	III	Brazil	1954
Pixuna BeAr 35645	IV	Brazil	1961

a/ Classification based on the kinetic hemagglutination-inhibition studies of Young and Johnson (2).

b/ Vaccine strain derived from Trinidad donkey strain.

The molecular weights of the structural proteins of each VEE prototype virus were determined by electrophoresis of dissociated virus on discontinuous polyacrylamide gels (8,9,17) (Fig. 1). Two envelope glycoproteins and a capsid protein are well resolved for most of the viruses (Fig. 1). The two envelope glycoproteins of subtype IV, Pixuna virus, migrated similarly but were separable from each other. Molecular weights of the structural proteins for each of the viruses were determined (Table 2). Nucleocapsid proteins of all strains were estimated to have molecular weights of 35×10^3 daltons and the E1 envelope glycoproteins 51×10^3 daltons. More size variation was observed in the molecular weights of the larger E2 glycoproteins of the different VEE subtypes ($53 - 56 \times 10^3$ daltons).

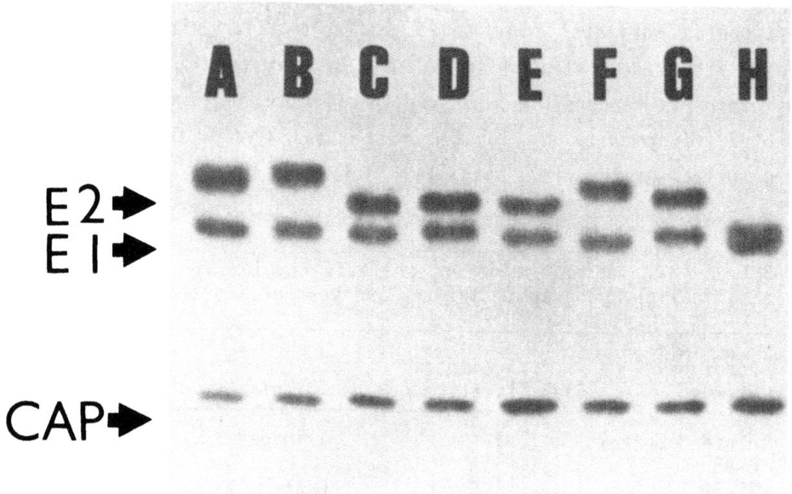

FIGURE 1. Polyacrylamide gel electrophoresis of proteins purified from [^{3}H] leucine-labelled VEE complex viruses. (a) TC-83, (b) PTF-39, (c) P676, (d) 3880, (e) Mena II, (f) Everglades Fe3-7c, (g) Mucambo BeAn 8, and (h) Pixuna BeAr 35645.

To determine the biological functions and antigenic specificities of VEE virus envelope glycoproteins, purified virus was solubilized with a nonionic detergent and the two glycoproteins separated by isoelectric focusing (8,9,10,17; Fig. 2). Typical isoelectric focusing profiles of all VEE virus glycoproteins revealed two peaks of radioactivity. The E1 glycoproteins focused with an approximate isoelectric point of 7.0 and the E2 glycoprotein with a pI of 9.0.

TABLE 2. Molecular weights of the structural proteins of prototype VEE complex viruses determined by SDS-PAGE.

Prototype Virus	Estimated mol. wt. (daltons x 10^3): E2	E1	Capsid
TC-83	56	50	35
PTF-39	56	51	35
P676	54	51	35
3880	54	51	35
Mena II	54	50	35
Everglades Fe3-7c	56	50	35
Mucambo BeAn 8	56	51	35
Pixuna[a] BeAr 35645	52	50	35

a/ Pixuna virus often shows a minor band at 57 x 10^3 daltons.

(From France *et al.*, 8; by permission of Journal of General Virology.)

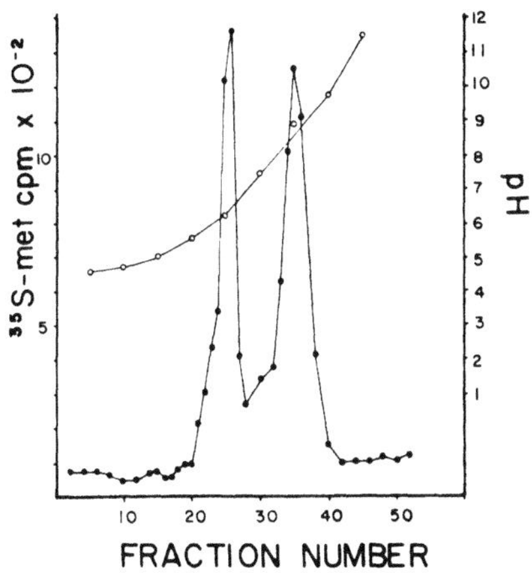

FIGURE 2. Isoelectric focusing of TC-83 glycoproteins. Purified, radioactive TC-83 virus was solubilized with Zwittergent, the nucleocapsid was removed by pelleting, and the glycoproteins separated in a pH 5-11 gradient. E1 (gp50) focused at pH 6.4, and E2 (gp56) focused at pH 9.0. ^{3}H-leucine labeled protein (closed circle); pH gradient (open circle). (From Roehrig *et al.*, 10; by permission of Virology.)

2.1.2. Antigenic and biological analysis. Antigenic and biological analysis of the purified glycoproteins was undertaken by preparing subunit antisera to each of the purified glycoproteins in rabbits (8,9,15; Table 3). Antiserum to the E1 glycoproteins of VEE virus reacted by radioimmunoassay and complement fixation with purified virus but did not neutralize virus or prevent viral agglutination of goose erythrocytes. The E2 antiserum neutralized virus and prevented virus agglutination of goose erythrocytes.

TABLE 3. Biological and immunochemical functions of VEE envelope glycoproteins.

Test	---- Antisera to: ---- E1	E2
Radioimmunoassay	25600	25600
Complement fixation	128	64
Neutralization	20	640
Hemagglutination inhibition	20	1280

VEE viruses representing subtypes I through IV and variants IA to IE were cross tested with the subunit antisera in the HAI test (Table 4). Antisera to virus E2 glycoproteins reacted specifically in the HAI test with those viruses which had previously been determined to be distinct in the kinetic HAI test (2). Subtype IB virus, PTF-39, reacted to equal titer with antisera to IA virus. The E2 anti-serum to IB virus was specific. Antisera to ID virus E2 glycoprotein reacted at equal titer with subtype I varieties A, B, and C suggesting a broadly reactive determinant is present in the hemagglutinin of subtype ID.

Virus testing with E2 subunit anti-serum confirmed the serologic relationships among the VEE viruses which had been established by HAI (Table 5). Subunit antiserum to the PTF-39 E2 glycoprotein was less specific by N than it was by HAI testing; however, the ID antisera to 3880 was more specific in the N test. The type II virus, Everglades, cross reacted to a significant extent with the antisera to type IAB viruses in both the N and HAI tests suggesting that these viruses share antigenic epitopes involved in both neutralization and agglutination to red blood cells.

On the basis of these serological results, it is clear that the E2 glycoprotein of VEE virus contains the biologically important critical N antigenic site responsible for VEE serologic specificity and eliciting protective host immunity (7-9). The VEE E1 glycoproteins contain complex reactive antigenic determinants which do not appear to be directly involved in the biological assays of N or HAI nor are they directly involved in the host protective immune response (15).

TABLE 4. Results of cross-testing VEE complex viruses by hemagglutination-inhibition using antisera prepared against the E2 glycoprotein.

Virus		Antisera to E2 Glycoprotein of:							
		TRD	PTF-39	P676	3880	Mena II	EVE	MUC	PIX
TC-83	(I-A)[a]	20480[b]	320	320	1280	40	40	40	ND[c]
PTF-39	(I-B)	20480	2560	320	2560	20	80	40	ND
P676	(I-C)	1280	640	5120	1280	ND	20	20	10
3880	(I-D)	320	160	80	1280	80	40	40	ND
Mena II	(I-E)	160	80	80	80	1280	160	80	ND
Everglades Fe3-7c	(II)	1280	320	40	40	20	2560	20	20
Mucambo BeAn 8	(III)	20	20	ND	40	40	40	640	80
Pixuna BeAr 35645	(IV)	20	20	20	20	20	ND	40	20480

a/ Short-incubation hemagglutination-inhibition subtype (2).
b/ Reciprocal of highest antiserum dilution exhibiting total inhibition of 4 to 8 units of hemagglutinin.
c/ ND - Not determined.

(From Kinney and Trent, 9; by permission of Journal of General Virology.)

2.2. Structural analysis of VEE virus envelope glycoproteins

2.2.1. Tryptic peptides. Serologic analysis of the glycoproteins strongly suggested that the antigenic and, therefore, primary structure of the VEE virus E1 proteins are highly conserved (15). To determine the relationship between the primary amino acid

TABLE 5. Results of cross-testing VEE complex viruses by N using antisera prepared against the E2 glycoprotein.

Virus	---------- Antisera to glycoprotein of: ----------							
	TRD	PTF-39	P676	3880	Mena II	EVE	MUC	PIX
TC-83 (I-A)[a]	10240[b]	320	160	320	40	40	80	ND[c]
PTF-39 (I-B)	10240	640	160	320	20	20	40	ND
P676 (I-C)	2560	640	20480	320	10	40	160	10
3880 (I-D)	80	80	80	1280	80	10	40	ND
Mena II (I-E)	10	10	ND	ND	640	ND	80	ND
EVE FE3-7C (II)	320	640	40	10	10	5120	ND	ND
MUC BeAn 8 (III)	ND	10	10	40	40	10	1280	80
PIX BeAr 35645 (IV)	ND	ND	ND	10	ND	ND	80	20480

a/ Short-incubation hemagglutination-inhibition subtype (2).

b/ Reciprocal of highest antiserum dilution (titer) which inhibited 80 percent or more of the p.f.u. used in the test.

c/ ND - not done.

(From Kinney and Trent, 9; by permission of Journal of General Virology.)

sequence of various VEE virus subtypes, virion structural proteins were purified, digested with trypsin, and mapped by high-performance liquid chromatography (HPLC) (17; Fig. 3). Tryptic maps of E1, E2, and C proteins of VEE IA virus revealed that the primary amino acid sequence structure of glycoproteins E1 and E2 were more similar to each other than they were to the capsid protein.

The HPLC tryptic map analysis of the E1 glycoproteins of VEE viruses representing each of the subtypes revealed that the tryptic peptides of IB, IC, and ID viruses were identical to those of IA virus (Table 6). Tryptic peptide maps of the E1 glycoproteins of IE, subtype III, and subtype IV viruses were very different from those of IA and from each other. The E1 glycoprotein tryptic maps of Everglades virus, subtype II, differed from that of IA by only four tryptic peptides.

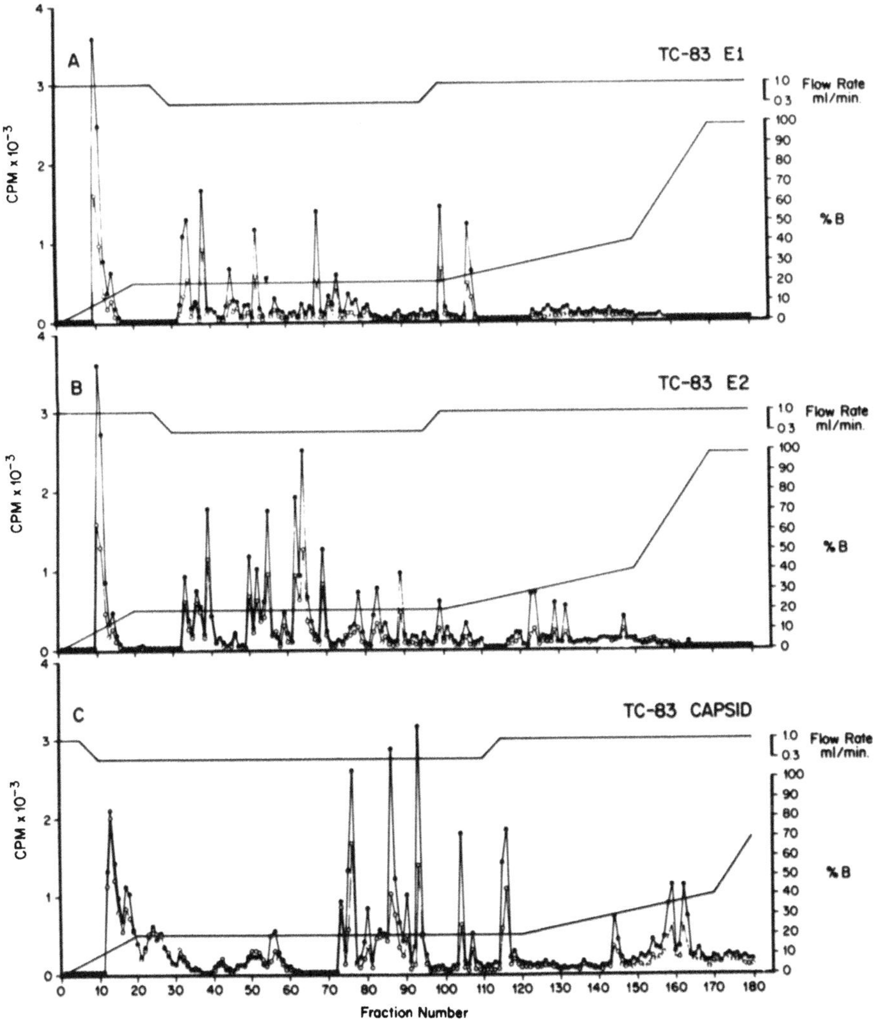

FIGURE 3. Tryptic maps of the purified structural proteins of TC-83 virus [^{14}C]. Lysine (open circles) and [^{3}H] lysine- (closed circles) labeled E1 envelope glycoprotein (A), E2 envelope glycoprotein (B), or capsid protein (C) were codigested with trypsin, and the resulting fragments were resolved by microprocessor-controlled high pressure liquid performance (HPLC). Fractions were collected every 0.5 min directly into scintillation vials, and the total radioactivity per fraction was determined. Percent B = percentage of acetonitrile in the mobile phase, which consisted of varying proportions of acetonitrile and sodium phosphate buffer (100 mM monosodium phosphate brought to pH 2.1 with phosphoric acid, 210 mM final phosphate concentration). (From Kinney and Trent, 17; by permission of Virology.)

TABLE 6. Tryptic peptide map analysis of VEE virus structural proteins by high performance chromatography.

Virus	Antigenic Subtype	Peaks Different From -- IA Virus TC-83 -- Capsid	E1	E2	--Similarity Ratio--a/ Capsid	E1	E2
TC-83	IA	0	0	0	100	100	100
PTF-39	IB	2	0	1	96	100	98
P676	IC	2	0	8	96	100	86
3880	ID	0	0	6	100	100	88
Mena II	IE	8	18	14	84	63	72
Everglades	II	6	4	9	88	92	82
Mucambo	III	9	16	24	79	58	52
Pixuna	IV	9	14	24	79	65	51

a/ 100 x (total ^{3}H peaks + total ^{14}C peaks) - (number of different peaks)/total ^{3}H peaks + total ^{14}C peaks.

(From Kinney and Trent, 15; by permission of Virology.)

The HPLC elution profiles of tryptic digests of the E2 glycoproteins of VEE viruses IA and IB viruses were similar, showing only one peptide difference (Table 6). The HPLC profiles of the E2 peptides of IC and ID viruses showed eight and six differences from that of IA virus, respectively. Digests of subtype IE contained at least 14 tryptic peptides which were different from the IA virus.

Analysis of E2 glycoproteins of subtype II, III, and IV viruses by tryptic peptide mapping revealed that subtype II virus E2 shared many more peptides with IA than did either subtype III or IV. The tryptic maps of type III and IV virus E2 glycoproteins were very distinct, containing very few peptides which eluted in coincidence with IA or each other (Table 6).

The nucleocapsid proteins of IA, IB, IC and ID viruses produced essentially identical tryptic peptide digestion products (Table 6). In contrast, the nucleocapsid tryptic peptide profiles of IE virus Mena II, Everglades virus subtype II, Mucambo subtype III and subtype IV virus Pixuna differed from that of IA virus by six to nine peptides (Table 6).

2.2.2. Amino-terminal sequence analysis. The amino-terminal amino acid sequence of the envelope glycoproteins of VEE and eastern encephalitis viruses were determined using a microsequenator automated Edman degradation (16). The amino terminal sequences of these viruses were compared to data published for Sindbis (19,20), Semliki Forest (21), and WEE viruses (22) (Fig. 4). As expected, there are no glycosylation sites or extensive hydrophobic regions in these portions of the VEE or other alphavirus glycoproteins. The extent of homology in the N-terminal sequences of WEE, Sindbis, VEE, EEE, and SFV viruses is very striking (Fig. 4). Although numerous amino acid changes have occurred in both E1 and E2 alphavirus glycoproteins, comparison of the sequence data shows that the proteins share regions and features of homology. Particularly striking is the conservation of cysteine residues involved in disulfide bridges which link distal regions of polypeptides. Their conservation implies that the overall three-dimensional conformation of the alphavirus glycoproteins is conserved. Of particular interest are the first few amino-acid residues of E1 and E2, because these are generated by proteolytic cleavage during the processing of the precursor polyprotein (23). The N-terminal amino acid sequences of E1 of the six viruses are not identical, although the differences observed are largely conservative. The first three N-terminal residues of the E1 glycoproteins of all seven viruses are tyrosine or phenylalanine, glutamic acid and histidine. Conservation of these N-terminal amino acids may indicate an important role in virus maturation or biological function. The aromatic amino acids, tyrosine or phenylalanine, have been conserved in several positions in E1 and E2 as have been proline and cysteine. Although all amino acid residues contribute to protein conformation, proline or cysteine residues confer major alterations in conformation. Antigenic reactive sites in native proteins can occur at bends in the polypeptide chain and antigenic reactivity is altered by conformational changes (24). The significance of these conserved or altered sequences in the antigenic and biological function of alphavirus glycoproteins must await further characterization.

The serologically-related SIN and WEE viruses show considerable homology in both E1 and E2 envelope glycoproteins. The E2 protein of these two alphaviruses has undergone a deletion of three amino acid codons at positions 8, 9, and 10 (Fig. 4).

```
E1
VEE/TRD     Y E H A T T M R ? Q A G I S Y N T I V N R A G Y A P L P I S I T P T K I K L I P T P N L E Y V T C H Y K T G V ? ? P T Y
VEE/TC-83   Y E H A T T M P S Q A G I S Y N T I V N R A ? Y A P L P I S I T ? T K I I L
SIN         Y E H A T T V P N V P Q I P Y K A L V E R A G Y A P L N L E I T V M S S E V L P S T N Q E Y I T C K F T T V V P S P K I
WEE         F E H A T T V P N V P G I P Y K A L V E R A G Y A P L N L E I T V V S S E L T P S T N K E Y V T C K F T T V I P S P Q V
SF          Y E H S T V M P N V V G F P Y K A H I E R P G Y S P L T L Q M Q V V E T S L E P T L N L E Y I T C E Y K T V V P S P Y V
EEE         Y E H T A V M P N K V G I P Y K A L V E R P G Y A P V H L Q I Q L V N T R I I P I T N L E Y I ? C K

E2
VEE/TRD     S T E E L F K E Y K L T R P Y M A R C I R C A V G - S C H T P I A I E A V K
VEE/TC-83   S T E E L F N E Y K L T R P Y M A R C P R C A ? (P - S P H Y P ? A)
SIN         S V I D D F T - - - L T S P Y L G T C S Y C H H T E P C F S P V K I E Q V W D E A D D N T I R I Q T S A Q F G Y D Q S G
WEE         S I T D D F T - - - L T S P Y L G F C P Y C R H S A P C F S P I K I E N V W D E S D D G T I R I Q V S A Q F G Y N Q A G
SF          S V S Q H F N V Y K A T R P Y I A Y C A D C G A G H S C H S P V A I E A V R S E A T D G M L K I Q F S A Q I G I D K S D
EEE         D L D T H F T Q Y K L A R P Y I A D C P N C M H S
```

FIGURE 4. N-terminal amino acid sequences of the E1 and E2 envelope glycoproteins of Venezuelan equine encephalitis viruses, strains Trinidad donkey (TRD) and TC-83, western equine encephalitis (WEE) virus, Sindbis (SIN) virus, Semliki Forest (SF) virus, and eastern equine encephalitis (EEE) virus, strain Ten Broeck. The single letter amino acid code is used: A, Ala; C, Cys; D, Asp; E, Glu; F, Phe; G, Gly; H, His; I, Ile; K, Lys; L, Leu; M, Met; N, Asn; P, Pro; Q, Gln; R, Arg; S, Ser; T, Thr; V, Val; W, Trp; Y, Tyr.

The N-terminal sequences of the E1 glycoproteins of VEE TRD and TC-83 viruses are nearly identical, exhibiting differences at only two positions; arginine and lysine of TRD have been substituted with proline and isoleucine at positions 8 and 37, respectively, in TC-83 virus. Although a lysine substitution has occurred in the E1 protein, no differences in the tryptic peptide maps of TRD and TC-83 E1 polypeptides have been detected (25). The E2 proteins of these two VEE viruses show more variability. Amino acids lysine (position 7) and isoleucine, position 20 in TRD have been substituted with asparagine (N) and proline (P), respectively, in TC-83. It appears that a glycine at position 25, cysteine at position 28, and threonine at position 30 in TRD have been substituted with proline, proline, and tyrosine respectively, in TC-83 virus. One or more of these differences may in part be responsible for the differences observed in the tryptic peptide maps of the E2 glycoproteins of TRD and TC-83 viruses (25).

2.3. Detailed antigenic characterization of VEE glycoproteins using monoclonal antibodies

2.3.1. Mapping antigenic epitopes. The antigenic structure of the gp56 (E2) and gp50 (E1) envelope glycoproteins of VEE virus has been mapped using monoclonal antibodies. From these studies we have been able to characterize the evolutionary antigenic drift among the VEE viruses, identify the critical antigenic sites involved with *in vitro* N and HA, and determine the epitopes responsible for eliciting a protective immune response (9).

The specificity of the monoclonal antibodies was determined using purified VEE structural proteins in their native configuration in an ELISA or protein A radioimmune precipitation. The class, subclass, and idiotype of each antibody was also determined. The characteristics of some of the monoclonal antibodies which were selected for further analysis are shown in Table 7. Most of the monoclonal antibodies isolated reacted with either gp50 or gp56 and were IgG subclass γ 2A.

2.3.2. Antigenic epitopes on VEE virus envelope glycoproteins. To understand the antigenic drift occurring within the surface glycoproteins of VEE alphaviruses, the binding characteristics of selected monoclonal antibodies with viruses in the VEE complex were analyzed by ELISA (Table 8).

Monoclonal antibodies to TC-83 virus glycoprotein could be clustered into one of seven distinct reactivity groups. Based on their cross-reactivity patterns, monoclonal antibodies binding with gp56 could be divided into at least three antigenic epitopes and those reacting with gp56 divided into four separate antigenic epitopes. Monoclonal antibodies specific for gp50[d] also reacted with Sindbis, eastern equine encephalitis, and Semliki Forest alphaviruses (data not shown). This suggested that less antigenic drift had occurred in the gp50 glycoprotein as indicated by the presence of a broadly cross-reactive epitope (10,14, unpublished data).

Monoclonal antibodies representing each of the gp50 and gp56 epitopes were tested for their ability to neutralize TC-83 virus infectivity or to inhibit virus mediated agglutination of goose erythrocytes (Table 9). Only those antibodies specific for the gp56^{c} epitope neutralized virus infectivity or inhibited hemagglutination to high titer indicating that this epitope is responsible for *in vitro* biological functions of N and HA. Epitope gp50^{b} also demonstrated HA and N activity, but to very low titer. These results indicate that the neutralization and hemagglutination sites on the VEE/TC-83 envelope glycoprotein gp56 must be equivalent.

TABLE 7. Characteristics of some TC-83 monoclonal antibodies.

Clone	Specificity	Class	Subclass
	gp50	IgG	γ1A
3B2A-9	gp50	IgG	γ2A
3B4C-4	gp56	IgG	γ1
3B6A-10	gp50	IgG	γ2A
3A3D-9	gp56	IgG	γ2A
3A2D-10	gp56	IgM	ND[a]
5B6A-6	gp50	IgG	γ2A
1A1D-4	gp56	IgG	γ2A
3B1A-5	gp50	IgM	ND[a]
1B2B-4	gp50	IgG	γ2A
1B1C-1	gp50	IgG	γ2A
5B4D-3	gp56	IgG	γ2A
4A1C-3	gp56	IgG	γ2A

a/ ND - not determined.

TABLE 8. Epitopes on gp50 and gp56 VEE virus glycoproteins determined with representative anti-TC-83 virus monoclonal antibodies using ELISA.

Representative Hybrid Cell Line	Epitope	Reaction of Monoclonal Antibody With Virus Subtype								
		IA	IA	1B	1C	1D	1E	2	3	4
5B4D	gp56^{a}	+	-	-	-	-	-	-	-	-
2A4B-12	gp56^{b}	+	+	+	-	-	-	-	-	-
3B4C-4	gp56^{c}	+	+	+	+	+	-	+	-	-
3B2D-5	gp50^{a}	+	+	+	±	+	-	-	-	-
3B2A-9	gp50^{b}	+	+	+	+	+	-	+	-	-
5B6A-6	gp50^{c}	+	+	+	+	+	+	+	+	-
3A5B-1	gp50^{d}	+	+	+	+	+	+	+	+	+
Epitopes shared with TC-83 virus		7	6	6	4-5	5	2	4	2	1

(From Roehrig *et al.*, 10; by permission of Virology.)

TABLE 9. Biological reactivities of anti-TC-83 monoclonal antibodies.

Antibody Source	Epitope	Titer of Ascitic Fluid[a/] HAI	N
5B4D-6	gp56[a]	<10	<10
2A4B-12	gp56[b]	<10	<10
3B4C-4	gp56[c]	2×10^6	2×10^6
3B2D-5	gp50[a]	<10	<10
3B2A-9	gp50[b]	160	160
5B6A-6	gp50[c]	<10	<10
3A1C-12	gp50[d]	<10	<10

a/ Ascites contained 7 mg/ml antibody. HAI was done against 4 HA units of antigen. Neutralization tests were performed with 60 plaque-forming units/test and 70 percent plaque reduction endpoints.

(From Roehrig *et al.*, 10; by permission of Virology.)

To determine the biological cross-reactivity of the gp50[b] and gp56[c] epitopes, HAI and N endpoint titrations were done using members of the VEE complex of alphaviruses (Table 10). Monoclonal antibody to IA epitope gp56[c] demonstrated HAI activity with VEE virus subtypes IA, IB, and IC. Virus subtypes ID and type II reacted with the IA gp56[c] antibody at lower titer; other subtypes were not reactive with this antibody. Antibody to the gp50[b] epitope did not react with the viruses in the HAI test. The infectivity of IA, IB, and IC viruses was neutralized to equal titer by the IA gp56[c] monoclonal antibody; however, ID and subtype II viruses were less efficiently N by this antibody. Subtype IE, III, and IV viruses were not N *in vitro* by this antibody. Antibody to the gp50[b] epitope did not efficiently N any of the viruses.

2.3.3. Spatial arrangement of epitopes on E2 (gp56) and E1 (gp50) glycoproteins. Competitive binding assays were done to determine the spatial arrangement of the antigenically defined epitopes. Competition was determined by comparing conjugate-binding-inhibition using immune and nonimmune ascites to inhibit the binding of various highly avid monoclonal antibodies (9) (Table 11). Competitive inhibition patterns indicated a close spatial

TABLE 10. Cross-reactivity of hemagglutinin and neutralization epitopes determined with monoclonal antibodies 3B4C-4 (gp56[c]) and 3B2A-9 (gp50[b]).

	TITERS OF ASCITES[a/]			
	---- HAI[b/] ----		---- N[c/] ----	
Subtype	gp56[c]	gp50[b]	gp56[c]	gp50[b]
1A (TC-83)	1×10^6	80	1×10^5	10
1A (Trd)	1×10^6	80	1×10^5	<10
1B	1×10^6	160	1×10^5	10
1C	1×10^6	160	1×10^5	10
1D	3×10^4	40	100	<10
1E	<10	40	10	<10
2	160	80	1×10^4	10
3	<10	<10	10	10
4	<10	40	<10	<10

a/ Ascites contained 7 mg/ml in the HAI test and 0.7 mg/ml in the N test.

b/ HAI was done against 4 HA units of antigen.

c/ N test was done by plaque assay with 20-60 plaque-forming units (PFU)/test and 70 percent plaque reduction endpoints.

(From Roehrig, *et al.*, 10; by permission of Virology.)

arrangement between most of the epitopes. Competition between gp56 epitopes a and c, and a and b, but not b and c indicates that gp56[b] and gp56[c] are most probably located on opposite sides to gp56[a]. Similarly, competition between gp50 epitopes b and c, and b and d, but not c and d suggests that gp56[c] and gp50[d] are probably located on sides opposite to the gp50[b] epitope.

Cross-competition between gp50[b] and gp56[c] monoclones was also observed (Table 11). These results could indicate that either these two epitopes are located proximal to each other within the glycoprotein spike structure or share amino acid sequences responsible for antigenic cross-reactivity. Because monoclonal antibodies directed against gp50[b] and gp56[c] will not precipitate purified isolated heterologous proteins; therefore, it is unlikely that sequence identities exist in gp56 and gp50 which would give antigenic homologies.

TABLE 11. Competitive binding assay mapping of gp56 and gp50 epitopes using representative monoclonal antibodies.

Epitope Specificity of Competitor[a/]	Competition Reaction with Enzyme-Conjugated Monoclonal Antibody					
	3A3D-9	3B4C-4	3B2D-5	3B2A-9	5B6A-6	3A5B-1
Controls						
Anti-VEE MHIAF	+	+	+	+	+	+
Control MHIAF	-	-	-	-	-	-
Tissue culture media	-	-	-	-	-	-
Anti-gp56[a] 3A3D-9	+	+	-	-	-	-
Anti-gp56[b] 2A4B-12	+	-	-	-	-	-
Anti-gp56[c] 3B4C-4	+	+	-	+	-	-
Anti-gp50[a] 3B2D-5	-	-	+	-	-	-
Anti-gp50[b] 3B2A-9	-	+	-	+	+	+
Anti-gp50[c] 5B6A-6	-	-	-	+	+	-
Anti-gp50[d] 3A5B-1	-	-	-	+	-	+

a/ All competing antibodies were purified from ascitic fluids and standardized to 1 mg/ml concentration.

(From Roehrig et al., 10; by permission of Virology.)

A hypothetical diagram of the organization of the antigenic epitopes on gp50 and gp56 within the viral surface spike structure is shown in Figure 5.

2.4. VEE glycoprotein epitopes involved in protective immunity

2.4.1. Passive protection. The availability of monoclonal antibodies with known specificities has provided an unique opportunity to investigate virus disease processes and the role of the humoral response in protective immunity. Three-week-old mice

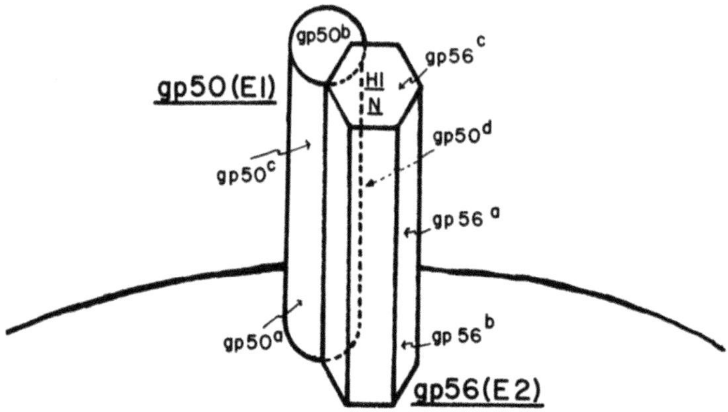

FIGURE 5. Hypothetical topological arrangement of the gp50 and gp56 antigenic epitopes on the virion spike based on competitive binding assays between monoclonal antibodies to these glycoprotein epitopes. (From Mathews and Roehrig, 11; by permission of Journal of Immunology.)

were passively immunized with various concentrations of monoclonal antibody to the critical N site $gp56^c$ and then challenged with increasing concentrations of virulent VEE virus (Table 12). Studies on clearance of passively administered monoclonal antibody in mice demonstrated that antibody was not cleared during the period of virus challenge (11). At concentrations of 5 and 20 μg of antibody to epitope $gp56^c$, most of the animals survived challenge with lower doses of viruses. Approximately 50 percent of the animals survived challenge with 10 million LD_{50} of virus when given as little as 5 μg of $gp56^c$ antibodies. Only 20 percent of those given 0.5 μg of antibody survived challenge with 100 LD_{50}. These results do not fit a typical dose response curve.

To determine whether the protection observed in animals passively immunized with monoclonal antibody was due to the passive antibody, and not the hosts own immune response, 3-week-old animals were immunosuppressed with cyclophosphamide (Table 13). Mice were treated with cyclophosphamide, passively immunized with 10 μg of monoclonal antibody to VEE virus epitope $gp56^c$, and then challenged with virulent virus. All animals were protected by the passive antibody. No survivors were observed among the cyclophosphamide-treated virus-infected animals or virus-infected control group.

TABLE 12. "In vivo" titration in 3-week-old mice of the protective capability of passively administered antibody against various VEE virus doses.

Passive[a] Antibody (μg)	------ I.P. LD_{50} of VEE Virus Challenge ------					
	10^2	10^3	10^4	10^5	10^6	10^7
20	10/10[b]	10/10	9/10	8/10	5/10	6/10
5	10/10	9/10	9/10	9/10	8/10	7/10
1	2/10	1/10	0/10	0/10	0/10	0/10
0.5	0/10	0/10	0/10	0/10	0/10	0/10
PBS	0/10	0/10	0/10	0/10	0/10	0/10

a/ Anti-gp56[c] (3B4C-4) was used as the passive antibody source given 24 hr before virus challenge.
b/ Survivors/total mice inoculated.
(From Mathews and Roehrig, 11; by permission of Journal of Immunology.)

TABLE 13. Effect of cyclophosphamide treatment on the immune response and in passive protection.

Immunosuppressive Treatment	Survivors	-- Antibody at Day 5 -- Post-Infection[a]		
		ELISA	HAI	N
Cyclophosphamide[b] + PBS + virus	0/10 [c]	ND[d]	ND	ND
PBS + virus control	0/10	ND	ND	ND
Cyclophosphamide + anti-gp56[c] + virus	10/10	+	+	ND
Anti-gp50[c] + virus	10/10	+	+	ND

a/ 20 μg antibody inoculated 24 hr before challenge.
b/ 400 mg/kg cyclophosphamide given i.v. 48 hr before i.p. inoculation of 100 i.p. LD_{50} VEE virus.
c/ Survivors/total mice inoculated.
d/ ND - not done.
(From Mathews and Roehrig, 11; by permission of Journal of Immunology.)

Animals which had received passive antibody had no measurable heterotypic antibody on the 5th day postinfection indicating that the host's humoral response was minimal during the first 5 days of infection. This indicated that passive antibody to the critical neutralizing site gp56[c] had protected the animals from disease.

Monoclonal antibodies to six of the seven VEE glycoprotein antigenic epitopes were used in passive immunization studies to determine which of the other epitopes might be involved and the role of antibody avidity in providing in vivo protection (Table 14). Antibody to gp50[a] did not protect mice when given as much as 100 μg of antibody protein. Only highly avid antibody to the critical neutralization site gp56[c] produced by clone 3B4C-4 protected in vivo although both 3B4C-4 and 4A1C-3 antibodies neutralized VEE virus in vitro. These two gp56[c] antibodies have different binding avidities. Previous studies with VEE glycoproteins indicated that gp50 plays no role in in vitro neutralization, nor does it induce protective immunity (26). In vivo passive immunization with monoclonal antibodies to gp50 support this idea; however, the highly avid antibody 3B2A-9 to gp50[b] protected mice at antibody concentrations similar to that which were observed for gp56[c]. Two other highly avid anti-gp50 antibodies (5B6A-6 and 3A5B-1) protected 40 and 80 percent of the mice when they were passively administered 100 μg of antibody. There was no isotype correlation with protection as both IgG1 and IgG2A subclass antibodies protected mice.

To further define the role of gp50 in the host immune response, 3-week-old mice were passively given polyvalent rabbit subunit antibody to VEE virus gp50 and gp56 proteins. Separate groups of mice were given antibody to gp50 and gp56 glycoproteins and challenged with 1,000 LD_{50} of virus. None of the animals receiving gp56 antibody died and 80 percent of the mice receiving anti-gp50 antibodies survived the virus challenge.

2.4.2. Dynamics of virus replication and humoral immune response in passively immunized animals. Although mice passively protected did not develop overt symptoms of disease, the spleen, brain, and serum were tested for the presence of virus during the first 5 days following virus inoculation and on the 17th day (Fig. 6). The control group of mice was administered virus but no

TABLE 14. Evaluation of TC-83 anti-epitope monoclonal antibodies in passive protection.

Hybrid Cell Line	Subclass of Antibody	Epitope	K[b/]	μg Antibody Inoculated i.v.[a/] 100	50	10
5B4D-6	IgG 2A	gp56[a]	10.1	0/10[c/]	0/10	0/10
3B4C-4	IgG 1	gp56[c]	11.1	10/10	10/10	10/10
4A1C-3	IgG 2A	gp56[c]	8.2	0/10	0/10	0/10
3B2D-5	IgG 2A	gp50[a]	10.1	0/10	1/10	0/10
3B2A-9	IgG 2A	gp50[b]	10.1	10/10	10/10	4/10
3B6A-10	IgG 2A	gp50[b]	9.2	0/10	0/10	0/10
5B6A-6	IgG 2A	gp50[c]	11.5	4/10	0/10	0/10
3A5B-1	IgG 1	gp50[d]	11.1	8/10	0/10	0/10

a/ Animals were inoculated with antibody 24 hr prior to challenge with 100 i.p. LD_{50} VEE virus. Results expressed as surviving animals.

b/ Log binding constant (M^{-1})

c/ Survivors/total

(From Mathews and Roehrig, 11; by permission of Journal of Immunology.)

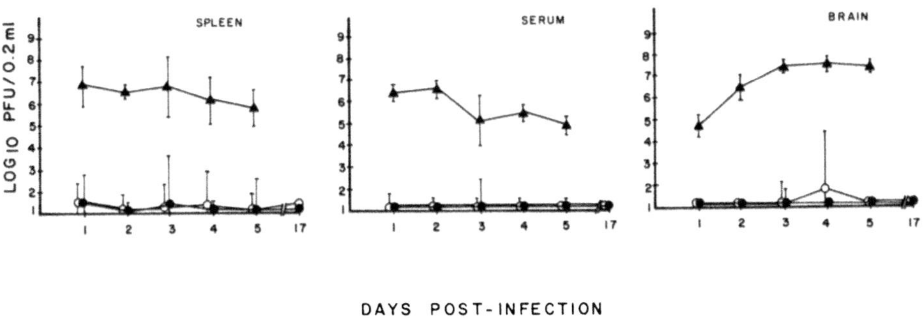

FIGURE 6. Viremia studies with VE (TRD) virus in various tissues in 3-week-old NIH white Swiss mice. Triangle, control mice; closed circles, mice given 20 μg of anti-gp56[c] 24 hr before virus challenge; open circles, mice given 40 μg of anti-gp50[b] 24 hr before virus challenge. N = 10 at each day. Points represent mean and standard deviation. (From Mathews and Roehrig, 11; by permission of the Journal of Immunology.)

passive antibody. In the control group, levels of virus in the spleen and serum peaked 1-2 days after infection; virus in the brain was highest in titer 3-5 days post-infection. Tissues of most of the mice given protective levels of anti-gp56^c (3B4C-4) and anti-gp50^b (3B2A-9) monoclonal antibodies and then challenged with 100 $IPLD_{50}$ of TRD virus did not have any detectable virus. Occasionally, however, there was 10^4 PFU/0.2 ml of virus present in the tissues of some mice which showed no overt signs of illness. The amount of passive antibody which was administered apparently did not neutralize all of the virus in the inoculum but limited the degree and rapidity of virus replication in the host. All of the animals which were given gp56^c antibody were devoid of virus on the 17th day following the infection.

Mice initially given gp56^c antibody and challenged with TRD virus were screened for anti-VEE antibody 25 days post-inoculation and given a secondary virus challenge. All of these animals survived the secondary virus challenge. Antibody levels were considerably higher in the passively immunized animals which were given the secondary virus challenge than they were in the control group which had been given passive antibody on day 1 but no primary virus challenge. Residual passive antibody, therefore, could not account for endpoint ELISA titers of $1x10^3$ to $1x10^4$ which were observed. Animals which were given large amounts of passive antibody appeared to have residual passive antibody which prevented virus replication so that no immune response occurred following secondary challenge. Therefore, unless high concentrations of passive antibody are administered to abrogate virus replication, secondary challenge results in limited virus replication, and the animals respond immunologically to the antibody restricted infection.

2.4.3. Challenge with other subtypes of VEE virus. Mice given protective amounts of monoclonal antibodies to gp50^b and gp56^c epitopes were challenged with 100 LD_{50} of VEE viruses representing subtypes IA through E and subtype II (Table 15). The *in vitro* ELISA binding values and N test cross reactivities of the gp56^c antibody correlated well with the protective effectiveness of this gp56^c antibody. Anti-gp50^b antibody did not neutralize virus *in vitro*.

However, all of the passively immunized animals challenged with IA, IB, or IC viruses survived the challenge. Epitope gp50^b protected animals also survived challenge by more distantly related VEE viruses 3880 (ID) and Mena II (IE). Animals challenged with Everglades virus (Fe3-7c) were not protected by antibody to IA epitope gp50^b. Mice given passive antibody to gp56^c were protected from challenge by viruses of subtypes IA, IB, IC, and subtype II virus Fe3-7c. Antibody to the critical neutralization site gp56^c did protect 50 percent of the animals from disease when challenged with ID virus but did not protect the animals when infected with Mena II virus (IE).

TABLE 15. Cross-protection of 3-week-old mice by passive transfer of monoclonal antibody followed by challenge with various VEE subtypes.a/

Virus Challenge (100 i.p.LD_{50})	Virus Subtype	ELISAb/	N	Survivors/ Totals/	PBS Control Survivors
		----- Anti-gp56^c -----			
Trd (20 PFU)c/	IA	77	10^5	10/10	0/10
PTF-39 (16 PFU)	IB	81	10^5	10/10	0/10
P676 (100 PFU)	IC	70	10^5	9/10	0/10
3880 (1 PFU)	ID	80	10^2	5/10	0/10
Mena II (10^6 PFU)	IE	19	10	0/10	0/10
Fe3-7c (330 PFU)	II	86	10^4	10/10	0/10
		----- Anti-gp50^b -----			
Trd (20 PFU)		77	<10	10/10	
PTF-39 (16 PFU)		78	10	10/10	
P676 (100 PFU)		66	10	10/10	
3880 (1 PFU)		86	<10	8/10	
Mena II (10^6 PFU)		19	<10	4/10	
Fe3-7c (330 PFU)		55	10	0/10	

a/ Passive immunization with 10 μg of either anti-gp56^c (3B4C-4) or anti-gp50^b (3B2A-9) antibody per mouse.

b/ ELISA values indicate percent of cross-reactivities where TC-83 reactivity = 100 percent. Cross-reactivities greater than 50 percent indicte relatedness to the TC-83 antigen.

c/ PFU=plaque forming units in Vero cell culture per 100 i.p. LD_{50}.

(From Mathews and Roehrig, 11; by permission of Journal of Immunology.)

3. DISCUSSION

The results presented in this report indicate that we have been able to assign specific biological and immunological functions to the VEE virus envelope glycoproteins. The N-terminal sequence data from the VEE glycoproteins have allowed us to make some limited comparisons with the corresponding proteins of SIN, SFV, EEE, and WEE alphaviruses (19-22). In particular, the cysteine residues at the N-termini of the alphavirus gp56 glycoprotein are invariant. The degree of homology in the amino acid sequence at the N-terminal of both glycoproteins lends support to the concept that the alphaviruses have descended from a common ancestor (22). The pattern of amino acid substitutions in the envelope proteins of the five alphaviruses provides evidence that SIN and WEE viruses arose from a common immediate ancestor, whereas SFV, EEE, and the VEE viruses evolved from a common ancestor but through a different lineage leading to the emergence of distinct virus complexes. Amino acid and nucleotide sequence homologies are in agreement with immunological studies of antigenic relationships (25). The VEE virus serologic subtypes I through IV appear to have evolved from a distant common ancestor with the variants I-A, I-B, I-C, and I-D representing descendents from a more recently evolved ancestor from which subtype II, Everglades virus has emerged. Mena II, variants I-E is the most divergent of the VEE subtype I species.

Attempts to assess the relationships of alphaviruses at the nucleic acid hybridization (29,30) or RNase T1 oligonucleotide map level (15,29,31) have been relatively uninformative probably due to the degeneracy of the genetic code. Thus, additional protein sequence information obtained from cloned viral cDNA will be necessary before the evolution and antigenic relatedness of the alphaviruses will be fully understood.

The envelope gp50 and gp56 glycoproteins of VEE viruses contain at least four and three epitopes respectively. Our data indicate that the $gp50^{b}$ and $gp56^{c}$ epitopes are closely associated within the intact virion spike. This proximal association of antigenic sites could explain the ability of anti-$gp50^{b}$ non-neutralizing antibody to inefficiently neutralize viral infectivity and block agglutination of red blood cells.

The site-specific N and HAI epitope $gp56^c$ is equivalent for subtype I varieties A, B, and C. The reactivity of ID virus with the IA epitope in the HAI test is less, and the virus is not neutralized by $gp56^c$ antibody This indicates that the N and HAI sites on ID virus are not identical. Everglades virus is neutralized by antiserum to the IA epitope $gp56^c$, but this antibody does not block hemagglutination of Everglades virus. This situation is a reversal of that observed with ID virus indicating that the HA and N sites on subtype II virus are split and not antigenically identical as they are for the IA, B, and C viruses.

Only highly avid antibody to the critical *in vitro* N and HAI epitope $gp56^c$ protects animals from challenge with virulent VEE IA virus. However, highly avid antibody to the $gp50^b$ epitope, which is located proximal to the $gp56^c$ epitope was nearly as effective in providing *in vivo* protection. Thus, both epitope specificity and antibody avidity are important in determining *in vivo* protection. Low-avidity antibodies to the critical neutralization site $gp56^b$ or the proximal $gp50^b$ provided little protection *in vivo*.

The passive administration of high titer antibody prior to infection prevented activation of the host own immune response to VEE virus. We postulate that this passive immunization provided antibody which could complex with virus with subsequent aggregation of virus particles, since the high-avidity anti-$gp56^c$ antibody-virus dose relationship was asymmetrical (32-34). Upon secondary virus challenge, the residual passive antibody modulated virus replication long enough for the host's own immune system to respond.

Monoclonal antibody to SIN virus gp53 (E1) which contains the HA but not N determinant has been observed to have N activity suggesting a proximal relationship between the N epitope on gp56 (E2) and the HA site on the gp50 protein (34). Schmaljohn *et al.* (35) have recently reported that non-neutralizing antibodies to SIN virus gp50 protein protected mice from virus challenge; however, no epitope assignments were described, and the amount of antibody administered was not reported. We have also found that administration of large doses of antibody to gp56 (E2) epitopes other than to the critical N site $gp56^c$ will protect. In our study the anti-gp50[b] antibody did bind to virus and protect animals from *in vivo* virus challenge although this antibody did not N *in vitro*. Subtype ID virus, which binds IA gp56[c] antibody in the HAI test but is not N *in vitro* by these antibodies, is not efficiently N *in vivo* by gp50[b] antibody as are variants IA, IB, and IC. Thus, it appears that antibody to the alphavirus hemagglutinin may be involved in providing protective immunity only when the critical N site and HA site are proximal to each other on the glycoprotein spike.

Currently we are cloning and sequencing the 42S genome of VEE strains TRD, TC-83, and Fe3-7c. These experiments should provide clearer insight into the genetic variation of the VEE complex and the molecular basis for changes in virus virulence. We are also using monoclonal antibodies specific for the various VEE subtypes to define antigenic drift within the VEE complex and its effect on resistance and recovery from virus disease. These studies should aid us in developing more effective disease control measures through the production of synthetic vaccines and improved rapid viral diagnosis for this important viral pathogen.

4. REFERENCES

1. Kubes, V. and Rios, F.A. Science 90:20-21, 1939.
2. Young, N. and Johnson, K.M. Am. J. Epidemiol. 89:386-307, 1969.
3. Shope, R.E., Causey, O.R., DeAndrade, A.H.P. and Theiler, M. Amer. J. Trop. Med. Hyg. 13:723-727, 1964.
4. Lord, R. Bull. Pan Amer. Hlth. Org. 8:100-110, 1974.
5. Yershov, F.I., Sokolova, T.M. and Titenko, A.M. Acta Virol. 21:213-221, 1977.
6. Mussgay, M. and Weibel, J. Virology 16:52-62, 1962.
7. Pederson, C.E., Jr. and Eddy, G.A. J. Virol. 14:740-744, 1974.
8. France, J.K., Wyrick, B.C. and Trent, D.W. J. Gen. Virol. 44:725-740, 1979.
9. Kinney, R.M., Trent, D.W. and France, J.K. J. Gen. Virol. 64:135- 147.
10. Roehrig, J.T., Day, J.W. and Kinney, R.M. Virology 118:269-279, 1982.
11. Mathews, J.H. and Roehrig, J.T. J. Immuno. 129:2763-2767.
12. Compans, R.W. Nature 229:114-116, 1971.
13. Dalyrmple, J.M., Teramoto, A.Y., Cardiff, R.D. and Russell, P.K. J. Immunol. 109:426-433, 1972.
14. Dalyrmple, J.M., Vogel, S.N., Teramoto, A.Y. and Russell, P.K. J. Virol. 12:1034-1042, 1973.

15. Trent, D.W., Clewley, J.P., France, J.K. and Bishop, D.H.L. J. Gen. Virol. 47:261-282, 1979.
16. Laemmli, U.K. Nature 227:680-685, 1970.
17. Kinney, R.M. and Trent, D.W. Virology 121:345-362, 1981.
18. Hunkapiller, M.W. and Hood , L.E. Science 207:523-525, 1980.
19. Bell, J.R., Hunkapiller, M.W., Hood, K.E., and Strauss, J.H. Proc. Natl. Acad. Sci. U.S.A. 75:2722-2726, 1978.
20. Rice, C.M. and Strauss, J.H. J. Mol. Biol. 150:315-340, 1981.
21. Garoff, H., Frichauf, A.-M., Simons, K., Lehrach, H. and Delius, H. Nature 288:236-241, 1980.
22. Bell, J.R., Bond, M.W., Hunkapiller, M.W., Straus, E.G., Strauss, J.H., Yamamoto, K. and Simiza, B. J. Virol. 45:708-714., 1982.
23. Schlesinger, M.J. and Kääriäinen, L. In: The Togaviruses (Ed. R.W. Schlesinger), Academic Press, Inc., 1980, pp. 371-392.
24. Wiley, D.C., Wilson, I.A. and Skehel, J.J. Nature 289:373-378, 1981.
25. Mecham, J.O. and Trent, D.W. Virology 121:251-261, 1982.
26. Pedersen, C.E. and Eddy, G.A. J. Virol. 14:740-744, 1974.
27. Calisher, C.H., Shope, R.E., Brandt, W., Casals, J., Karabatsos, N., Murphy, F.A., Tesh, R.B., and Wiebe, M.E. Intervirol. 14:229-232, 1980.
28. Cancedda, R., Villa-Komaroff, L., Lodish, H.F. and Schlesinger, M. Cell 6:215-222, 1975.
29. Wengler, G., Wengler, G. and Filipe, A.R. Virology 78:124-134, 1977.
30. Rentier-Delrue, F. and Young, N.A. Virology 106:59-70, 1980.
31. Trent, D.W. and Grant, J.A. J. Gen. Virol. 43:365-381, 1980.
32. Massey, R.J. and Schochetman, G. Virology 115:20-32, 1981.
33. Cooper, N.R. In: Comprehensive Virology (Eds. H. Fraenkel-Conrat and R.R. Wagner), Plenum, N.Y., 1979, pp. 123-158.
34. Chanas, A.C., Gould, E.A., Clegg, J.C.S., and Varma, M.G.R. J. Gen. Virol. 58:37-46, 1982.
35. Schmaljohn, A.L., Johnson, E.D., Dalrymple, J.M. and Cole, G.A. Nature 297:70-72, 1982.

22

ISOLATION AND CHARACTERIZATION OF THE VIRAL RNA DEPENDENT RNA POLYMERASES FROM CELLS INFECTED WITH EITHER THE MAHONEY OR SABIN ATTENUATED(LSc,2ab) STRAINS OF POLIOVIRUS TYPE 1

JAMES B. FLANEGAN AND CONCEPTA A. CLIFFORD
Department of Immunology and Medical Microbiology
University of Florida; Gainesville, Florida 32610;USA.

Summary

The poliovirus RNA-dependent RNA polymerase was isolated from cells infected with either the Sabin (LSC,2ab) or Mahoney strains of type 1 poliovirus. Both polymerases had similar levels of activity on all templates tested. Increasing the reaction temperature decreased the activity of both polymerases. No significant difference was found in the temperature sensitivity of the two enzymes. The size of the product RNA was measured by electrophoresis in CH_3HgOH-agarose gels. Both enzymes were shown to synthesize full-sized product RNA and to have similar elongation rates at 30°C. Thus no significant difference was found between the *in vitro* activities of the two polymerases.

Introduction

Poliovirus is an enterovirus and is one of a large number of human pathogenic viruses that are members of the picornavirus family. Three strains of poliovirus have been identified, and all three types are known to cause paralytic poliomyelitis. Attenuated strains of poliovirus were developed for vaccine production , but the molecular basis for their loss of pathogenicity has not been determined. To approach this problem, we have been studying the molecular mechanisms involved in the replication of poliovirus RNA. Poliovirus has a single-stranded RNA genome of positive polarity that has a 3' terminal poly(A) sequence and a small viral protein (VPg) covalently linked to its 5' terminal end (1,2). The replication of the viral RNA requires an RNA-dependent RNA polymerase that is found only in the cytoplasm of infected cells. To study the mechanism of viral RNA replication *in vitro*, we have purified the polymerase as a soluble and template dependent enzyme. We and others have shown that only one virus-specific protein, p63(also designated as NCVP4 and P3-4b), copurifies with the polymerase activity(3,4). In the presence of an oligo(U) primer, the purified polymerase will synthesize full-sized copies of poliovirion RNA(5) as well as several other types of viral and cellular RNA that contain 3'terminal poly(A) sequences (6). In the absence of oligo(U), the completely purified polymerase requires a cellular protein fraction or host-factor to initiate RNA synthesis _in vitro_(7).

The complete nucleotide sequences for both the virulent Mahoney strain of type 1 poliovirus and the Sabin attenuated strain, LSc-2ab, have been determined(8-10). Because the LSc strain was derived from the Mahoney strain, a comparison of the two sequences should indicate what amino acid changes have taken place during the attenuation process. The sequences indicate that 21 amino acid replacements have taken place with 12 of these in the capsid proteins and 4 in the polymerase protein(8). At the present time, it is not known which of these changes are responsible the attenuation of the virus. In this study, we have isolated the polymerase from cells infected with either the Mahoney or LSc strains of the virus and have compared the _in vitro_ activities of the two enzymes.

Polymerase Isolation

Stocks of the Mahoney strain virus and the LSc-2ab strain virus were prepared by infecting HeLa cell monolayers at 36°C. The yield of infectious virus with the LSc strain was about ten-fold less than the yield obtained with the Mahoney strain (Table 1). The virus stocks were concentrated by centrifugation and titered at 36° and 40°C. As expected, the titer of the LSc strain at 40° was about two logs lower than the titer obtained at 36°.The titer of the Mahoney strain virus remained constant at both temperatures. The concentrated virus stocks were then used to infect HeLa cell suspension cultures as previously

TABLE 1. Recovery of infectious virus and polymerase activity from poliovirus infected cells.

Strain	Virus Yield (pfu/cell)	Polymerase activity (units*/10^8 cells)
Mahoney	600	7.3
LSc,2ab	40	8.8

*One unit of activity is equivalent to the incorporation of 1 nmole of UMP into product RNA in 30 min.

described(11). Cytoplasmic extracts of the infected cells were prepared at 5 hr pi , and the polymerase was purified as described(3). Polymerase activity was assayed using a poly(A) template and oligo(U) primer. The purification involved the preparation of a 200,000 x g supernatant of the cytoplasmic proteins (Fraction I), precipitation in ammonium sulfate (Fraction II), and chromatography on phosphocellulose (Fraction III). Although the yield of infectious virus was much lower with the attenuated strain of the virus, the total units of Fraction III polymerase recovered was about the same with both strains of the virus(Table 1).

Comparison of Polymerase Activities

The activities of the LSc strain polymerase and the Mahoney strain polymerase were measured on poliovirion RNA *in vitro*. About the same level of activity was found with both polymerases(Table 2). As previously described with the Mahoney strain polymerase, the addition of oligo(U) to the *in vitro* reaction stimulated the activity of the LSc polymerase 5-10 fold (Table 2).In a 1 hr reaction, both polymerases catalyzed the linear incorporation of [32P]UMP into product RNA in the primer-independent reaction (Fig 1). Thus, the LSc strain polymerase was able to carry out both the primer-dependent elongation reaction and the primer-independent initiation

TABLE 2. Activity ofpoliovirus polymerase(Fraction III) on poliovirion RNA at 30°C.

Strain	oligo(U)	[^{32}P]UMP incorporated cpm
Mahoney	+	164,200
"	-	18,900
LSc	+	166,300
"	-	29,600

reactions that were previously described for the Mahoney strain polymerase(5,7).

Effect of Temperature on Polymerase Activity

The LSc strain when compared to the Mahoney strain is temperature-sensitive for the production of infectious virus particles when grown at 40°C (see above). To determine if the LSc strain polymerase was more sensitive to temperature inactivation than the Mahoney strain polymerase, we compared the activities of the two enzymes as a function of temperature. When poliovirion RNA was used as a template with an oligo(U) primer, the elongation activity of both enzymes decreased by about the same amount as the assay temperature was increased between 30° and 45°C (Fig 2). To measure the effect of temperature on initiating polymerase activity, the activity of both enzymes was measured on poliovirion RNA in the absence of an added primer. The activity of both polymerases again decreased by about the same amount as the tempereature was increased(Fig 3). As an alternate approach to checking temperature-sensitivity, both polymerases were heat-inactivated at 40°C and then assayed for initiating activity at 30°C. Both enzymes were rapidly inactivated during the first min at 40°, and then at a much slower rate between 1 and 12 min (Fig 4). These results showed that the inactivation kinetics were similar for both enzymes, and suggested that the LSc strain polymerase may in fact be slightly less sensitive to heat inactivation than the Mahoney strain polymerase. These results suggested that the LSc polymerase was not temperature-sensitive and were consistent with the observation that similar amounts of labeled RNA were synthesized in cells infected with either strain of the virus at 40°C (Fig 5).

Size of Product RNA

The size of the product RNA was characterized by electrophoresis in 1% agarose gels containing 5mM CH_3HgOH. Under these conditions, the product RNA is completely denatured and migrates in the gel as a function of its size. We have previously shown that the Mahoney strain polymerase can synthesize full-sized copies of

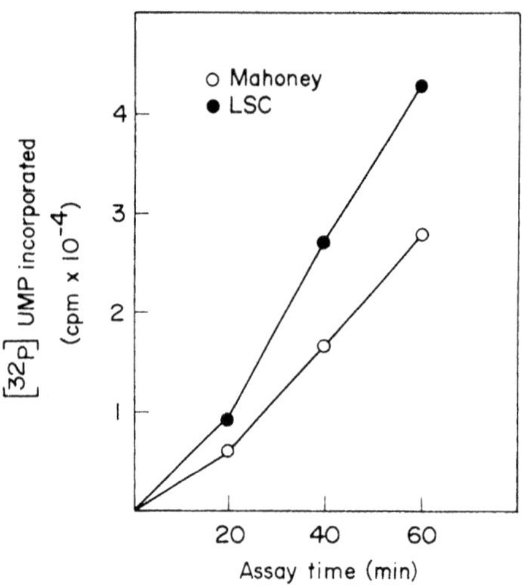

Fig 1. Activity of each polymerase on poliovirion RNA at 30°C.

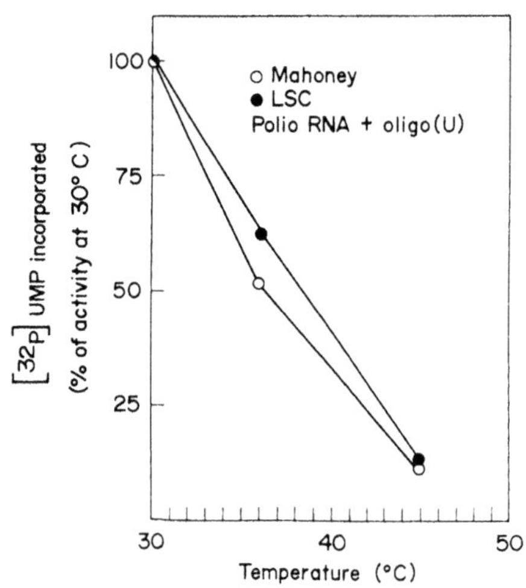

Fig 2. Effect of temperature on the activity of each polymerase.

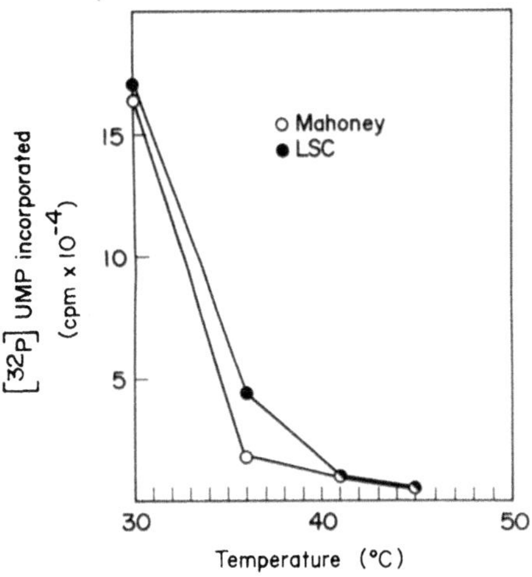

Fig 3. Effect of increasing temperature on the initiating activity of the polymerase. Poliovirion RNA was used without oligo(U).

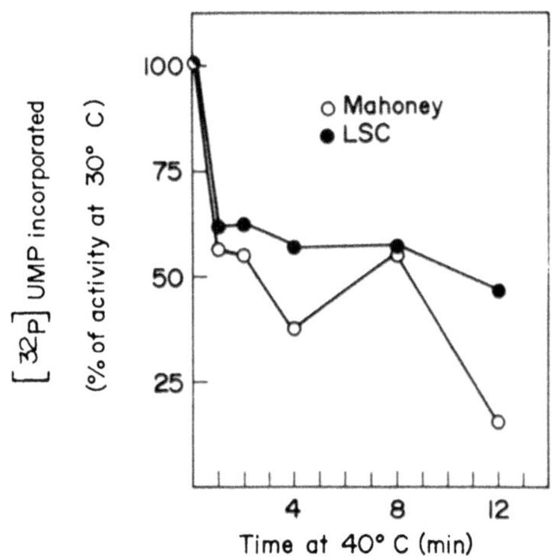

Fig 4. Effect of heat inactivation on the initiating activity of the polymerase. Each polymerase was heated at 40° as indicated and then assayed at 30°.

poliovirion RNA when oligo(U) is used as a primer(5). When tested in this study, the LSc strain polymerase was also found to synthesize full-sized copies of poliovirion RNA in the presence of oligo(U) at both 36° and 40°C (Fig 6). At 30°, full-sized product RNA was not synthesized in the 30 min reaction by either the LSc or the Mahoney strain polymerase (Fig 6). This was consistent with our previous results which indicated that more than 30 min are required under the reaction conditions used in this study (i.e. pH 8.0 and 3mM Mg^{+2}) for the synthesis of full-sized product RNA (5). In the 30° reaction, the largest product RNA synthesized by each polymerase was about the same size (Fig

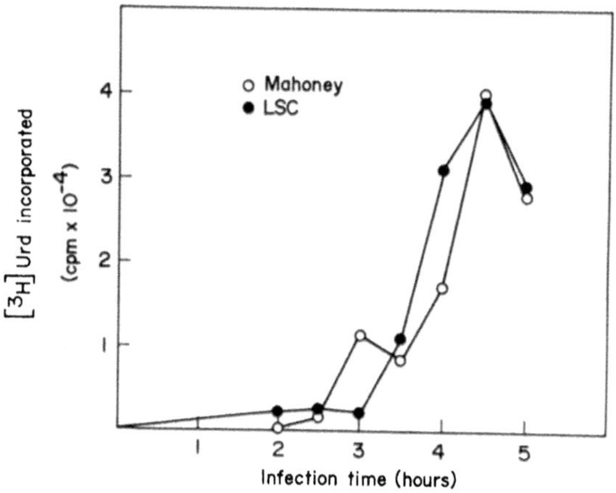

Fig 5. Synthesis of poliovirus RNA at 40° in cells infected with either the Mahoney or LSc strains of the virus.

6). This suggests a similar elongation rate for each polymerase.

Discussion

In this study, we were not able to detect any significant difference between the *in vitro* activities of the RNA polymerases that were coded for by the LSc and Mahoney strains of poliovirus type 1. The results suggest that the amino acid changes that were found in the attenuated strain polymerase had little effect on its enzymatic activity. This is consistent with the previous suggestion that the large number of mutations in the capsid region of the viral genome results in the attenuated phenotype of the LSc strain(8). It would be interesting to determine if the LSc strain could be further attenuated by specifically inserting mutations in the polymerase region of the viral genome which have an effect on the enzymatic activity of the polymerase.

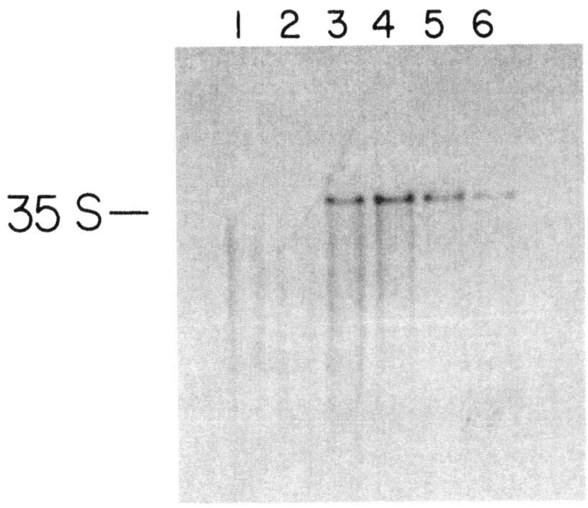

Fig 6. Electrophoresis on CH_3HgOH-agarose gel of labeled product RNA synthesized by LSc and Mahoney strain polymerases on poliovirion RNA template and oligo(U) primer. Gel contains the product RNA synthesized by the Mahoney polymerase at 30°(lane 1), 36°(lane 3), and 40°(lane 5), and the LSc product RNA at 30°(lane 2) at 36°(lane4), and 40°(lane 6).

References

1. Flanegan,J.B.,R.F.Petterson, V. Ambros, and D. Baltimore 1977. Proc. Natl. Acad Sci. USA 79:961-965.

2. Lee Y.F., A. Nomoto, B. Detjen, and E. Wimmer 1977. Proc.Natl. Acad. Sci. USA 72:4157-4161.

3. VanDyke,T.A., and J.B.Flanegan 1980. J.Virol 35:732-7450.

4. Baron,M.H., and D. Baltimore 1982. J.Virol 43:969-978.

5. VanDyke, T.A., R.J. Rickles, and J.B.Flanegan 1982. J.Biol.Chem. 257:4610-4617.

6. Tuschall,D.M., E.Hiebert, and J.B.Flanegan 1982. J.Virol. 44:209-216.

7. Dasgupta,A., P.Zabel,and D.Baltimore 1980. Cell 19: 423-429.

8. Nomoto,A., T. Omata, H. Toyoda, S.Kuge, H.Horie, Y.Kataoki, Y.Genba, Y. Nakano, and N.Imura 1982. Proc.Natl.Acad.Sci.USA 79:5793-5797.

9. Kitamura,N., B.Semler, P.G.Rothberg, G.Larson, C.Adler,

A.Dorner,E.Emini, R.Hanecah, J.Lee, S.van der Werf, C.Anderson, and E.Wimmer 1981. Nature(London) 291:547-553.

10.Racaniello,V.R. and D.Baltimore 1981.Proc.Natl.Acad.Sci. USA 78:4887-4891.

11.Villa-Komaroff,L., M.McDowell, D.Baltimore, and H. Lodish 1974. Methods Enzymol. 30F:709-723.

23

HERPES SIMPLEX VIRUS TYPE 1 THYMIDINE KINASE GENE CONTROLS VIRUS PATHOGENESIS AND LATENCY IN THE NERVOUS SYSTEM

Y. BECKER, Y. SHTRAM, J. HADAR, T. BEN-HUR, A. BARASOFSKY, Y. ASHER, E. TABOR, A. SCEMAMA and D.H. GILDEN*
Department of Molecular Virology, Faculty of Medicine, Hebrew University, P.O.B. 1172, 91 010 Jerusalem, Israel

SUMMARY

The pathogenicity of herpes simplex virus type 1 (HSV-1) was investigated in mice of different ages, using virus isolates that differed in the level of thymidine kinase (TK) activity. The TK^+ isolate from wild type virus was pathogenic in mice aged four to 28 days when inoculated onto scarified corneas. A mutant strain (designated TK¼) that had 25% of the TK activity of the TK^+ isolate was pathogenic for mice up to 10 days of age but was less pathogenic in older mice than the TK^+ virus. The TK^- virus with no TK activity killed mice that were four to seven days old,but mice older than 10 days survived. Infant mice with an undifferentiated nervous system are thus susceptible to the TK^- virus which replicates in the neurons. Latent virus could be isolated from the ganglia of adult mice infected with the TK^+ and TK¼ virus strains but not with the TK^- virus. The TK genes of the TK^+ and TK¼ virus strains were cloned in plasmid pBR322 and the degree of expression of these genes was investigated. A sequence of 800 bp upstream to the BglII-BamHI 2800 bp fragment containing the viral TK gene was found to affect the expression of this gene in the TK¼ virus strain.

A working hypothesis is presented for the molecular processes that may be involved in HSV-1 infection of the skin, trigeminal ganglia and brain.

INTRODUCTION

Herpes simplex virus type 1 (HSV-1) is a human pathogen that causes localized lesions in the skin and invades the nervous system by infecting nerve endings. The virus replicates in epithelial cells and in neurons and non-neural cells in the ganglia, causing ganglionitis. Subsequently, the

* Present address: Department of Neurology G1, University of Pennsylvania, Philadelphia, Pennsylvania

virus penetrates into the central nervous system (CNS), causing encephalitis, and establishes latent infections by residing in neurons in the form of naked viral DNA (1). Primary infections of the skin with HSV-1, as well as ganglionitis and encephalitis, involve a lytic interaction between the infecting virus and the cells, leading to virus replication, the production of virus progeny and cell death. The latent phase of the virus infection is an outcome of the lytic infection: the viral DNA interacts with the host cell (e.g. the neurons in the ganglia) in such a way that viral expression is controlled by the neuron. The latent infection is reversible, and the viral genome can be reactivated to replicate in some neurons in the ganglion, followed by transport of the virus through the nerve axons to the skin, where a lytic virus-cell interaction takes place and a lesion is formed (recurrent infection).

Marked advances have been made in recent years in understanding the nature of the lytic infection with HSV-1 in permissive cells. A number of viral genes have been identified and some were characterized at the level of the nucleotide sequences. Although not all of the events leading to the synthesis of virus progeny in permissive cells are known, it is clear that three classes of viral genes (α, β and γ) are involved in the regulation of viral synthesis (2). Far less is known about the molecular events that lead to the establishment of the latent state of the viral DNA in neurons in ganglia (and possibly in the brain). The exact mechanisms involved in the inflammatory process at the sites of virus infection (skin, ganglion and brain) that control the response of macrophages and lymphocytes to virus infection and lead to natural inhibition of the development of the lesion are not yet fully understood.

The interaction between HSV-1 and host cells depends on the nature of the viral glycoproteins. Thus, HSV-1 strains with mutations in the glycoprotein genes may be affected in the initial stage of the virus-cell interaction. The nature of the virus infection may also be controlled by the amounts of the glycoproteins produced by the virus mutants. Although mutations in genes involved in viral DNA replication could prevent the formation of virus progeny, it appears that the wild type HSV-1 strains affecting humans contain all the necessary enzymatic systems for virus replication, and they are able to replicate in permissive cells in the body. However, a different situation exists with neurons in the ganglia and CNS, since these cells are terminally differentiated and have no capacity to divide and duplicate their genomes. Such a biochemical situation poses difficulties for the herpesvirus which invades the neurons. If the viral genes are intact and able to supply the necessary building blocks for DNA and protein synthesis, then virus replication is possible. If, however, the invading virus is a mutant that lacks a gene (or genes) essential for the biosynthesis of viral DNA and proteins, then the viral replicative cycle will be affected or prevented.

A group of HSV-1 mutants with the ability to replicate in permissive cells,but not in differentiated cells like neurons, has been studied in recent years. These are the thymidine kinase-deficient (TK^-) mutants of HSV-1. IUdR-resistant mutants of HSV-1 were found to have lost their virulence for rabbits when inoculated into the eye (3). Similarly, TK^- mutants of HSV-1 and HSV-2 selected by bromodeoxyuridine (BUdR) were less virulent than the parental TK^+ virus (4). A detailed study by Tenser et al. (5) showed that TK^-, $TK^{\pm}$ and TK^+ HSV-1 strains that were able to replicate in the mouse eye to the same extent differed in their ability to infect the ganglia. The TK^- mutants of HSV-1 were rarely isolated from the trigeminal ganglia of mice. A relationship was found between the virus titer in the trigeminal ganglia and the ability of the virus to produce TK (5-7). Studies on infection of the superior cervical ganglia (SCG) revealed that TK^- mutants were unable to infect the ganglia, whereas the TK^+ virus strains were able to do so (8-10). It was not possible to reactivate TK^- mutant strains in the SCG after explantation and incubation with permissive cells in-vitro. Similarly, TK^- HSV-1 strains resistant to acyclovir were found to be unable to cause latent infections (11).

In our studies (12, 13), we have confirmed the above reports that TK^- virus mutants are apathogenic for mice and do not cause latent infections in the trigeminal ganglia after inoculation onto scarified corneas. It was possible to reactivate the virus from trigeminal ganglia of mice infected with our TK^+ virus strain isolated from wild type (w.t.) virus, but not with our TK^- mutant isolated in the presence of BUdR.

In addition, we studied the properties of a mutant of HSV-1 that had 25% of the TK activity of the TK^+ isolate (13). This mutant (designated TK¼) was able to establish a latent infection in the trigeminal ganglia of mice after inoculation onto scarified corneas. This and other studies (5, 7) showed that mutants of HSV-1 expressed the viral TK gene at levels ranging from 1% to 80% of that of the w.t. virus with 100% TK activity (ref. 4; R. Tenser and J. Jones, personal communication). Tenser et al. (5) established that there is a relationship between the ability of the virus to cause latent infections in the trigeminal ganglia and the level of TK activity, and showed that a low expression of TK was sufficient for the virus to establish a latent infection. The TK^- virus mutants were unable to cause latent infections.

These studies drew attention to the fact that HSV-1 needs to have an active TK gene for its neuropathogenicity. Since the TK is an enzyme that phosphorylates thymidine to thymidine monophosphate (TMP), differentiated neurons in the ganglia and the CNS probably cannot provide the virus with TMP; thus, only a virus that has an active TK gene will be able to replicate in neurons. According to the level of viral TK gene expression in the neurons, the virus will either be able to produce enough progeny to reach the CNS or there will only be sufficient virions to establish a latent infection. Another aspect is the nature of the mutation in the TK gene that causes reduced expression of the gene. Reduced expression means a range of 1% to 80% of the TK activity of the w.t. TK^+ virus (5). Since such mutants were obtained from natural virus populations without the use of mutagens, the HSV-1 strains with reduced TK expression are probably defective in a control mechanism of TK gene expression.

As opposed to the possible role of the viral TK gene in neurovirulence, far less is known about the expression of the cellular TK gene in the neurons in the ganglia. Studies done in rats revealed that TK activity in the CNS of newborn rats reached maximal levels on day 6 after birth and markedly decreased during the next 10 days (14, 15). We found no information regarding cellular TK gene expression in neurons in the ganglia. The relationship between the sensitivity of mice of different ages to infection with a TK^- mutant was studied in our laboratory (16), and the results showed that mice up to the age of about one week were sensitive to the TK^- mutant of HSV-1, whereas older mice (three to four weeks old) were resistant to the virus.

The role of the immune system, interferon, antibodies, as well as macrophages, in the host defense mechanism against the virus has been studied (1). The immune cell system response is essential in limiting the development of skin lesions and ganglionitis, but the mechanisms involved in the recruitment of lymphocytes at the lesion site and the development of the defensive response (interferon, anti-HSV-1 antibodies, cytotoxic T cells, macrophages) are not known.

The present communication summarizes our studies on the importance of the HSV-1 TK gene in the neuropathogenicity of the virus for mice and analyzes the nature of the mutations in the expression of the viral TK gene. A working hypothesis for the mechanisms involved in the response of the host defense system to the viral lesion is provided.

Neurovirulence of HSV-1 TK^+, $TK^{1/4}$ and TK^- strains

Virus strains

The large plaque variant of an NIH strain of HSV-1 was used in the study (12, 13). This virus (designated TK^+) has an active TK gene (Vmax 8.44×10^{-6} M/Min and Km 9.47×10^{-7} M). A TK^- mutant (designated TK^-) was isolated from BUdR-treated virus; no TK activity was found in mouse L(TK^-) cells infected with this virus. A virus strain which expresses 25% of the TK activity (designated $TK^{1/4}$), as compared to the TK^+ virus (Vmax 2.07×10^{-6} M/Min and Km 9.68×10^{-7} M) was also isolated. This mutant produces small plaques in BSC-1 monolayers, as compared to the TK^+ and TK^- virus strains that produce large plaques, but the plaque morphology is independent of the TK gene activity (13). The specific activity of the TK induced in mouse L(TK^-) cells by the TK^+ virus was 4.1 pmol of ^{3}H-TMP/μg protein/hr, while that of the $TK^{1/4}$ virus was 0.96 pmol of ^{3}H-TMP/μg protein/hr.

Pathogenicity of the virus strains for mice of different ages

The results presented in Fig. 1 demonstrate the pathogenicity of the three virus strains for mice of different ages. The mice were inoculated onto scarified corneas with one drop of virus suspension with a titer of 10^7 pfu/ml and observed for three weeks, after which percent mortality was determined. The TK^+ virus was highly pathogenic for mice of all ages (four days to four weeks). The $TK^{1/4}$ virus was also highly pathogenic for mice up to 10 days of age, but its pathogenicity decreased in mice aged three to four weeks. The TK^- mutant was pathogenic in four to seven-day old mice, but the pathogenicity decreased with age. Mice older than 10 days were fully resistant to the virus. These results indicate that HSV-1 strains expressing the viral TK can replicate in the neurons of the ganglia and CNS in adult mice with a differentiated nervous system, even though they have a fully developed immune system. The TK^- mutant, on the other hand, cannot replicate in mice with a differentiated nervous system in which there is no TK activity. However, when the nervous system is

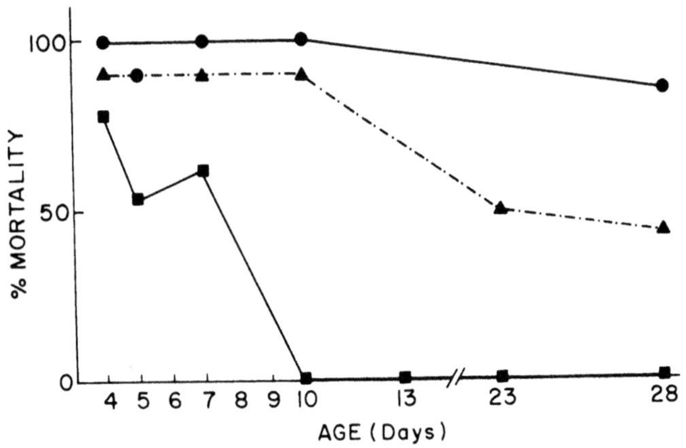

FIGURE 1. Neurovirulence of HSV-1 strains in mice of different ages. Mice were inoculated onto scarified corneas with 10^7 pfu/ml of the TK^+ (●), TK¼ (▲) and TK^- (■) virus strains.

undifferentiated, as in young mice, the TK^- virus probably multiplies by obtaining the necessary TMP through the activity of the cellular TK enzyme.

Latency of the HSV-1 strains

The three HSV-1 strains were found to differ in their ability to establish latent infections in the trigeminal ganglia of mice inoculated as above. Of the mice that survived infection with the HSV-1 TK^+ virus (30% at an inoculation dose of 10^6 pfu/ml), 75% were found to have latent virus. In adult mice inoculated with the TK¼ strain, it was possible to reactivate latent virus from the trigeminal ganglia of 80% of the surviving mice three weeks after infection. Four months later, only 25% of the survivors retained reactivable latent viral DNA in the ganglia (13). TK^- mutant virus was not found in the trigeminal ganglia of four-week-old infected mice and could not be reactivated from the trigeminal ganglia after explantation and in-vitro incubation.

Thus, latency of HSV-1 in the trigeminal ganglia can only be achieved by virus strains able to express the TK gene. However, virus strains with a reduced level of TK expression are better able to establish latent infections than the highly pathogenic TK^+ strains which, after infecting the ganglia, migrate to the CNS where they cause encephalitis and death. Our

Table 1. Recombinant plasmids containing the HSV-1 TK gene.

Plasmid	Cloned fragment	HSV-1 strain	Size of cloned DNA (bp)
pBRTK[a]	BamHI-BamHI	F	3600
pBY16	"	TK^+	"
pBY20	"	TK^+	"
pBYAB53-12	"	TK¼	"
pBY718	BglII-BamHI from pBRTK	F	2800
pBYAB719	" from pBY16	TK^+	"
pBYAB720	" from pBY20	TK^+	"
pBYAB53-12-16	" from pBYAB53-12	TK¼	"

The BamHI/Q fragment of viral DNA was cloned into pBR322 at the BamHI site.

[a] Provided by Dr. H. Cedar (24)

studies showed that latency of the TK¼ strain follows a stage of ganglionitis when virus replication takes place (12, 13).

Organization of the HSV-1 TK gene

Cloning of the TK gene in pBR322 and expression in E. coli

To study the TK gene of HSV-1, it is possible to clone the BamHI/Q fragment of the viral DNA in plasmid pBR322. Reports from several laboratories (17-23) provided information on the cloning and expression of the HSV-1 TK gene in bacteria. In our studies (12, 23), the TK gene of the TK^+ virus was cloned and expressed in *E. coli*, using the tetracycline resistance promoter for the expression of the viral gene (plasmids pBY16 and pBY20, Table 1). Removal of 800 bp upstream to the viral TK gene (BamHI-BglII fragment), followed by recloning of the BglII-BamHI fragment of 2800 bp in pBR322 at the BamHI site, enhanced the expression of the TK gene fourfold in *E. coli* (23).

These results suggest that the 800 bp fragment (containing the TATTA box of the TK gene) has a sequence that reduces expression of the viral TK gene in *E. coli* (plasmids pBY718, pBY719 and pBY720, Table 1). To study the nature of such a sequence, as well as the reason for the reduced expression of the TK gene of the HSV-1 TK¼ strain, the BamHI/Q fragment of the TK¼ viral DNA was cloned in pBR322 (Table 1, pBYAB53-12). The plasmids were introduced into *E. coli tk*⁻ (Ky895) and the expression of the viral TK

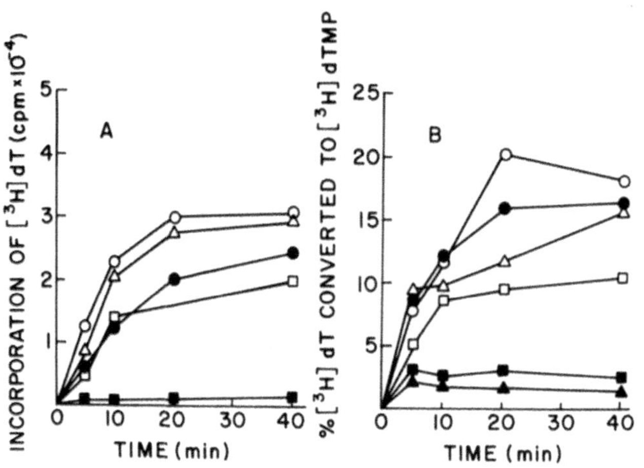

FIGURE 2. Expression in E. coli of HSV-1 TK isolated from HSV-1 strains and cloned in pBR322. The recombinant plasmids (Table 1) were transferred to E. coli tk^- (Ky895) and expression of the TK gene was determined by measuring the incorporation of [^{3}H]dT into cellular DNA (A) and by measuring the conversion of [^{3}H]dT to [^{3}H]dTMP (B). Symbols: (●) pBY718; (△) pBYAB719; (□) pBY16; (■) pBYAB53-12; (o) pBYAB53-12-16; (▲) pBR322

was determined by [^{3}H]dT incorporation and by determining the conversion of [^{3}H]dT to [^{3}H]dTMP. Fig. 2 shows that recombinant plasmid pBYAB53-12 did not express the TK gene obtained from the HSV-1 TK¼ mutant. However, cloning of the BglII-BamHI 2800 bp fragment containing the viral TK gene without the upstream 800 bp (plasmid pBYAB53-12-16) led to good expression in E. coli. This indicates that the 800 bp upstream sequence of both TK$^+$ and TK¼ virus strains contains a sequence which affects the expression of the TK gene when added to the BglII-BamHI 2800 bp fragment.

A possible attenuation sequence upstream of the TK gene

Fig. 3 provides information on the anatomy of the HSV-1 TK gene, which is a β viral gene. It was noted from the nucleotide sequence reported by McKnight (18) and Wagner et al. (22) that a palindrome could be identified on either side of the BglII site in the viral DNA. This palindromic sequence resembles an attenuator sequence in the gene trp of E. coli (25-27; reviewed in 28). It may be possible that an attenuator sequence is present half-way between the TATTA box and the codon of the first

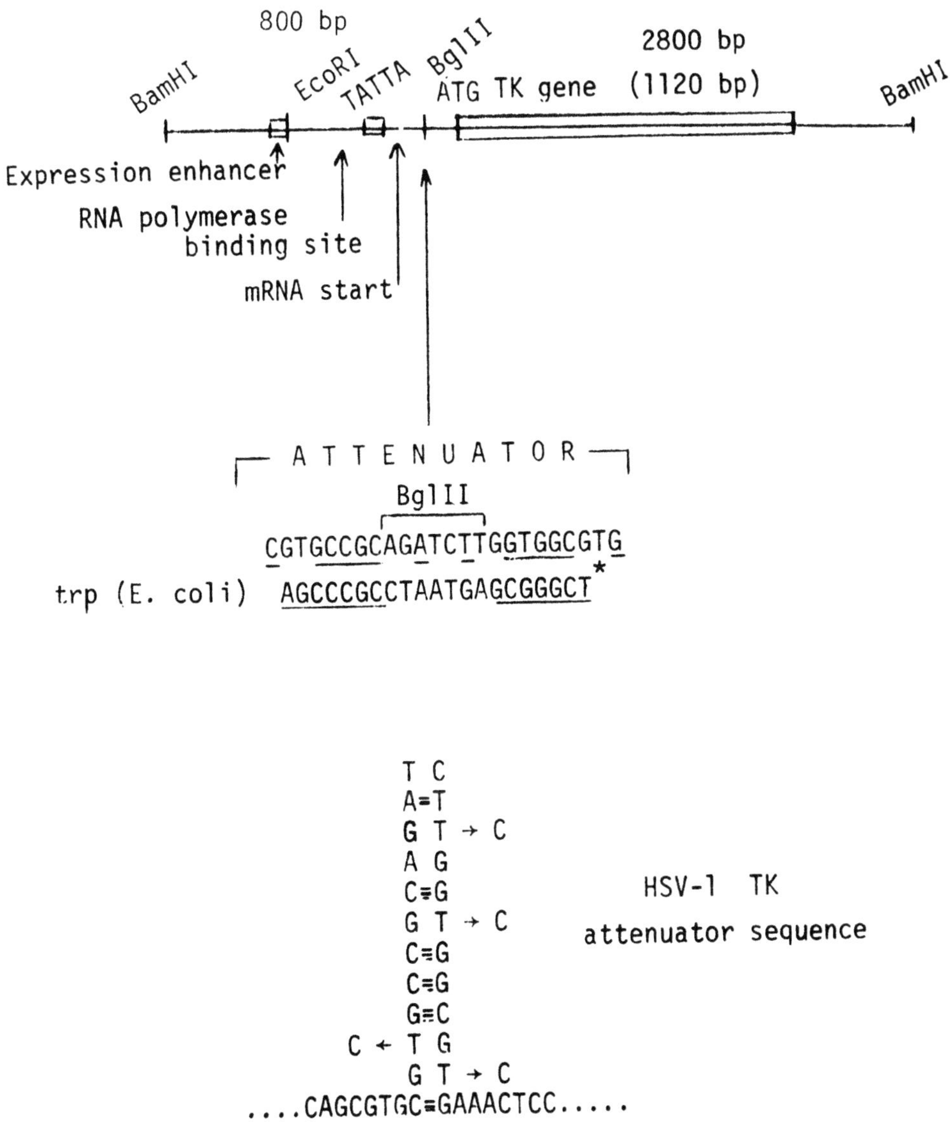

* See refs. 25-28.

FIGURE 3. Schematic representation of the HSV-1 TK gene and the upstream regulatory sequences.

methionine (ATG) of the viral TK gene. Mutations of thymidines to cytidines or vice versa in the attenuator sequence could change the efficiency of the attenuator sequence. A complete palindromic sequence may be a better attenuator, and expression of the viral TK gene could thus be markedly regulated. Sequencing of the attenuator sequence in HSV-1 TK½ is under way.

It is of interest that a similar attenuator sequence has been identified in the HSV-1 glycoprotein sequence, centrally located between the TATTA box and the codon for the first methionine (R.J. Frink, R. Eisenberg, G. Cohen and E.K. Wagner, personal communication).

DISCUSSION

The neurovirulence of three strains of HSV-1 that differ in the level of TK activity (TK^+, TK½ and TK^-) was studied in young mice with undifferentiated nervous and immune systems and in adult mice. Our studies confirm and extend the results obtained by other groups (4-7, 11) that the viral TK gene is essential for the establishment of a neuronal infection. In the absence of such a gene, HSV-1 TK^- mutants are unable to replicate in the trigeminal ganglia, although they replicate well at the site of infection (the eye) in adult mice. In young mice with undifferentiated nervous tissue, the HSV-1 TK^- mutant is highly pathogenic, suggesting that the cellular TK provides the needed TMP for the synthesis of the viral DNA and virus progeny in the nervous system.

A major problem is the identification of the neurons in the ganglia in which the virus DNA resides in a latent form. Identification of such neurons may provide a clue to the mechanism of virus reactivation. Recent studies on type C fibers and pain neurons reported that the undecapeptide substance P (SP) (Fig. 4) is the neurotransmitter in the pain neurons (29). It was also reported that SP resembles tuftsin (Fig. 4) in its ability to activate macrophages and human polymorphonuclear leukocytes (30). Studies by Openshaw et al. (31) provided information that HSV-1 migrates through myelinated fibers to the neurons, but no information is available as to whether HSV-1 can infect type C fibers and travel to the pain neurons in the ganglia. To study this possibility, we used mice treated with capsaicin (Fig. 4) that prevents the development of type C fibers and possibly the pain neurons (32). Mice aged two to three days were injected intraperitoneally with capsaicin, and six weeks later the mice were infected by inoculation onto scarified corneas with the TK^+ virus (10^7 pfu/ml). Fewer mice (50%) succumbed to the virus infection after capsaicin treatment, as compared with untreated

Tuftsin Thr-Lys-Pro-Arg

Substance P Arg-Pro-Lys-Pro-Gln-Gln-Phe-Phe-Gly-Leu-Met$(NH)_2$

Capsaicin

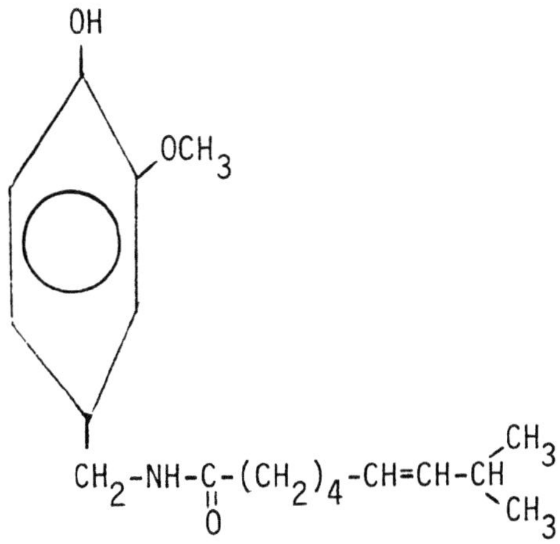

FIGURE 4. Structure of tuftsin, substance P and capsaicin

infected mice, of which 85-90% died (J. Hadar and Y. Becker, unpublished data).

These results suggest that mice lacking type C fibers are more resistant to virus infection, and that, therefore, the pain neurons might be involved in hosting the viral DNA. If this is the case (and more experiments need to be done), then activation of the virus from the pain neurons could occur by biochemical triggers like prostaglandins, that function in the skin at the site of the type C fiber nerve endings. In addition, substances that trigger the release of substance P internally, like fever and hormonal changes, might also reactivate the virus. Although little is known about the interaction of type C fibers and pain neurons with herpesviruses, the possibility of herpes simplex virus residing in such neurons in the ganglia may be worthwhile investigating.

Table 2 provides a working hypothesis for the possible molecular processes involved in HSV-1 infection of the skin, the ganglia and the CNS.

Table 2. Molecular processes involved in HSV-1 infection

A. Primary infection with the virus: lesion in epithelium

Primary virus infection of epithelial cells

↓

Virus replication in permissive cells
(viral TK gene expression unimportant)

↓

Expression of viral glycoprotein genes
determine lesion size and pathogenicity

↓

Destruction of epithelial cells
Axon endings revealed

↓

Infection of myelinated and unmyelinated (type C) fibers → Migration to the ganglion (see B)

Release of substance P (SP) to lesion

↓

Migration of lymphocytes to the lesion

Activation by SP (tuftsin-like)

↓

Phagocytosis of infected cells

Antibody production: virus neutralization

↓

Healing: activation of connective tissue fibroblasts

Regeneration of epithelium

Table 2 (cont'd)

B. Infection of neurons in ganglia and brain: ganglionitis, encephalitis and latency.

Migration of virions by axoplasmic flow

↓

Uncoated virions reach neurons in the ganglion

↓ ↓

Neurons with myelinated fibers → ↓ Release of neurotransmitter

Neurons (pain) with unmyelinated fibers (type C) → ↓ Release of neurotransmitter

(Both types of neurons →) Entry of viral DNA into nuclei of infected neurons (viral TK gene expression essential)

↓ ↓ ↓

Virus strains

Mice (days)	TK^+	TK^-	TK½
4	+	+	+
10	+	-	+
23	+	-	±
28	+	-	±
	Replication in ganglion	No replication	Low level of replication in neurons

TK^+: Replication in ganglion → ↓ Migration through axons to brain → ↓ Infection of neurons → ↓ Encephalitis

TK^-: No replication

TK½: Low level of replication in neurons → ↓ Latency of viral DNA (see C)

Release of neurotransmitter (neurons with unmyelinated fibers) → ↓

Lymphocytes
Migration to ganglia

↓

Activation of macrophages
Interferon production

↓

Removal of infected neurons and virus

↓

Healing of ganglion

Table 2 (cont'd)

C. Latency and reactivation of HSV-1

Neurons in ganglia

↓ Pain and heat neurons (Unmyelinated type C)

↓ Neurons with myelinated axons → unchanged (?)

Damage to skin

↓

Release of prostaglandins → (Pain and heat neurons)

Fever → (Pain and heat neurons)

Hormonal changes → (Pain and heat neurons)

↓

Effect on nerve endings

↓

Release of SP from neurons
(Expression of gene coding
for SP precursor protein)

↓

Activation of latent viral DNA

↓

Virus replication in neurons
(Viral TK gene expression necessary)

↓

Migration of virions to
epithelium through axons

↓

Lesion formation
(Recurrent infections)

↓

Recovery as in A

Further studies are needed to throw more light on the biochemical processes involved in latency and reactivation of the virus.

ACKNOWLEDGMENTS

These studies were supported by grants from the Chief Scientist, Ministry of Health, Israel, the Israel Academy of Sciences and Humanities, Basic Research Commission, and the Joint German-Israeli Research Program sponsored by the Gesellschaft fuer Strahlen- und Umweltsforschung, Muenchen.

The studies on cloning of the HSV-1 TK gene in pBR322 were done in collaboration with Drs. A. Cohen and A. Honigman, Department of Molecular Genetics, our medical school. Studies on the neurovirulence of HSV-1 strains were done in collaboration with Dr. Y. Gordon, Department of Ophthalmology, Hebrew University-Hadassah Medical Center, Jerusalem.

REFERENCES

1. Nahmias, A.J., Dowdle, W.R. and Schinazi, R.F. (Eds.) The Human Herpesviruses: An Interdisciplinary Perspective. Elsevier North Holland, Inc., New York, 1981.
2. Mackem, S. and Roizman, B. Proc. Natl. Acad. Sci. USA 77:7122-7126, 1980.
3. Marcialies, M.A., La Colla, P., Schivo, M.L., Flore, O., Firinu, A. and Loddo, B. Experientia 31:502-503, 1975.
4. Field, H.J. and Wildy, P. J. Hyg. Camb. 81:267-277, 1978.
5. Ressel, S. and Dunstan, M.E. Virology 112:328-341, 1981.
6. Tenser, R.B., Miller, R.L. and Rapp, F. Science 205:915-917, 1979.
7. Tenser, R.B. and Dunstan, M.E. Virology 99:417-422, 1979.
8. Price, R.W. Science 205:518-520, 1979.
9. Price, R.W. and Schmitz, J. Infec. Immun. 23:373-383, 1979.
10. Price, R.W. and Kahn, A. Infec. Immun. 34:571-580, 1981.
11. Klein, R.J., Friedman-Kien, A.E. and Destefano, E. Infec. Immun. 34:693-701, 1981.
12. Becker, Y., Shtram, Y., Snipper, D., Asher, Y., Tabor, E., Gordon, Y., Gilden, D., Wellish, M., Hadar, J., Becker, O., Cohen, A., Laban, A., and Honigman, A. In: Herpesvirus: Clinical Pharmacological and Basic Aspects (Eds. H. Shiota, Y-C. Cheng and W.H. Prusoff), Int. Congress Series 571, Excerpta Medica, Amsterdam-Oxford-Princeton, 1982,pp.57-73.
13. Gordon, Y., Gilden, D.H., Shtram, Y., Asher, Y., Tabor, E., Wellish, M., Devlin, M., Snipper, D., Hadar, J. and Becker, Y. Arch. Virol. (in press).
14. Weichsel, M.E. Jr. Brain Res. 78:455-465, 1974.
15. Yamagami, S., Mori, K., and Kawakita, Y. J. Neurochem. 19:369-376, 1972.
16. Ben-Hur, T., Hadar, J., Shtram, Y., Gilden, D.H. and Becker, Y. (Submitted).
17. Colbere-Garapin, F., Chousterman, S., Horodniceanu, F., Kourilsky, P. and Garapin, A-C. Proc. Natl. Acad. Sci. USA 76:3755-3759, 1979.
18. McKnight, S.L. Nucl. Acids Res. 8:5949-5964, 1980.
19. Garapin, A-C, Colbere-Garapin, F., Cohen-Solal, M., Horodniceanu, F

and Kourilsky, P. Proc. Natl. Acad. Sci. USA 78:815-819, 1981.
20. Hayward, G.S., Reyes, G.R., Gavis, E.R. and McKnight, S.L. In: Herpesvirus DNA (Ed. Y. Becker). Developments in Molecular Virology, Vol. 1, Martinus Nijhoff Publishers, The Hague-Boston-London, 1981, pp. 271-306.
21. Kit, S., Otsoka, H., Qavi, H., and Kit, M. Gene 16:287-295, 1981.
22. Wagner, M.J., Sharp, J. and Summers, W. Proc. Natl. Acad. Sci. USA 78:1441-1445, 1981.
23. Becker, Y., Shtram, Y., Honigman, A., Laban, A. and Cohen, A. Gene (in press).
24. Pollack, Y., Stein, R., Razin, A. and Cedar, H. Proc. Natl. Acad. Sci. USA 77:6463-6467, 1980.
25. Lee, F. and Yanofsky, C. Proc. Natl. Acad. Sci. USA 74:4365-4369, 1977.
26. Lee, F., Bertrand, K., Bennett, G. and Yanofsky, C. J. Mol. Biol. 121:193-217, 1978.
27. Squires, C., Lee, F., Bertrand, K., Squires, C.L., Bronson, M.J. and Yanofsky, C. J. Mol. Biol. 103:351-381, 1976.
28. Rosenberg, M. and Court, D. Ann. Rev. Genet. 13:319-353, 1979.
29. Bacaglini, P.I. and Hogan, P.B. Proc. Natl. Acad. Sci. USA 80: 594-598, 1983.
30. Fridkin, M. and Gottlieb, P. Mol. Cell. Biochem. 41:73-97, 1981.
31. Openshaw, H., Sekizawa, T., Wohlenberg, C. and Notkins, A.L. In: The Human Herpesviruses: An Interdisciplinary Perspective (Eds. A.J. Nahmias, W.R. Dowdle and R.F. Schinazi). Elsevier North Holland, Inc., New York, 1981, pp. 289-296.
32. Wall, P.D., Fitzgerald, M., Nussbaumer, J.C., van der Loss, H. and Devor, M. Nature 295:691-693, 1982.

24

ROUND TABLE DISCUSSION

chaired by

Stuart A. Aaronson, M.D.
National Cancer Institute
Bethesda, MD 20205, USA

Dr. Aaronson: We have observed at this meeting the power of viruses as tools to understand basic mechanisms by which genes in cells are regulated. Recent concepts that have emerged from studies of viruses, include activator/enhancer sequences as discussed by Dr. Gruss, attenuator sequences described by Drs. Aloni and Becker, and translational control mechanisms as summarized by Dr. Kaempfer. Virology has moved to the point where the replication process is pretty well understood in most viral systems. We are now starting to look at the interaction of viruses with the cell in terms of their pathogenicity.

In this round table discussion I thought it would be worthwhile to devote the time to discuss some of the strategies by which it has been possible to intercede in the processes leading to pathogenicity. We ought to start with one of the great successes of modern virology, the poliovirus system. Dr. Flanegan, in retrospect, why was it relatively easy to develop vaccines for poliovirus?

Dr. Flanegan: There are several reasons why it was possible to develop effective vaccines against poliovirus. In the first place, there are only three serotypes of the virus. This is an obvious advantage when compared to other viruses that have a large number of different serotypes. A second fact that helped in the production of these vaccines is the ease with which the virus can be grown in culture. This made it possible to grow large amounts of virus that could be used in vaccine production. Finally, the fact that poliovirus has no known animal reservoir has helped in the control of poliomyelitis by the widespread use of the poliovirus vaccines.

There are still some problems that are associated with the use of the live-attenuated and killed virus vaccines. Both vaccines are still used in different countries around the world. The killed-virus vaccine has proven to be effective in controlling poliomyelitis when 90% or more of the population in a given country receives a complete set of injections with the vaccine. The killed-virus vaccine, however, is more difficult and costly to administer than is the live-attenuated (oral) vaccine. The oral vaccine also produces longer lasting immunity and the secondary spread of the virus also results in the immunization of a larger percentage of the population. On rare occasions, however, the secondary spread of the virus has resulted in cases of vaccine-associated paralytic diseases. A clear example of this is when a non-immunized parent has been infected with the vaccine virus from their recently vaccinated infant. A challenge for the future is to further attenuate the strains used in the oral vaccine to eliminate all cases of vaccine-associated paralytic disease. Still another problem that has been associated with the use of oral vaccine has been its failure to induce satisfactory immunity when administered to populations in some underdeveloped warm-climate areas of the world. This may result from interference by other enteroviruses that are already present in the intestinal tract. Thus, another challenge for the future lies in the development of a vaccine that is also effective in these situations.

Dr. Aaronson: When a vaccine strain causes disease is it due to reversion in the vaccine strain or because it is not attenuated enough?

Dr. Flanegan: The virus can mutate as it passes through the human population, and it may in certain cases revert to a more virulent form. In addition, adults are more susceptible to developing paralytic disease than are young children. Thus, the vaccine virus itself may cause some paralytic disease in adults that would not be observed in children.

Dr. ter Meulen: Does vaccination with killed vaccine provide life-long immunity or do we have to repeat vaccination at intervals?

Dr. Flanegan: Four doses of the killed-virus vaccine are given to children. Adults are not usually revaccinated. Antibody titers in individuals vaccinated with the killed-virus vaccine drop to very low levels after about five years. It has been argued, however, that individuals without detectable serum antibody titers may still be protected from clinical illness by "immunological memory." The antibody titers produced by the killed-virus vaccine are lower than those produced by the live (oral) vaccine, but both have proven to be effective in preventing poliomyelitis.

Dr. Aaronson: Is the protection elicited humoral or cell mediated?

Dr. Berns: It is humoral.

Dr. ter Meulen: The observation of polio vaccination failures raises the question about the function of the immune system in these recipients. In children with immune defects it has been observed that enteroviruses can persist causing a chronic CNS infection if the immune system fails to mount a normal immune response. It is of interest, that to my knowledge, only Echo virus strains have been isolated. Enteroviruses do not normally persist, but cause a lytic infection.Other mechanisms must play a pathogenetic role in these disorders.

Dr. Bishop: In Munich Dr. Herbert Mayer-Ewert has been studying a case of an immunocompromised individual who is continuously excreting vaccine virus.

Dr. Choppin (to Dr. Flanegan): You mentioned that there were a number of mutations in the attenuated poliovirus capsid protein. If I remember correctly one of the first cleavages of the polio polyprotein is by a host protease, and the other cleavages are by virus protease. Are any of the changes in the capsid protein in the region of the cleavage by a host protease?

Dr. Flanegan: For the vaccine-strain of poliovirus type 1 the complete nucleotide sequence has now been determined. The cleavage pattern for the viral proteins has also been studied in some detail. The cleavage of the viral polyprotein is apparently carried out by a virus-specific protease. It has been argued that this protease is autocatalytic and may be able to cleave itself out of the original polyprotein. The alternative is that a host protease carries out the initial cleavage of the polyprotein. In any event, the protease coded for by the vaccine-strain virus has not been characterized. Future studies on this enzyme might be interesting since VP1 in the vaccine-strain virus is different in size (as measured by migration in a gel) from VP1 in the Mahoney-strain.

Dr. Aaronson: Would Dr. Choppin be willing to comment on the problems that exist in measles virus vaccine and what approaches are utilized?

Dr. Choppin: Measles live virus vaccine has been very effective where it has been used extensively. CDC has been attempting to eliminate measles from the USA by extensive vaccination. Had we known about SSPE before the introduction of live measles vaccine, there might be hesitation in its wide use. We now know that one does not get SSPE from the live vaccine. Early in the development of the vaccine Katz & Enders actually injected the attenuated strain of measles directly into the CNS of monkeys and showed that it did not replicate. Thus the live measles virus vaccine does not produce persistent infection that is observed in some cases of wild measles infection. With regard to paramyxoviruses in general, they all can produce persistent infection in one type of cell or another and many of them in some animals. Since we now know why previous inactivated vaccines for other paramyxoviruses failed and we know to which antigens immunity must be stimulated, we should re-evaluate the question of vaccines for respiratory syncytial virus and other paramyxoviruses. Some of us feel that the use of purified viral proteins will prove to be useful.

Dr. ter Meulen: With reference to the CDC program, I would like to remind you of the historical incidence of the Faröe Islands 140 years ago. At that time about 7,000 inhabitants were living on this island and acute measles had been absent for about 60 years. However, when measles virus was introduced by a visitor, 80% of the inhabitants came down with severe acute measles within a short time. Only those who had had measles more than 60 years ago escaped the infection probably as a result of a long persisting immunity against measles which lasted for more than 60 years in the absence of reinfection. However, to obtain such a lifelong immunity, antigenic restimulation has to occur and it has therefore been suggested that measles virus probably persists after acute measles infection. Yet, no information is available about the mechanisms by which measles virus persists or how it is reactivated. However, in respect to the vaccines discussed at the present time, consisting of purified viral proteins only, it has to be taken into consideration that long-lasting immune responses are necessary in order to obtain lifelong protection. Otherwise a new outbreak of acute measles by introduction of this contagious agent could occur.

Dr. Aaronson: What type of immunity does one obtain after measles virus vaccination?

Dr. ter Meulen: One obtains both, humoral and cellular immunity. Patients who have a T cell deficiency cannot respond with a cell-mediated immunity and normally come down with a severe acute measles infection associated with the variety of complications.

Dr. Aaronson: We are now learning more about molecular mechanisms in addition to the ability to prevent disease by vaccines. Now we should move to areas where knowledge has been gained concerning the complex interactions of the infecting virus with the host. Would Dr. Bishop be willing to comment about pathogenesis in relation to molecular mechanisms?

Dr. Bishop: I should like to address two virus systems: bunyaviruses and arenaviruses. In the bunyavirus group although there are many viruses only few are associated with diseases of economic importance to man. Some of them are serious diseases, such as La Crosse encephalitis, and Rift Valley Fever. Over the last few years, Rift Valley Fever virus found in Egypt a suitable vector and devastated both domestic animals as well as humans. There were some 600 - 700 people who died during the initial viral epidemic.

As to bunyavirus pathogenesis, we know very little. We have determined by genetic methods that the viral glycoproteins are the major determinants of viral pathogenesis. This appears to be true for both viscerotropic and neurotropic bunyaviruses. What cell receptors are addressed by the viral glycoproteins to allow invasion are not known.

The other virus family which is extremely important in terms of a viral disease is the arenavirus family. These viruses include Argentine hemorrhagic fever and the African Lassa fever. Attempts to produce vaccines against these viruses have so far not been particularly successful. Genetic studies indicate that the pathogenic abilities of these viruses are related to their S-RNA gene products (glycoproteins and nucleoproteins). It has yet to be elucidated how the interaction of these gene products with cells relate to cell invasion.

Dr. Rott: Any change in the viral genome will affect virulence or pathogenicity. One has to bear in mind, however, that reassortment in viruses *in vitro* might be artifacts; gene constellations are produced that make the viruses unable to grow in natural hosts. The studies with enveloped viruses indicate that glycoproteins are important determinants essential for pathogenicity. In the avian influenza system, cleavage is really necessary for the rapid spread of the virus in the organism. This situation might not be true for localized influenza infection in mammals, where the virus grows in epithelial cells of the respiratory tract and spreads and dissemination does not occur. It is not clear in this case what is the pathogenic factor in such an infection, is it hemagglutinin or neuraminidase, or perhaps are these other factors such as contaminant bacteria that provide the cleaving enzyme in normal non-permissive cells.

In other viruses the glycoproteins are cleaved but nevertheless there are differences in pathogenicity, and in such cases the receptors may be of importance. But our knowledge of the receptors as determinants of pathogenicity is very limited. It should be a goal of future investigations to define the relations between receptors and the viral glycoproteins.

Dr. Catravas: As a biochemist among virologists, I should like to ask whether one could use isotopically labeled viruses to find the fate of the viruses that become latent (e.g. herpes), or how a latent virus changes to a virulent one. It is known that radiation affects glycoproteins by splitting carbohydrates, so perhaps the use of radiation would throw some light on the function of viral glycoproteins?

Dr. Skehel: For the majority of viruses one would not unfortunately know whether one follows isotopically labeled virus that is biologically important or biologically irrelevant viruses. The particle/pfu ratio is therefore of great importance in such studies.

Dr. Catravas: What about double-label techniques?

Dr. Skehel: It is not the problem of labeling, but it is a problem of stability of the virus. The problem is not how to obtain radioactively labeled but biologically active virus.

Dr. Choppin: Richard Price of the Memorial Sloan Kettering Institute, working with herpes simplex virus has made use of the fact that only infected cells can phosphorylate certain antiviral drugs, by using a radiolabeled drug to find the infected cell in the nervous system, because only the infected cell will phosphorylate the drug and retain it in the cell. With computer assisted tomography and radioautography they can detect individual infected cells, and thus can follow in an animal the progression of the infection. This method may be useful for early diagnosis of herpes encephalitis without the necessity to perform brain biopsy.

Dr. Becker: In order to obtain the expression of the viral thymidine kinase gene, one requires the replication of the virus in the cell; if one speaks about silent viral DNA in the neurons, then one would not expect the TK gene to be expressed. What is the form of the latent DNA in the ganglion cells? As a result of our studies over the years, it was proposed that the infecting viral DNA molecules undergo intramolecular recombination (Becker, Y., J. Theoret. Biol. 75:339, 1978). Dr.N. Fraser (personal communication) studied the latent DNA in the brain stem in infected mice. He isolated the total DNA from the brain stem cells, cleaved it with restriction enzymes, separated the DNA fragments on agarose gels, and used radiolabeled viral DNA probes for hybridization. He found that the ends of the DNA molecule were not in the correct position in the gel, meaning that there was an intramolecular recombination event taking place in the viral DNA; this may have been the reason for the silent viral DNA not being expressed. The reactivation event must open up the integrated molecular ends of the viral DNA and release the DNA from arrest, in order to allow expression of the TK gene; only then would one be able to detect cells in which the virus replicates. Using DNA:DNA hybridization techniques, it is difficult to identify the cells in which the DNA is latent.

Now a few words about treatment of herpes simplex virus infections. One of the major problems in the treatment of herpetic encephalitis with araA is that it is usually given too late to the patient because of the delay in diagnosis. Only very early treatment of encephalitis patients with araA would have a therapeutic effect. As to the possibility of vaccination against herpes infections, Epstein, who discovered the Epstein-Barr virus (EBV), has suggested that, in order to afford protection against EBV infections, it is necessary to develop a vaccine against the virus. It is known that antibodies to EBV develop in cases of nasopharyngeal carcinoma, but whole virus vaccines cannot be used, even if the virus is inactivated. This is because the possibility always exists that the DNA in such inactivated virus may still be active and potentially oncogenic, or act as a helper virus for a cellular oncogene. It is, therefore, necessary to develop vaccines made of purified antigens, free of nucleic acids. There are attempts to develop a vaccine of this type against HSV-2 and to use it for the vaccination of prepubertal girls, and thus afford protection against the virus and reduce the incidence of genital herpes. As to HSV-1, which is pathogenic and infects practically all children before the age of 3, one may have a situation in immunocompromised children in which a normally innocuous herpes infection becomes fatal. Acyclovir is not used in such infections, since the virus settles in the ganglia and is not sensitive to this drug. Acyclovir is used to cure children with generalized herpes simplex virus infections. In view of all this, the current idea is to produce an efficient vaccine based on viral glycoproteins: it is necessary to isolate the appropriate gene from the viral DNA, clone it, and make it express itself in bacterial cells or mammalian cells. The product will be purified on columns with monoclonal antibodies to obtain a subunit protein vaccine. At present glycoprotein

D vaccine of HSV-1 has been shown to afford protection in mice against HSV-1 and HSV-2. The same idea may be applied to varicella zoster virus (VZV) and EBV; critical experiments in humans, however, have not yet been done.

Dr. Aaronson: Does anyone have any experience with peptide vaccines that can elicit good antibodies? I believe that one problem so far is that peptides don't seem to produce good immunologic "memory."

Dr. Kohn: As far as I know that is only one group in Israel working in this direction. The group of Ruth Arnon studies vaccines based on synthetic peptides corresponding to isolated loop peptides important in the expression in viral antigens (lysosome, bacteriophages, influenza virus). They have now an article in "La Recherche", and in February 1983 there was a review on the subject by Lerner in Scientific American.

Dr. Malkinson: I wonder what is the state of art in the gene products of retroviruses. How important are these gene products in pathogenesis as expressed by tumors in animals? Should the strategy for vaccination against tumors be based on the use of glycoproteins?

Dr. Aaronson: With respect to RNA tumor viruses, such as avian, murine and bovine leukemia viruses, the immunity elicited against the envelope proteins can be effective in preventing infection. There is now evidence that at least some lymphomas may be associated with a retrovirus as an aetiologic agent. Gallo and coworkers have provided strong evidence for the association of HTLV in clusters where adult T cell lymphoma is prevalent. In order to obtain protection against such an agent one approach might be to use viral components or synthetic peptides as vaccines. It would be very difficult for any virus associated with human tumors to justify use of an attenuated live virus vaccine. With respect to the use of modern technology for virus diagnosis, Dr. David Ward at Yale has developed nucleotides conjugated to biotin, which can be incorporated into appropriate DNA probes. Hybridization with viral nucleic acids in infected cells can then be detected by peroxidase labeled avidin. Indeed, hybridization probes derived from conserved regions of adenovirus DNA genome have been used by Ward and colleagues for *in situ* hybridization to detect adenovirus infected cells. Similar technology could perhaps be used for detection of herpesvirus infection. This is just one aspect of molecular biology that may be useful in the future for diagnosis of virus infection.

Dr. Flanegan: In the diagnosis of paralytic poliomyelitis it is important to know if the case resulted from an infection by a vaccine strain or wild strain of the virus. Oligonucleotide fingerprint analysis can now be used to distinguish between vaccine and wild strains of poliovirus.

Dr. Wagner: In this conference on pathogenesis, it is proper to ask what is the definition of pathogenesis. When we infect cells with viruses, we may look at what is happening following the infection on DNA and RNA protein synthesis, on memberane permeability, on lysosome effects, on syncytia formation, and then determine that all these phenomena are cytopathological and may be related to cell death. In any research on viral infection, one may observe any of these cellular phenomena, occurring either early or late in the infection. Some of these effects may be primary, some secondary. In measles infection, before one reaches the stage of fusion following virus replication (defined as fusion from within), there may occur many stages that lead to this effect, and there may be many factors involved. So we really don't know the exact progress of the changes occurring following infection. In fact we are just nibbling around the edges in our research.

Dr. Choppin: VSV does not kill cells in the same way as measles virus. In each virus:cell system the primary event that leads to cell death may be different. In some cases this may be the interruption of macromolecular synthesis, in others, damaging of the cell membrane. For example in SV5 infection, in some cells there is no effect on cellular macromolecular synthesis, and the cells continue to divide, cell fusion may progress to formation of very large giant cells that are still viable. The only event that causes cell death eventually is damage to the cell membrane, so that the giant cells disintegrate. If however, one prevents infected cells from fusing the cells remain viable. VSV and poliovirus induce in the infected cells the shut-off of macromolecular syntheses. This shut-off kills the cells more rapidly than the shut-off induced by metabolic inhibitors such as puromycin and actinomycin. This is not the shut-off *per se* that kills the cells. Enucleated cells infected with VSV permit virus multiplication which eventually leads to disruption of the cell. Even in such "pure" system there are many factors which lead to the death of cells, and there would be more in a whole organism. In the case of paramyxovirus the primary factor has been shown to be membrane damage induced by the F protein.

Dr. Kaempfer: A cell that is actively producing a virus cannot be called a dead cell: its mechanisms are intact and it is operating full blast to produce virus. One has to be careful with semantics of killing and pathogenesis, because the healthy reproductive act of virus production leads eventually to the death of the host cells. We should be careful on what we concentrate. Is it on the lysis of cell that is equivalent to killing, or should we concentrate on the events that lead up to lysis? What are the viral signs that lead to the death of the cell? One should examine the events that begin the chain of events leading to death.

Dr. Kohn: The definition of pathogenesis can be made on several levels, on a molecular level - when one observes the shut-off of macromolecular syntheses, or on cellular level when one speaks of the death of a cell but the common use of the term pathogenesis concerns the effects of the virus on the host as a whole. Perhaps the term virulence would be more appropriate. (Turning to Dr. Rott): There are virulent and avirulent strains of Newcastle Disease Virus. Serologically they are practically identical. Can one put the finger on a molecular difference between these velogenic and lentogenic strains? Can one define the correlation between the structure of these viruses and their behavior in nature?

Dr. Rott: There is indeed a very good correlation between the pathogenicity of NDV, and the structure of the viral glycoproteins. In the highly virulent velogenic strains both glycoproteins, the HN and the F are cleaved in all cells that were tested for it. In mesogenic strains of moderate virulence only HN is cleaved and in the nonpathogenic (lentogenic strains) only few cells are permissive for cleavage of both glycoproteins. Thus there is a very good correlation between the susceptibility of both glycoproteins to cleavage by proteolytic enzymes. Why is it so? We don't know, since we don't have yet the exact sequence of aminoacids in these glycoproteins, and therefore we don't know whether there is any difference in the cleavage sites.

Dr. ter Meulen: I would like to comment on the virus effect on host cell function. It has been shown that in a state of persistent infection cellular dysfunction can occur. As shown for example with measles or rabies virus infection in neuroblastoma cells. In addition, in Aleutian Mink Disease excess production of gammaglobulin is observed to such an extent that the animal dies. This is another example of how a virus infection interferes with a lymphocyte function. Certainly, this aspect has been neglected in past studies of virus-induced disease process and associated pathogenic mechanisms.

Dr. Choppin: In the case of Sendai virus, Andy Scheid isolated a number of mutants that were cleaved by different proteases. He showed that one could change the host range in these mutants: The N-terminus in these mutants was the same, meaning that the mutation had to be situated to the "left" of the cleavage site, and the change on the mutated protein had made it susceptible to different proteases and thus caused the change in host range. One can made the prediction that the same occurs in NDV; there is a change in cleavability due to the change in that region of the protein.

Dr. Wagner (answering a question by Dr. Fuchs): The strategy of VSV, myxoviruses and the paramyxoviruses for virus transcription and translation is fairly close. The main difference is evidence for the presence of a leader sequence in VSV. This sequence is presumably present in the paramyxoviruses, but absent in myxoviruses. One has yet to determine whether there is a general mechanism relating to virulence which would be the same in VSV as well as paramyxoviruses, but surely it is different for myxoviruses, who have different mechanisms of replication, including a nuclear phase.

Dr. Becker: We agree that cellular genes may control viral genes. Can the same virus, therefore, cause different diseases? One of the examples was mentioned by Dr. Choppin. Reye's syndrome starts with influenza B or varicella in a child; this proceeds in a predictable manner, but then the disease takes a different course, and after apparent convalescence, there is fatty degeneration of the liver, increase in intracranial pressure, and death. The question is then, why does the disease in some children become fatal in this manner? There must be a host or genetic factors involved.

Dr. Wagner: There is wide variation in reaction of cells to virus infection. For example, pre-B myeloma cells are much more susceptible to VSV than other cells. The difference is not in adsorption, but in the actual events inside the cell. We have to be concerned about what type of cells are involved when we speak of cytopathogenicity. The cells modulate the viral infection to considerable but varying effect.

Dr. Aaronson: We should now like to turn to tumor viruses. One of the most successful persistent viral infections which has been in process for millions of years in a variety of animals has been that in which chronic retroviruses have undergone a stable germ line infection of vertebrates. It is clear that such events have affected humans too. Groups at the NIH, headed by Malcolm Martin and Jeffrey Schlom have cloned from human DNA retrovirus related sequences, which have all of the structural features of retrovirus genome including long terminal repeats. The number of copies of endogenous retroviral sequences in animals and in man can vary. In the mouse, for instance, Robert Callahan (NIH) has calculated that 0.1 - 1% of mouse cellular DNA is composed of retroviral sequences.

Eli Canaani (Weizmann Institute) has shown that an endogenous retroviral sequence in the mouse can be transposed and be placed in a position so as to activate a cellular oncogene.

There have also been genetic studies in mice in which genetic changes involving disruption of a particular gene (coeat color) is associated with the reintegration of an endogenous virus. (Turning to Dr. Canaani): What can be done to prevent pathogenesis associated with endogenous retroviral sequences?

Dr. Canaani: If it is only a transposition event, then one cannot prevent it or do anything about it. But if one knows what is the type of oncogene that is activated, then one can think about some drug that would specifically inhibit or change the product of the gene without affecting the function of the cells as a whole. There accumulates now the new evidence based on translocation - the movement of a gene from a position where it is silent to a position where it is changed in its composition or lacks a part and where it is active in transcription. More examples of hematological tumors in humans could also fall into such category. If one knows what oncogenes are activated in hematological tumors, one may find a way to modify these tumor proteins, perhaps one could find drugs that modify this protein.

Dr. Becker: I should like to comment about translocation and activation of oncogenes. In ataxia-telangiectasia there is a clastogenic (DNA-breaking) factor in the blood that causes chromosomal breaks in peripheral lymphocytes. A child who developed leukemia had a typical translocation in a clone of lymphocytes which were regarded as preleukemic. Five years later leukemia became manifest and the tumor cells exhibited the same type of translocation. In this case there was a chromosomal translocation in preleukemic cells without expression of a cancer state.

The triggering of the appearance of the tumor must have been due to some additional events beyond translocation, perhaps by additional viral information or activation of the translocated oncogene by an additional, unknown event.

Dr. Aaronson: There may perhaps be more than two steps necessary for the development of a tumor.

Dr. Malkinson: I understand that the cell oncogenes are linked to the differentiation process in the embryonal life of the vertebrates, whilst the endogenous leukosis virus came from the outside into the genome. In fact, Astrin & Heywood have bred chickens without endogenous retrovirus genes, yet the differentiation sequences remain in these chickens. Is there any information about what the cell oncogenes are doing when they are not being stimulated by viral oncogenes?

Dr. Aaronson: This is the central question to research in this field. The functions of the normal counterparts of the viral oncogenes are not yet known. What changes these gene products to make them transform cells may in fact require knowledge of the normal gene functions.

Dr. Kohn: In the bladder carcinoma studied by Weinberg's group, there is a protein, p 21, where a change in a single amino acid is sufficient to make the cell harboring it malignant; There is a change from a normal protein product to a cancer product. How does this finding correlate with the oncogenes and their translocation as initiators of the cancer state?

Dr. Aaronson: Your question really is how can a single point mutation occur and not create a problem earlier in the development of the organism. If this single change were indeed sufficient for the cell to become malignant, it should have, on statistical grounds, occurred relatively early in development. It is unlikely therefore that one change in a gene would be sufficient, and be the entire explanation of how a normal cell becomes malignant.

We have covered a number of topics that relate to virus pathogenesis. I should now like to invite questions from the audience.

Dr. Flanegan: One clear challenge that remains for virologists dealing with viral diseases is the development of new antiviral drugs. To accomplish this, we need to know more about the specific proteins involved in viral pathogenesis. For the time being, there are very few effective drugs. Vaccines alone will not be able to control all viral diseases.

Peter Howley (NIH) has successfully used interferon to cure papilloma virus infections and transformation of NIH/3T3 cells in culture.

Dr. Kohn: Also here in Israel throat papilloma in children has been successfully treated with interferon, and saved the patients from frequent operations. Although these few cases provide only anecdotal evidence for the efficacy of interferon in these cases, it seems to be working.

Dr. Catravas: Some years ago I worked with dimethylbenzathracene (DMBA) and found that this reagent labeled with radioisotopes bound strongly to Z-DNA (left-handed). Has Z-DNA something to do with oncogenicity?

Dr. Canaani: Z-DNA has a structure that resembles some of the enhancer elements found in SV40 and in the LTR of A-particles, which we cloned. Alexander Rich told us that Z-DNA sequences fit very well with portions of inserted element in c-mos. Thus Z-DNA may be functioning as an enhancer of transcription.

Dr. Shalitin: Concerning Z-DNA, Shostak claims that cloned yeast telomers have Z-DNA structure as deduced from the sequence and therefore Z-DNA may have something to do with origin of replication.

Dr. Aaronson: I would like to thank all of you for contributing to a stimulating discussion.

AUTHOR INDEX

SUBJECT INDEX

GPSR Compliance

The European Union's (EU) General Product Safety Regulation (GPSR) is a set of rules that requires consumer products to be safe and our obligations to ensure this.

If you have any concerns about our products, you can contact us on ProductSafety@springernature.com

In case Publisher is established outside the EU, the EU authorized representative is:

Springer Nature Customer Service Center GmbH
Europaplatz 3
69115 Heidelberg, Germany

Batch number: 10583998

Printed by Printforce, the Netherlands